27th SNAME Offshore Symposium 2022

AA005472

Houston, Texas, USA
22 February 2022

ISBN: 978-1-7138-5750-1

Printed from e-media with permission by:

Curran Associates, Inc.
57 Morehouse Lane
Red Hook, NY 12571

Some format issues inherent in the e-media version may also appear in this print version.

Printed with permission by Curran Associates, Inc. (2022)

For permission requests, please contact Society of Naval Architects and Marine Engineers (SNAME) Texas Section at the address below.

Society of Naval Architects and Marine Engineers (SNAME) Texas Section
99 Canal Center Plaza, Suite 310
Alexandria, VA 22314
USA

Phone: +1-703-997-6701

www.sname.org/texassection

Additional copies of this publication are available from:

Curran Associates, Inc.
57 Morehouse Lane
Red Hook, NY 12571 USA
Phone: 845-758-0400
Fax: 845-758-2633
Email: curran@proceedings.com
Web: www.proceedings.com

TABLE OF CONTENTS

SESSION: 1

SESSION: 2

SESSION: 3

THE INTERNATIONAL COMMUNITY
FOR MARITIME AND OCEAN PROFESSIONALS

Application of a Comprehensive Design Method to Counter-Rotating Propellers

Kyungjung Cha
Ocean Engineering Group, Department of Civil, Architectural and Environmental Engineering,
The University of Texas at Austin

Spyros A. Kinnas
Ocean Engineering Group, Department of Civil, Architectural and Environmental Engineering,
The University of Texas at Austin

Proceedings of the 27th Offshore Symposium, February 22nd, 2022, Houston, Texas
Texas Section of the Society of Naval Architects and Marine Engineers

ABSTRACT

A comprehensive design method is applied to design of counter-rotating propellers (CRP). A numerical nonlinear optimization algorithm is first used for design of each propeller. This approach represents a blade by B-spline geometry and the design variables at the location of the vertices of the B-spline polygon determine the optimal blade shape. The nonlinear optimization method aims at minimizing the torque for a given target thrust with constraints, e.g. the minimum pressure constraint for a fully wetted propeller or maximum allowed cavity area for a cavitating propeller. Then the interaction of the designed propellers and a given pod including the viscous flow field around the two propellers is analyzed by coupling a vortex-lattice method (VLM) with a Reynolds-Averaged Navier-Stokes (RANS) solver. The analysis determines a new inflow for a new design of propellers. The procedure of the design and interaction analysis finishes when the propeller thrust converges within a certain tolerance. Finally, the designed propellers are compared with the original propellers.

Keywords: Counter-rotating propellers, Propeller/pod interaction, Podded CRP, Optimization, VLM, RANS

Introduction

A traditional method to design marine propellers first determines the optimum circulation distributions by lifting line model subject to the circumferentially averaged inflow and then applies an inverse method to design the blade producing the determined circulations (Greeley and Kerwin 1982, Kerwin 2016). The designed propeller is analyzed by vortex-lattice method (VLM), panel methods, finite volume methods (CFD) in order to predict the propeller performance in the actual non-axisymmetric inflow.

A nonlinear optimization algorithm known as CAVOPT-3D for propeller design was developed by Mishima and Kinnas (Mishima 1996, Mishima and Kinnas 1997). The algorithm is coupled with the VLM (code name:

MPUF-3A) to consider cavitating propeller subject to non-axisymmetric inflow (Kinnas 1983, Du et al. 2018), and designs the blade surface directly, without the need of knowledge of the circulation distribution. CAVOPT-3D uses a B-spline polygon which accurately defines the blade camber surface with a relatively fewer number of parameters, and uses second-order Taylor expansions with respect to those parameters (Mishima and Kinnas 1997). The method can be applied to design propellers working under high loading subject to given non-axisymmetric inflow, by specifying the amount of allowable cavitation. In addition, the method can be applied to design propellers to achieve the highest thrust for a target power for given operation conditions. Griffin and Kinnas (1998) later proposed the minimum pressure constraint and skew optimization in the code. The algorithm showed an improved performance of propellers by incorporating the constraints. Du and Kinnas (2018) developed the optimization method to couple with a panel method, PROPCAV, and included the under-relaxation factors to improve the stability of the optimization process. This method was recently applied in order to design open or ducted propellers to move a given axisymmetric body at given speed, where the full interaction of the hull with the propeller was included, by coupled VLM with RANS, without the need of assumed wake fraction or thrust deduction coefficients (Kinnas et al. 2021). In the same work the effects of propeller or outer duct diameter and of RPM on the required optimum power were also studied.

This paper starts by describing details of propeller analysis by the VLM, propeller design by the nonlinear optimization method, and propeller/pod interaction by the VLM/RANS interactive method, and then presents a comprehensive design method that includes the above-mentioned techniques. The comprehensive design method is applied to counter-rotating propellers for given axisymmetric pod as well as design conditions. The major advantage of this systematic design method is that the wake fraction and thrust deduction factors are not required. The designed propellers are eventually compared with the original propellers.

METHODOLOGY

Analysis of Propeller

A vortex lattice method (VLM) in a code known as MPUF-3A is adopted for propeller analysis. The presence of propellers is represented by singularities of vortices and sources on the blade mean camber surface and its assumed wake surface. The strength of singularities is determined by applying the kinematic and dynamic boundary conditions at some control points. The linearized blade thickness is applied and the blade thickness to chord ratio is thus assumed to be small. The suction and pressure sides of the blade are collapsed into the mean camber surface. The linearized boundary conditions are satisfied on the blade mean camber surface. In this method, viscous force is obtained from the frictional drag coefficient C_f which is uniformly applied on the wetted surface of the blade (Kinnas 1983, Du et al. 2018).

Design of Propeller

A nonlinear optimization method by CAVOPT-3D, CAVitating blade OPTimization-3D, is applied to design propellers. In the code, the propeller is modeled by B-spline polygon of 4×4 control points to represent the camber surface. The B-spline polygon ensures the propeller blade is uniquely defined with fewer parameters. Cubic B-spline surfaces are defined as follows

$$\begin{aligned} P(u,w) &= [x(u,w), y(u,w), z(u,w)] \\ &= \sum_{i=0}^{N_u-1}\sum_{j=0}^{N_w-1} d_{ij} N_{i,4}(u) N_{j,4}(w) \end{aligned} \qquad (1)$$

where d_{ij}=B-spline control points, $N_{i,4}(u), N_{j,4}(w)$=B-spline basis of order 4, u, w=parameters for B-splines, and N_u, N_w=number of B-spline control points in u and w-direction. The cubic B-spline surface with 4×4 vertex polygon and 10×9 propeller grid in MPUF-3A is presented in Figure 1. The B-spline polygon defines the blade geometry without skew, and the skew is then included independent of the blade geometry.

Design variables x_i exist at locations of vertices of B-spline polygon to determine a shape of blade. The vertices move at each optimization iteration to provide an improved blade geometry. The direction of movement of each design variable is decomposed into two directions (Figure 2) such that the direction normal to an initial leading edge-trailing edge line on a horizontal plane and the chordwise direction. The number of design variables n depends on an option of chord length distributions; $n = (N_u - 1) \times (N_w - 1) + 1$ for a fixed chord length and $n = N_u \times (N_w - 1) + 1$ for a determined chord. The vertices are thus not allowed to move in the chordwise direction when the radial chord length distributions are given. The vertices of the leading edge and trailing edge at a particular radius move the same amount in the opposite direction, therefore, the skew and rake cannot be included.

The method starts with an initial blade geometry represented by the vector of design variables. The first step is the linear approximations of the objective and constraint functions. MPUF-3A runs to get the coefficients of the linear functions in terms of the design variables. Then the minimizing problem is numerically solved for the linear objective and constraint functions to get an improved geometry. The next step is the quadratic approximations of the objective and constraint functions and then the problem is solved. In this algorithm, the thrust and torque are approximated by the second-order Taylor expansions with respect to the parameters that define the B-spline surface from the most recent values determined by the VLM because the evaluation of gradient of the objective function for three-dimensional propeller problem, e.g. torque of propeller, is computationally expensive than approximation by polynomial expansions.

The nonlinear optimization method needs to specify the design conditions such that

- Advance ratio: $J_S = V_S/(n_P D_P)$
- Cavitation number based on n_P: $\sigma_n = (P - P_V)/(0.5\rho n_P^2 D_P^2)$
- Froude number: $Fr = (n_P^2 D_P)/g$
- Number of blades: Z
- Hub radius: R_{HUB}/R
- Required thrust coefficient: K_{T0}

where V_S is the ship speed, n_P is the propeller rotational frequency (RPS), and D_P is the propeller diameter. P_V is the vapor pressure, ρ is the fluid density, and g is the gravitational acceleration.

The optimization method minimizes the torque for a given target thrust. This problem is defined as follows

$$\begin{aligned} &\text{minimize} && f(x) = K_Q(x) \\ &\text{subject to} && h(x) = \frac{K_T(x) - K_{T0}}{K_{T0}} = 0 \end{aligned} \tag{2}$$

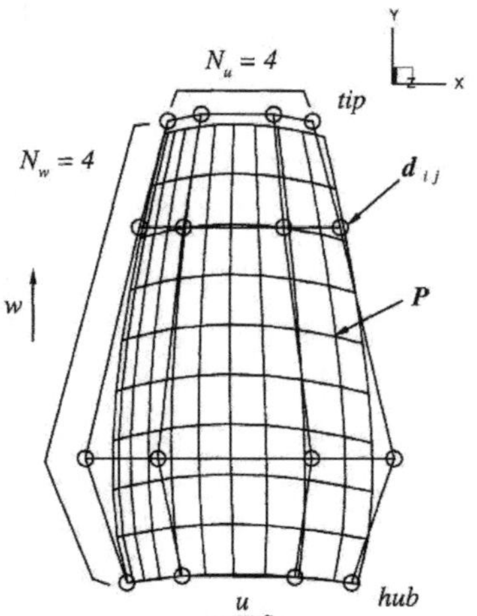

Figure 1: Cubic B-spline surface with 4×4 vertex polygon and 10×9 propeller grid in MPUF-3A (From Mishima and Kinnas 1997)

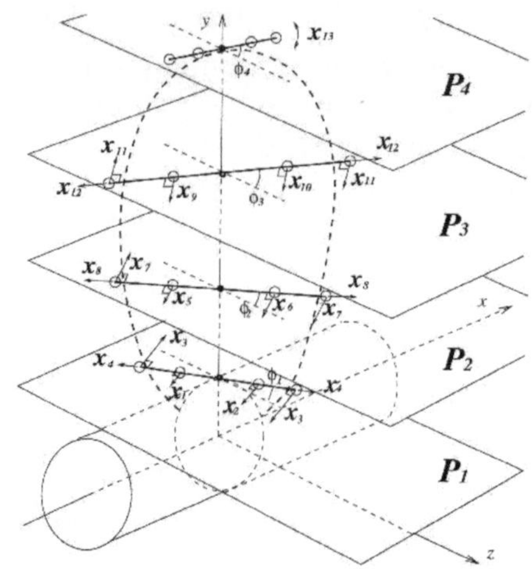

Figure 2: Initial B-spline vertices and the vertex movements (From Mishima and Kinnas 1997)

where $K_Q(x)$ is the torque coefficient, $K_T(x)$ is the thrust coefficient, and x is the design variable vector to define the blade geometry. The objective function $f(x)$ represents the torque and the equality constraint $h(x) = 0$ is given in terms of the thrust. Also the inequality constraints $g_i(x) < 0$ are able to be imposed to minimize the objective function. The inequality constraints that can be incorporated are defined as follows

$$\partial^2 C_i / \partial x_{4i}^2 < 0 \quad (3)$$

$$DPV > BTOL \quad (4)$$

$$DCP > CTOL \quad (5)$$

$$CA < CAMAX \quad (6)$$

The left-hand side of the inequality constraints of Equation (3)~(6) shows instantaneous values at a particular iteration whereas the right-hand side of the constraints presents the limit of the value on the left-hand side.

The constraint of the second derivative of chord $\partial^2 C_i / \partial x_{4i}^2$, where $C_i = C_{0,i} + x_{4i}$ $(i = 1, 2, 3)$, imposes more reasonable shape of the blade when the chord length distribution is determined (Equation (3), Figure 3). The design variable at tip to determine the chord length does not exist because the tip vertices simply rotate.

Equation (4)~(5) impose limits of the pressures. Figure 4 shows definitions of the pressure related constraints. DPV constrains the minimum pressures by the difference between a standard to occur cavitation and a point of minimum pressure; $DPV = \sigma_n/2 - max(-C_p)$ where C_p is the pressure coefficient that $-C_p = \left(P_0 - P\right)/\left(\rho n_p^2 D_p^2\right)$. This constraint excludes the tip to avoid over restriction. Thus $BTOL = 0$ shows the minimum pressure should be above the vapor pressure, $P > P_V$. Also, DCP controls the pressure difference between the suction and pressure sides of the blade with a definition of $DCP = \left(-C_p^+\right) - \left(-C_p^-\right) = \Delta P$, excluding the leading and trailing edges and the tip. $CTOL = 0$ requires the pressure difference to be positive, $\Delta P > 0$.

The back cavity area constraint $CAMAX$ represents the maximum allowable back cavity area over the blade area.

The final solutions of the nonlinear design method are the propeller geometry PG, thrust coefficient K_T, torque coefficient K_Q, and propeller efficiency η_p.

- Thrust coefficient: $K_T = T/(\rho n_P^2 D_P^4)$
- Torque coefficient: $K_Q = Q/(\rho n_P^2 D_P^5)$
- Propeller efficiency: $\eta_P = (K_T/K_Q)(J_S/2\pi) = (TV_S)/(Q\omega)$

where T and Q are the propeller thrust and propeller torque, respectively. $\omega = 2\pi n_P$ is the propeller angular velocity. The propeller efficiency denotes the useful power (TV_S) for the power given to the propeller ($Q\omega$).

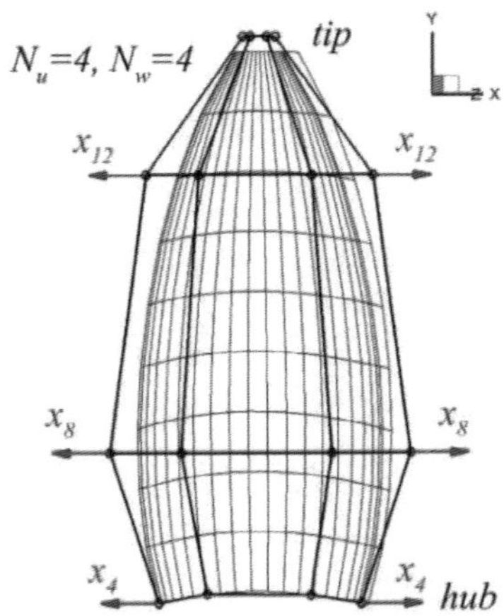

Figure 3: Design variables to determine the chord length

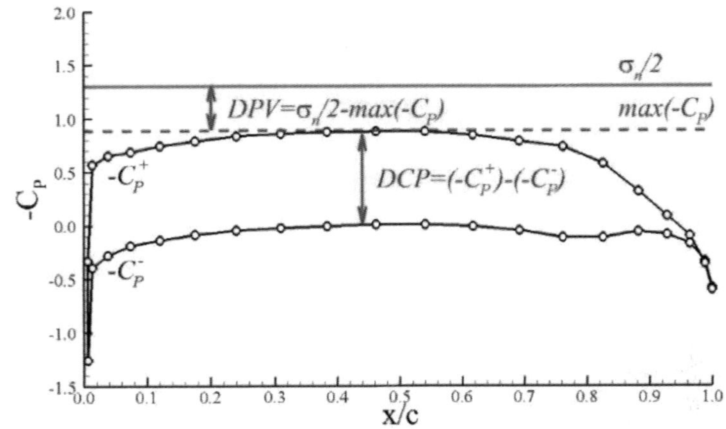

Figure 4: Pressure constraints

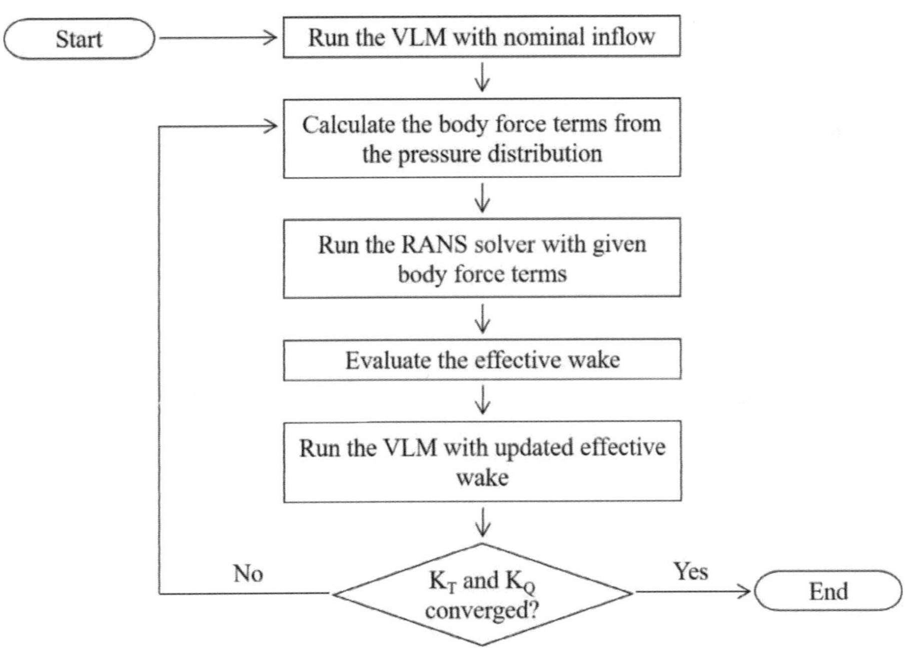

Figure 5: Numerical algorithm of the VLM/RANS interactive scheme

Propeller/Pod Interaction

The VLM is applied for the analysis of propeller and the RANS solver, ANSYS FLUENT, is used for the prediction of viscous flow field around the propeller and drag force including the friction force on the pod. Figure 5 describes the procedure to consider propeller and pod interaction. Propeller is represented by the body forces obtained from the pressure data over the blades by the VLM, and the body forces are then interpolated to the RANS grids. The effective wake EW_i is calculated by difference between total velocity from the RANS, U_{total}, and propeller induced velocity in the VLM, $U_{prop\ induced}$. The propeller/pod interaction is considered without the need for wake fraction and thrust deduction factors but advance ratio and thrust coefficient for propeller are necessary in early stage. In addition, a scaled propeller is able to be modeled by scaling the coordinates, body forces, and mass sources when the RANS solver couples with the VLM with a scale factor k (Equation 7).

$$k = \frac{R_P}{R} \tag{7}$$

where R_p is the scaled propeller radius and R is an initial propeller radius in the code (=1m).

Comprehensive Design Method for CRP

Figure 6 shows a flowchart of the comprehensive design method. This method starts by assuming the propeller effective wake as uniform inflow and the design conditions (e.g. J_S, σ_n, Fr) should be determined.

- Step 1: Forward and aft propellers are designed using the optimization method with given pod geometry and the effective wake. The target thrust is assumed as the thrust of original propellers at determined design conditions. In this step, locations of each propeller are taken into account but the size of propellers is not included.
- Step 2: Interaction of the pod and designed propellers from Step 1 is analyzed by the VLM/RANS coupling method. The effect of propeller size is considered. The propeller thrusts and the effective wakes are determined based on the effect of the pod to the blade.
- Step 3: The revised effective wakes from Step 2 are updated as inflow for each propeller, and the swirl velocity is considered for the aft propeller.

A combination of Step 1 and 2 is counted as one outer iteration. The procedure continues until the propeller thrust from Step 1 and 2 converges with a specified tolerance.

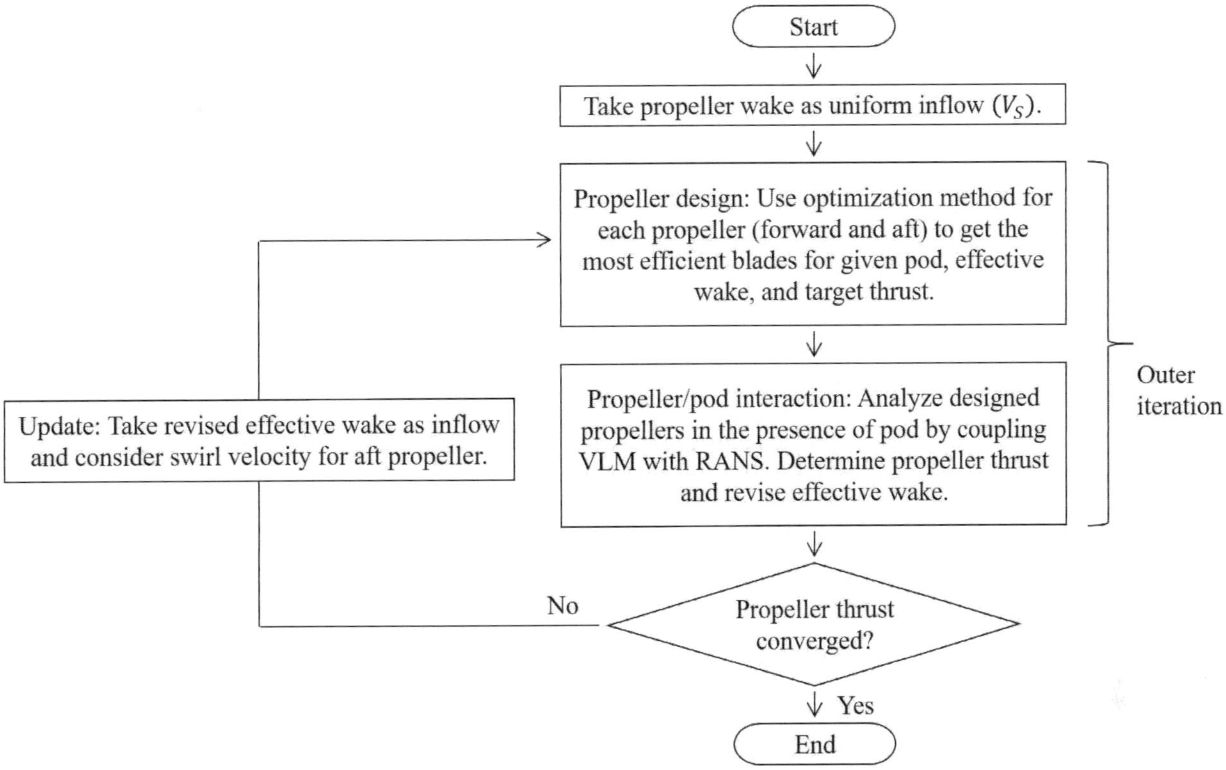

Figure 6: Flowchart of the comprehensive design method

Results

Counter-rotating podded propellers (podded CRP)

A typical CRP is shown in Figure 7, and the predicted performance by BEM/RANS interactive method (Kim 2021) is shown in Figure 8, together with the measured values.

Figure 7: A typical set of podded counter-rotating propellers

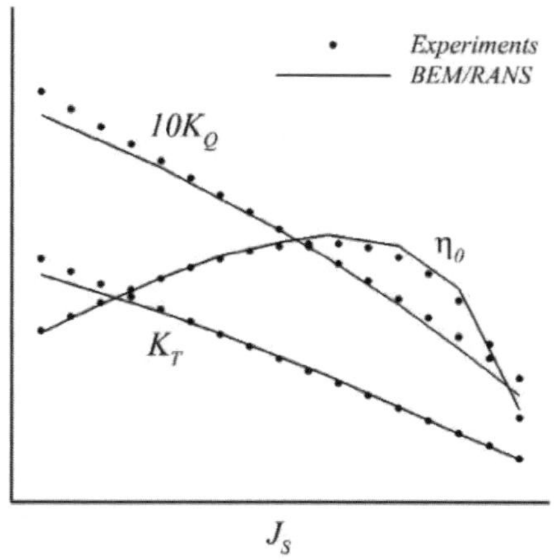

Figure 8: Performance predicted the BEM/RANS method and measured for a CRP (the scales are omitted due to propriety issues) (From Kim 2021)

Design conditions and details

In this paper, we consider an existing set of podded counter-rotating propellers (CRP), and the present method is applied in order to redesign the propellers to produce the same total thrust (and the same thrust division between the two propellers), with higher efficiency, i.e. smaller total torque. This study thus takes the target thrusts (not only the total) to be the same as those of the original design. The force on the pod is not considered in this study. The forward propeller is right-handed propeller with 4 blades and the aft propeller is left-handed one with 5 blades, and 20×9 propeller grid in the VLM is used. The thrust ratio between the two propellers is T_F: T_A=0.66:0.34. Design conditions are K_{TT}=0.3222, J_S=0.8000, σ_n=1.4418, and Fr=8.4306. The subscripts F, A, and T denote the forward and aft propellers, and the total.

Unsteady wetted (non-cavitating) flow is only considered in this study. The viscous pitch correction which approximates the viscous effect by an empirical frictional coefficient over the blades and empirical correction of the pitch angles (Kerwin and Lee 1978) is considered. The wake alignment is applied so that the wake geometry is generated based on the mean inflow and the propeller induced velocity (Greeley and Kerwin 1982). The chord and thickness distributions take as given in this problem, but the pitch and camber are determined by the optimization method. The movement of design variables is constrained by 0.01 at each optimization iteration and the camber range is constrained between 0.0 and 0.1. The forward propeller is designed in constrained conditions of $P > P_V$ and $\Delta P > 0$, and the aft propeller is designed by unconstrained run.

Pod without propellers

Figure 9 shows the RANS domain of size -1.50<x/L_H<2.00 and 0.00<y/L_H<1.25. The cell number of this domain is 30k. Reynolds number based on the pod length is $Re = V_S L_H/\nu = 4.07\times10^7$ where ν is the kinematic viscosity of the fluid. The force on the pod without the propellers was evaluated by RANS (ANSYS/Fluent) as R_{H0}=2038 N, and the pressure distribution over the pod is shown in Figure 10.

Figure 9: 2D axisymmetric RANS domain

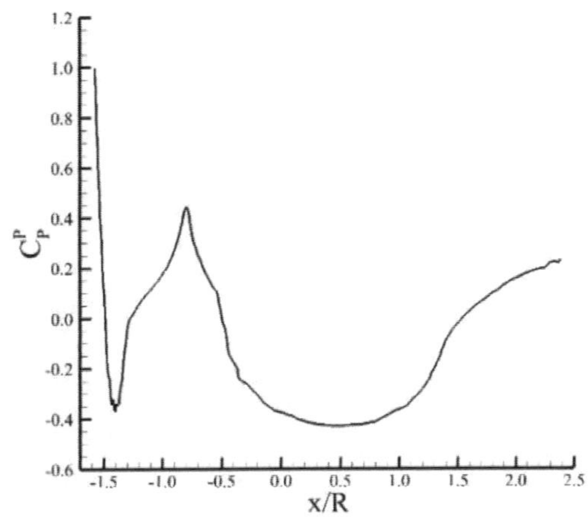

Figure 10: Pressure distribution over the pod from RANS simulation without propeller; $C_P^P = (P - P_0)/(0.5\rho V_S^2)$

Converged results and comparison

The convergence history of propeller thrusts from design and analysis steps are presented in Figure 11. The procedure finishes at the 4th outer iteration within 1.5% tolerance that 1.45% and 0.59% for the forward and aft propellers, respectively.

Figure 12 shows the axisymmetric VLM model with designed propellers and wake geometry. Figure 13 presents the convergence history of the forces and propeller efficiency, and of the constraints. The constraints of forward propeller are almost satisfied with the desired conditions, the minimum pressure of DPV>0 and the positive loading of DCP>0, by the fact that the final values of DPV=-0.0876 and DCP=-0.0466. Figure 14 presents the history of design variables for the optimization iterations and the converged design variables for both propellers correspond well to the propeller geometry shown in Figure 12.

Figure 15 presents the designed propeller geometry, pitch (P/D) and camber (f/c). The camber distributions are well designed in that the imposed range is 0.0<f/c<0.1. Figure 16 and Figure 17 present favorable circulation distributions of the designed propellers, and the axial body force zone to represent scaled propellers in the RANS. Figure 18 shows the axial effective wake for both propellers which is the decelerated flow that could affect the propeller thrust of the VLM/RANS method. Figure 19 presents the swirl velocity around (before, between, and after) the designed propellers such that the tangential velocity is cancelled out behind the after propeller, which is an indication of a successful counter-rotating propeller design.

Figure 20~21 show the pressure distributions of designed propellers. The forward propeller slightly violates the desired pressure distributions based on the final DPV and DCP such that the pressures occurrence above $\sigma_n/2$ at r/R =0.858 and almost invisible negative loading over all blade sections. The aft propeller is designed by unconstrained condition, but it is worth noting that the pressures seem to have reasonable distributions; the maximum of $-C_p$ is well below cavitation occurrence ($\sigma_n/2$) and no visible negative loading over the blade sections except the leading edge which is excluded in the algorithm to avoid over restriction.

Table 1 compares the total final solutions between the original propellers and designed propellers. The final propeller thrust well enough to satisfy the desired propeller thrust. The total propeller efficiency shows the designed propellers are 9.36% more efficient than the original propellers by the fact of 6.70% beneficial torque coefficient as well as 3.63% lower propeller thrust coefficient.

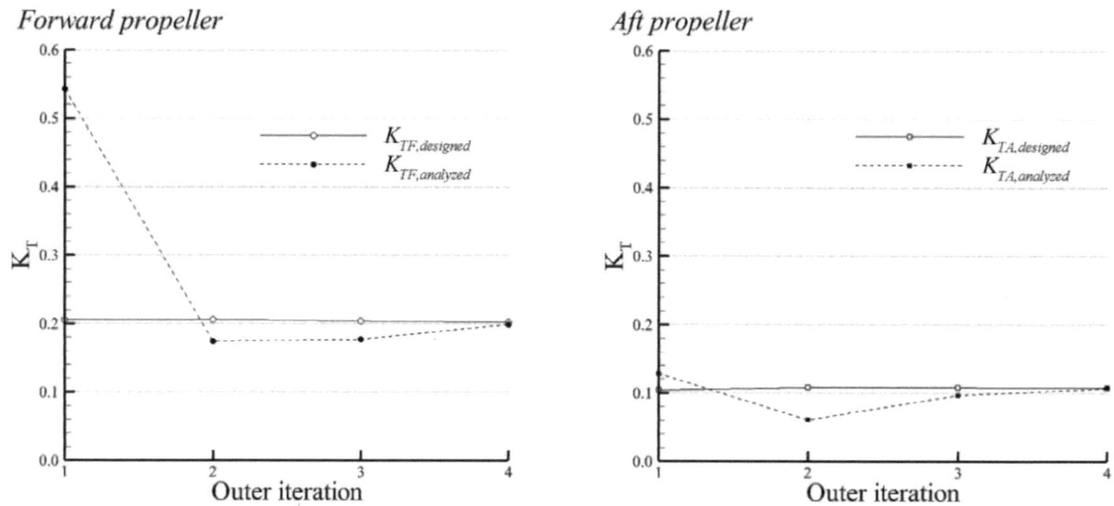

Figure 11: Propeller thrust history from the optimization method and VLM/RANS interactive method (K_T converges with 1.5% tolerance); forward propeller (left) and aft propeller (right)

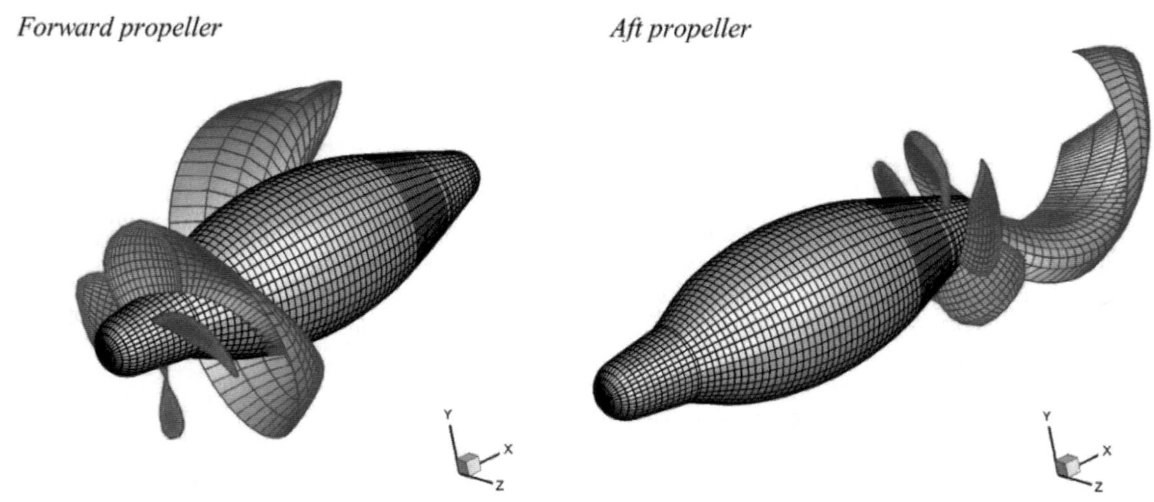

Figure 12: Axisymmetric VLM model with the designed propellers; forward (left) and aft (right) propellers

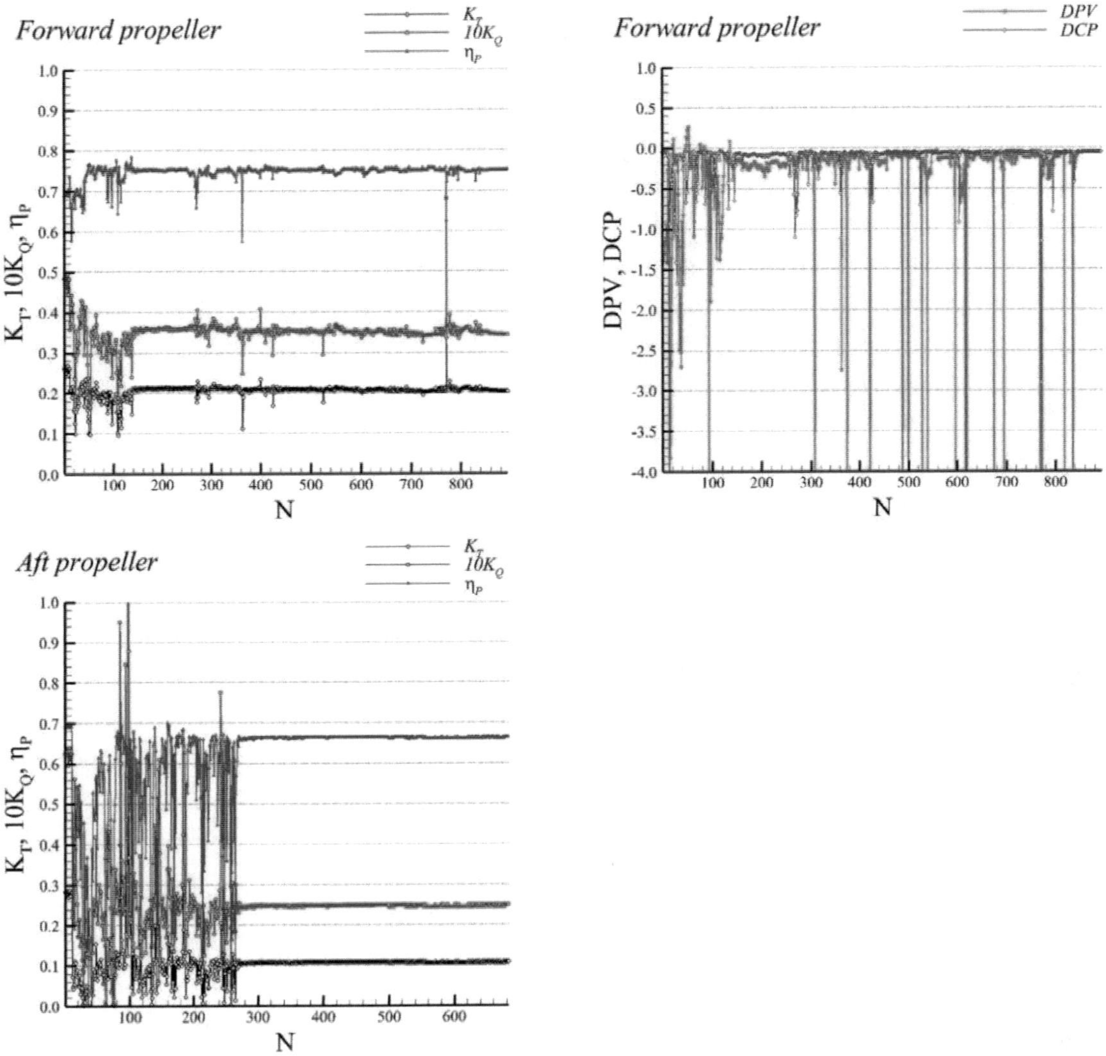

Figure 13: Convergence history; forward propeller (top) and aft propeller (bottom)

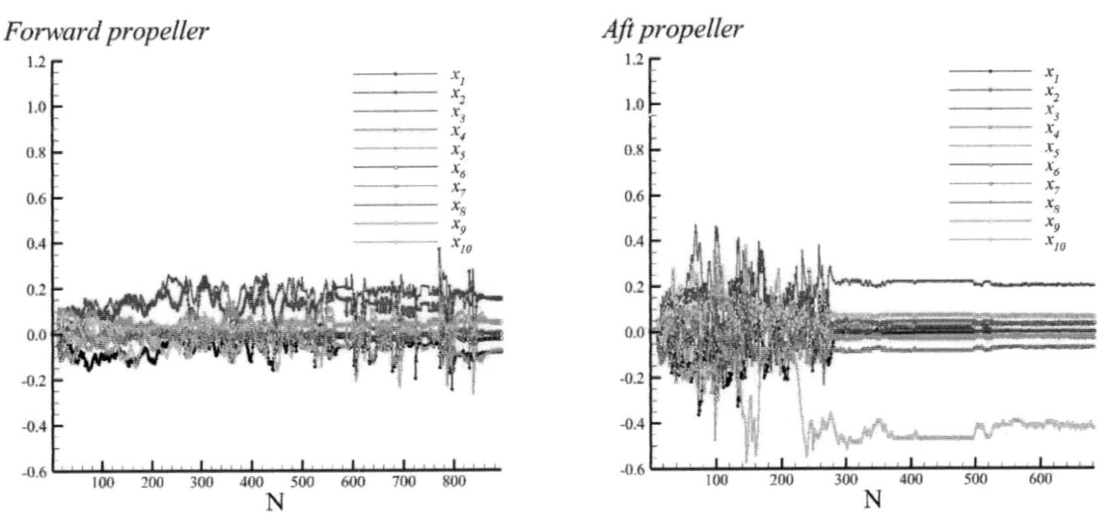

Figure 14: Convergence history of the design variables; forward propeller (left) and aft propeller (right)

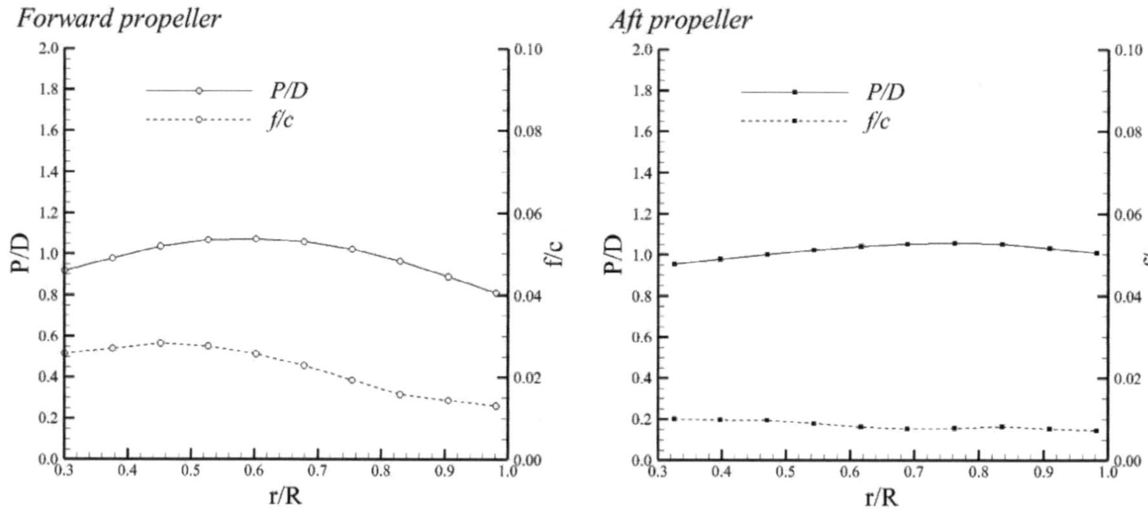

Figure 15: Designed pitch (P/D) and camber (f/c) distributions; forward propeller (left) and aft propeller (right)

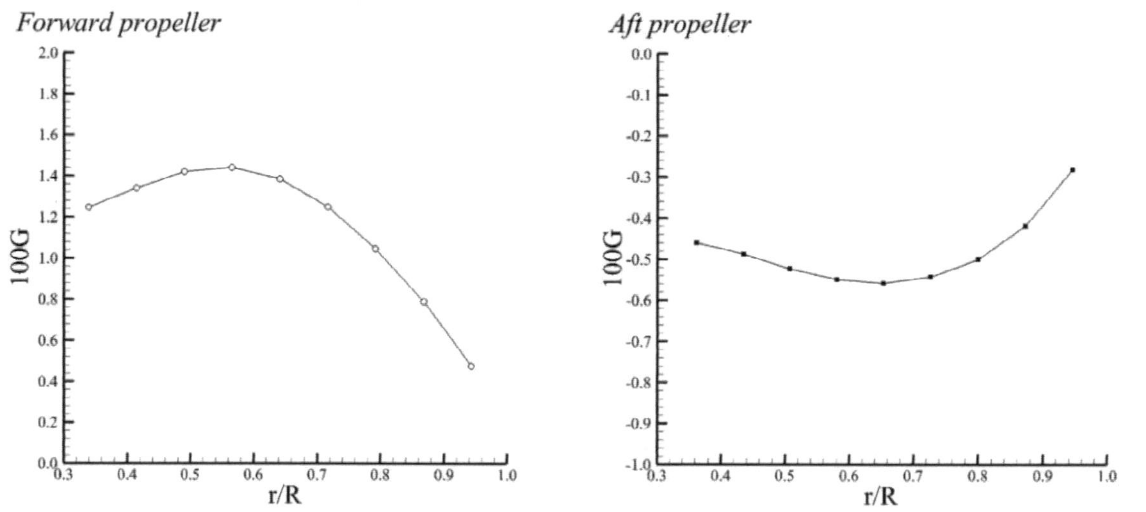

Figure 16: Circulation of the designed propellers; forward propeller (left) and aft propeller (right);
$G = 100\Gamma/(2\pi R_p V_R)$**: non-dimensional circulation,** Γ**: circulation,**

$V_R = [V_S^2 + (0.7 n_p \pi D_p)^2]^{1/2}$ **: reference velocity**

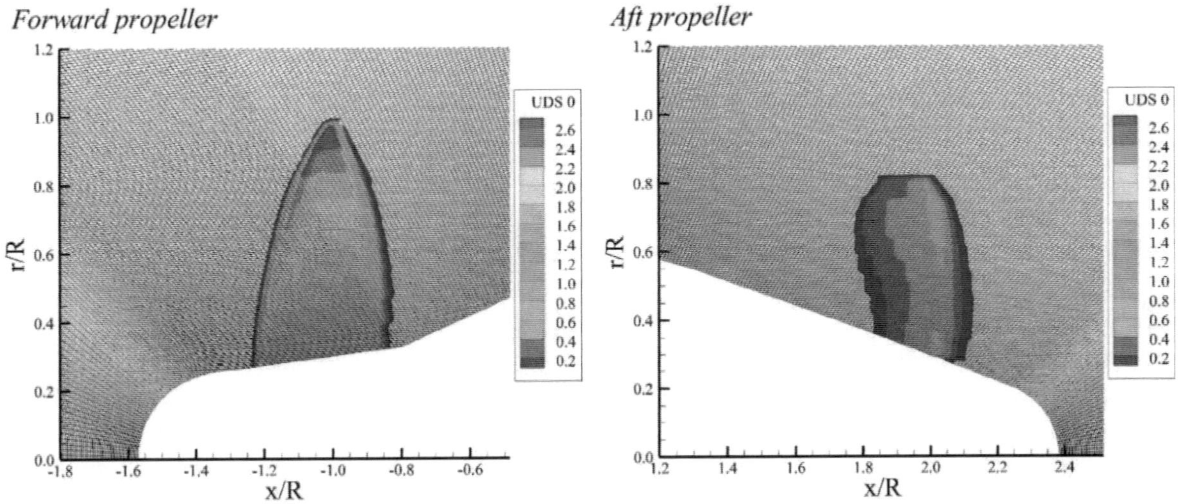

Figure 17: Axial body force distribution in the RANS; forward propeller (left) and aft propeller (right)

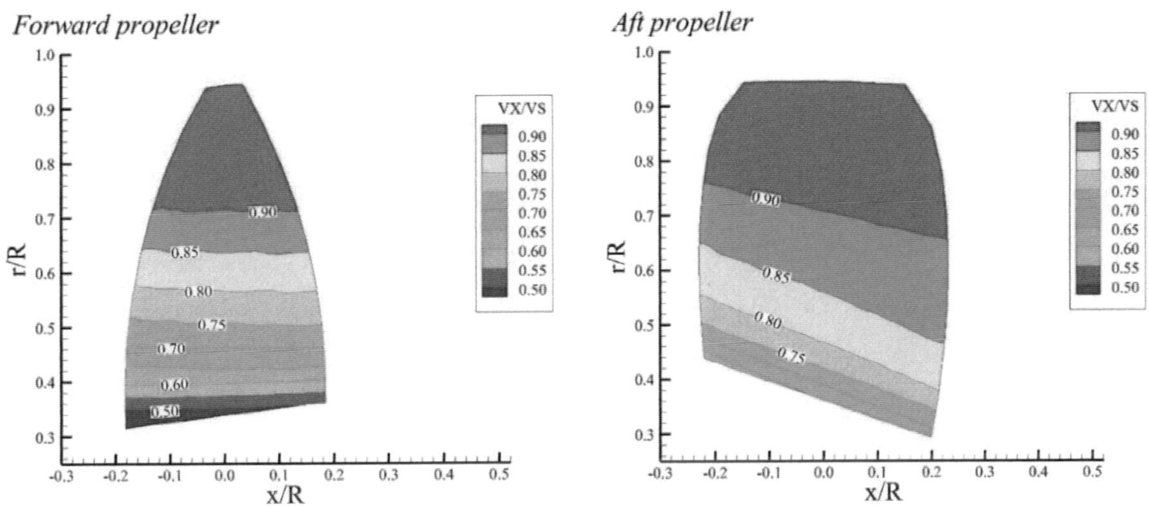

Figure 18: Axial effective wake; forward propeller (left) and aft propeller (right)

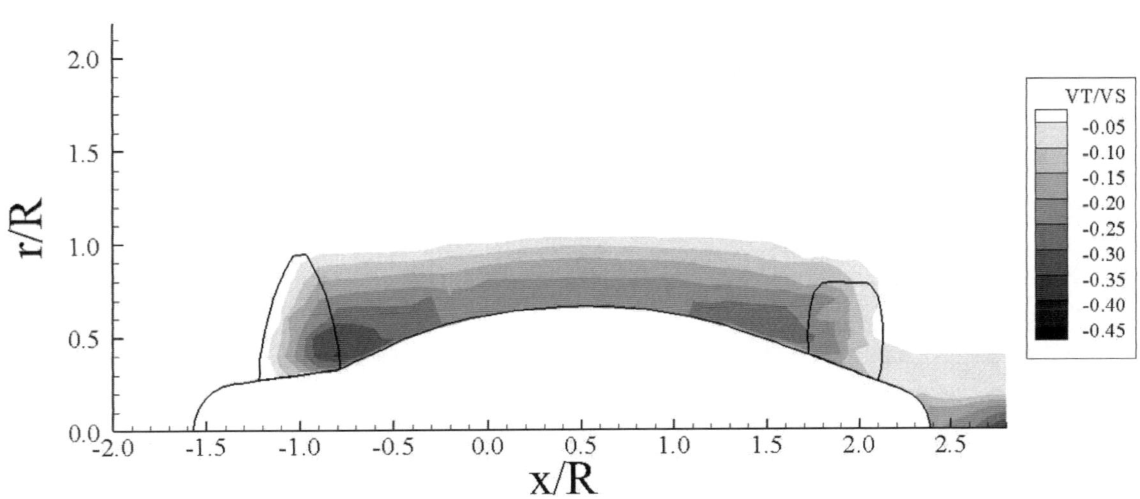

Figure 19: Swirl velocity around the designed propellers (Note the swirl cancelation downstream of the aft propeller)

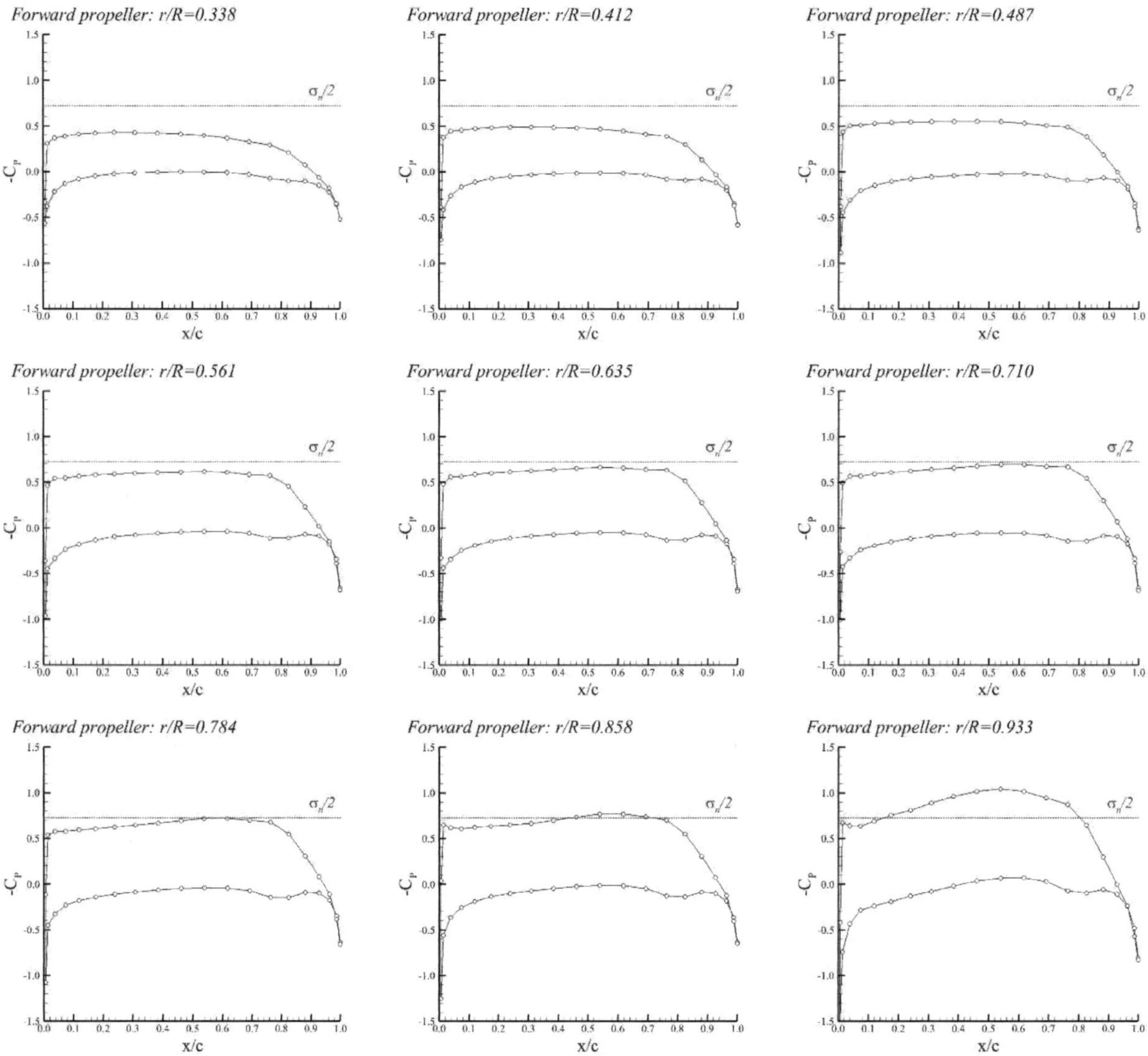

Figure 20: Pressure distributions of the designed forward propeller from the 1st strip around the hub to the 9th strip around the tip (r/R=0.338, 0.412, 0.487, 0.561, 0.635, 0.710, 0.784, 0.858, 0.933)

Table 1: Comparison of the total final solutions from original propellers and designed propellers

Propeller	K_{TT}	$10K_{QT}$	η_{PT}
Original	0.3222	0.6354	0.6696
Designed	0.3105	0.5928	0.7323

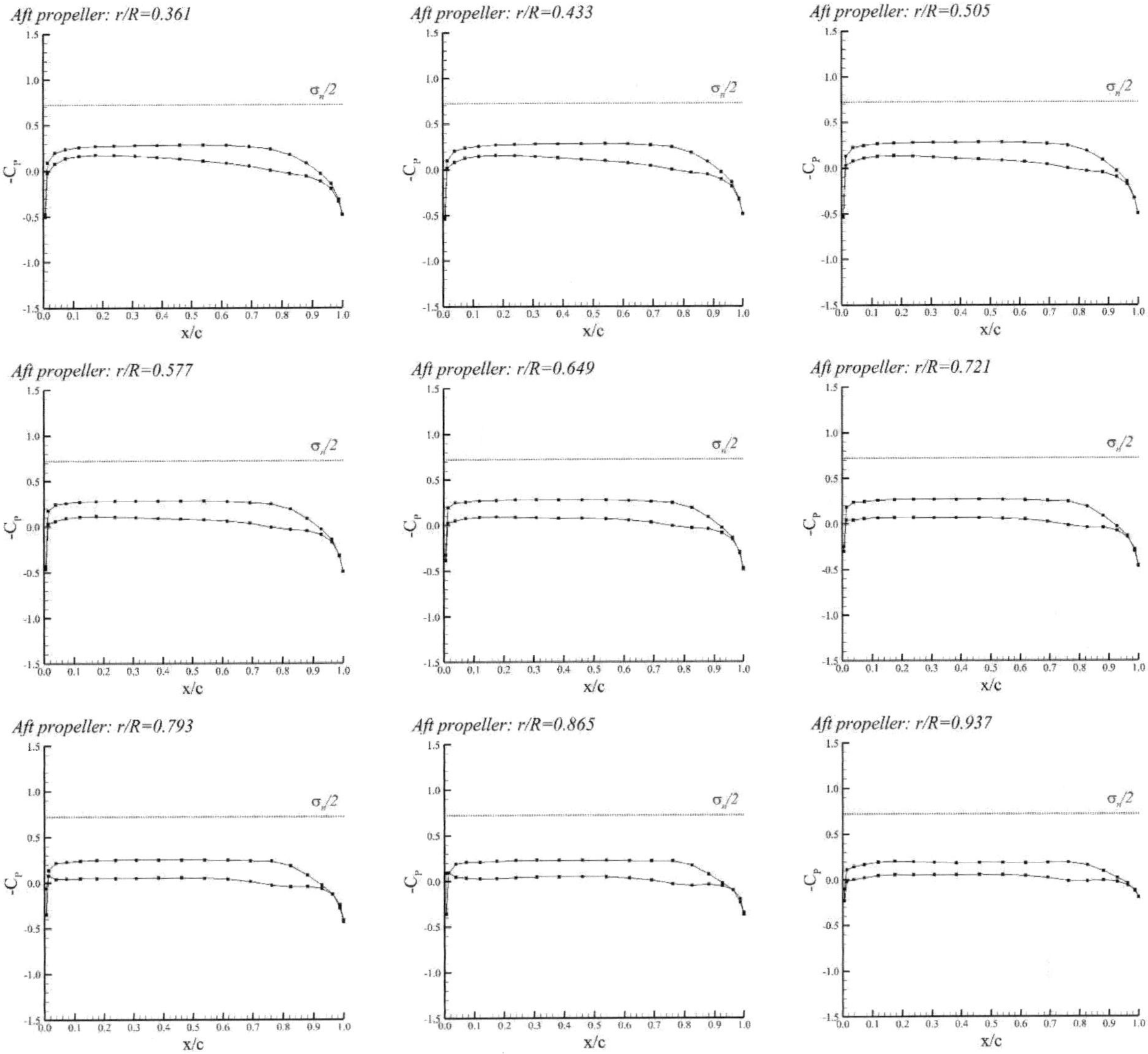

Figure 21: Pressure distributions of the designed aft propeller from the 1st strip around the hub to the 9th strip around the tip (r/R**=0.361, 0.433, 0.505, 0.577, 0.649, 0.721, 0.793, 0.865, 0.937**)

Conclusions And Future Work

A comprehensive design method is presented. The nonlinear optimization method is applied to design counter-rotating propellers with the maximized propeller efficiency for a given axisymmetric pod with specified design conditions and constraints. The VLM/RANS interactive technique is then used to analyze the propeller/pod interaction. The analysis should be included because the propeller/pod interaction significantly changes the propeller thrust and provides the effective wake which could be applied as a new inflow to new design of propeller. Also the swirl velocity for the aft propeller would be included for the CRP design. The procedure of the design and interaction analysis continues until the propeller thrust converges. The final solutions of the comprehensive design method are the optimal propeller geometry, thrust and torque coefficients, and propeller efficiency.

The results of this study indicate that the present algorithm is a useful tool for propeller design. This method is particularly convenient to design propellers for desired conditions by imposing constraints, e.g. the minimum pressure constraint and pressure difference constraint, with less expensive computations. The propeller RPM and cavitation number should be properly considered for a design of non-cavitating propeller. In addition, the designed propellers present more efficiency for the total required power compared to the original ones. The method could be extended to include the effect of the strut for podded CRP as well as of the hull, so that the propellers are designed for best performance (least required power) to move the hull at a desired speed.

Acknowledgements

Support for this research was provided by the U.S. Office of Naval Research (Grant Numbers N00014-18-1-2276 and N00014-21-1-2488; Dr. Ki-Han Kim) and by Phases VIII and IX of the "Consortium on Cavitation Performance of High Speed Propulsors".

References

Du, W. and Kinnas, S.A. "Coupling a numerical optimization technique with a panel method or a vortex lattice method to design cavitating propellers in nonuniform inflows." Proceedings of the 10th International Symposium on Cavitation, 2018, Baltimore, Maryland, USA.

Du, W., Kim, S. and Kinnas, S.A. "MPUF-3A Version R2018, Report No. 18-1." User Manual and Documentation, UT Austin, 2018.

Greeley, D.S. and Kerwin, J.E. (1982) Numerical methods for propeller design and analysis in steady flow. SNAME Transactions, 90, 416–453.

Griffin, P.E. and Kinnas, S.A. "A design method for high-speed propulsor blades." Journal of Fluids Engineering, 120(3):556–562, 1998.

Kerwin, J.E. "A Rapid Design/Analysis Method for Open, Ducted or Waterjet Propulsors with Single or Multiple Blade Rows" 31st Symposium on Naval Hydrodynamics, Monterey, CA, USA, September 11–16, 2016.

Kerwin, J.E. and Lee, C.-S. "Prediction of steady and unsteady marine propeller performance by numerical lifting surface theory." SNAME Transactions, 86, 1978.

Kim, S. "Numerical prediction of underwater noise induced by non-cavitating/cavitating propellers" Ph.D. thesis, Ocean Engineering Group, UT Austin, 2021.

Kinnas, S.A. "A Numerical Method for the Analysis of Cavitating Propellers in Nonuniform Inflow." MIT-PUF-3 Program Documentation, MIT, 1983.

Kinnas, S.A., Cha, K. and Kim, S. "Comprehensive Design Method for Open or Ducted Propellers for Underwater Vehicles", SNAME Maritime Convention 2021, 27-29 October, Providence RI

Mishima, S. "Design of Cavitating Propeller Blades in Non-Uniform Flow by Numerical Optimization." Ph.D. thesis, MIT, 1996.

Mishima, S. and Kinnas, S.A. "Application of a Numerical Optimization Technique to the Design of Cavitating Propellers in Nonuniform Flow." Journal of Ship Research, 41:93-107, 1997.

DESIGN AND ANALYSIS OF TIDAL TURBINES VIA A NONLINEAR OPTIMIZATION TECHNIQUE COUPLED WITH A BEM

SEUNGNAM KIM
OCEAN ENGINEERING GROUP, THE UNIVERSITY OF TEXAS AT AUSTIN

THOMAS S. WU
OCEAN ENGINEERING GROUP, THE UNIVERSITY OF TEXAS AT AUSTIN

SPYROS A. KINNAS
OCEAN ENGINEERING GROUP, THE UNIVERSITY OF TEXAS AT AUSTIN

Proceedings of the 27th Offshore Symposium, February 22nd, 2022, Houston, Texas
Texas Section of the Society of Naval Architects and Marine Engineers

ABSTRACT

In this paper, a nonlinear optimization method is coupled with a boundary element method (BEM) to design optimum ocean current turbines, and some preliminary results are presented. The optimum turbine geometry is designed to produce the highest power output under several initial constraints. The coordinates of the blade are defined using a B-spline geometry with 4×4 control points being the parameters to be optimized for a given thrust/inflow in the absence of any cavitation. The design process is repeated in an iterative manner by a fully automated interface, which interacts between the hydrodynamic BEM and nonlinear optimization method until the blade geometry becomes fully stabilized, producing maximum power. The present method is first applied to ocean current turbines in open water condition to investigate the influence of design constraints and wake alignment models on the optimal efficiency of the designed blades. full-blown RANS simulations are conducted separately to validate the predicted analysis results using the optimal blade geometry. Comparisons show satisfactory agreement between the results from the present method and those from viscous simulations, and importantly a significant impact of the wake alignment model on the turbine blade design.

Keywords: Tidal turbine, Turbine design, Numerical optimization method, Boundary element method, Lifting-line optimization method

INTRODUCTION

In recent years, the interest in ocean current energy harvesting has been growing in accordance with the needs of clean and renewable energy. Among many different types of tidal current generators, axial tidal turbines have been of great interest and implemented in many different places (Charlier, 2003, Grabbe et al., 2009) because of their similarity to wind turbines and higher density of water than air. The high density of water allows tidal turbines to be manufactured with a much smaller size than wind turbines, and the tip rotational speed of tidal turbines is usually too slow to be lethal to marine lives dwelling near a tidal turbine farm in the ocean. In order to assess and utilize the tidal energy via optimization, the accurate prediction of tidal turbine performance on- and off-design condition in a single- or multi-turbine configuration is crucial.

There have been many different approaches for the analysis of tidal turbine performance. Batten et al. (2007) and Allsop et al. (2017) used the blade element momentum theory (BEMT) to design marine current turbines. Malki et al. (2013), Belloni et al. (2017), and Turnock et al. (2011) coupled BEMT with a Reynolds-averaged Navier-stokes (RANS) solver to analyze the power performance from an array of current turbines. Kim et al. (2021) used a similar approach where BEMT is replaced by a boundary element method (BEM), results of which given as body forces are interpolated into the RANS domain to solve for the total flow around an axial tidal turbine farm. They also applied BEM in a single turbine problem with/without yaw in non-cavitating/cavitating conditions. Xu & Kinnas (2010) used a vortex-lattice method (VLM) to predict the performance of a horizontal axis marine current turbine. A similar approach can be found in Baltazar & Falcão de Campos (2009) and Salvatore et al. (2018). Otto et al. (2012) studied the flow over a marine current turbine using a viscous flow RANS approach over a broad range of inflow conditions.

The literature review commonly reveals that the flow field around tidal turbines involves viscous effects, such as flow separation, which becomes more significant at lower tip-speed ratios (TSRs) affecting the overall turbine performance. Viscous flow solvers should be the best option in such cases over potential flow solvers to analyze the turbine performance more accurately. In optimization problems, on the other hand, the analysis part is called a number of times to search for the optimum blade based on the intermediate turbine performance; therefore, it *must be* computationally efficient. In such cases, potential flow solvers with numerical corrections on the viscous effects could be preferable alternatives to viscous flow solvers since the former has been proven to be satisfactory in capturing boundary layers on the blade in a time-efficient way.

In marine propeller problems, one of the preliminary design methods is based on the optimal radial distribution of circulation using a lifting-line method, as described in Kerwin & Hadler (2010). This approach does not require knowledge of the propeller series. The method is applied to turbine/propeller design problems, and the output geometry is used for the analysis using a vortex-lattice method (VLM) or BEM (Kinnas et al., 2012, Arán et al., 2019). As an improved alternative approach to the design method, Mishima & Kinnas (1996,1997) developed a numerical optimization technique coupled with VLM to design non-cavitating/cavitating propellers in non-axisymmetric inflow. This method utilizes a 4×4 B-spline polygon net to represent the blade geometry instead of control points used in VLM. The optimum objective and the constraints in terms of the propeller forces and cavity quantities are approximated via first- or second-order Taylor expansions. The main advantage of this method is to allow a constrained amount of propeller forces or cavity volume in the design process considering either axisymmetric or non-axisymmetric inflow. The method is later further developed by Griffin & Kinnas (1998) such that the minimum pressure constraint and skew optimization process are incorporated in the case of propeller problems. Very recently, Kinnas et al. (2021) coupled the nonlinear optimization method with a viscous flow solver to determine the most efficient propeller blade shapes for a given axisymmetric hull considering their viscous interaction.

Beyond the propeller problem, the objective of this paper is to couple the numerical optimization technique with BEM to design ocean current turbines for their optimum blade geometries such that they produce the highest power performance under specified design constraints. This paper briefly introduces the fundamentals behind the lifting line optimization method (D. H. M. Arán & Kinnas, 2014), BEM (or panel method) (Kim et al., 2018, 2021, Kim & Kinnas 2021), and numerical optimization method modified to turbine problems. The lifting line method is first implemented to obtain the optimal turbine blade, which is then analyzed by the hydrodynamic BEM. Results from both approaches are compared to see the optimum circulation can be reproduced by the analysis method. The hydrodynamic performance is then validated against the results from viscous RANS simulations. In this part, the importance of considering the viscous effect in BEM calculations will be discussed. The numerical optimization method will be implemented next with the optimum turbine geometry as the base (initial) geometry to further increase the turbine power performance under a specified thrust constraint. Ideally, the turbine design has to be free of any thrust constraint and released to achieve the maximum power output. In the current level of development, however, the thrust constraint is present such that the optimal performance is achieved producing a prescribed thrust. BEM is selected in this paper for the evaluation of hydrodynamic quantities because of its fast and accurate predictions. Different wake alignment models, initially devised for propellers, are tested to understand how they affect the turbine design. This paper will present some preliminary results out of the numerical optimization technique applied to tidal turbine design.

Methodology

Lifting line optimization method

In the lifting line optimization method, the turbine can be represented by a lifting line model, wherein each blade is represented by a radially concentrated bound vortex strength $\Gamma(r)$. The pitch of horseshoe vortices was originally aligned with the flow only at the lifting line, and the free vortex sheet behind the turbine is in a helical shape, known as the Lerbs-Wrench formula. Recently, the full wake alignment of the trailing wake has been incorporated in the lifting line model by Menéndez Arán & Kinnas (2014) to represent the wake geometry based on the local flow downstream, improving the accuracy of the predicted propeller efficiency. This paper adopts the latter for best results since it is reported that the circulation distributions obtained through the Lerbs-Wrench formulas produce significantly lower power performance than the Betz limit (Arán et al., 2019). The wake induced velocities in tangential $\left(u_t^*\right)$ and axial $\left(u_a^*\right)$ are computed numerically based on the Biot-Savart law and relative locations of the discretized blade and vortex segments in the wake. Figure 1 shows the combined velocity and force diagram for a turbine blade of length R at a section radius r.

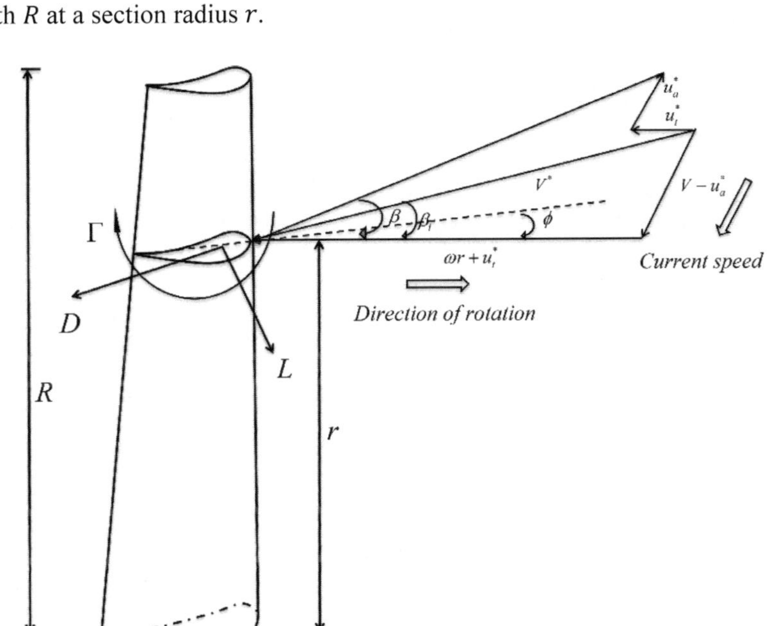

Figure 1 Schematics showing combined velocity and force diagram on a blade section at radius r, taken from Kinnas et al. (2012)

Once the induced velocity on the lifting line is obtained, the total torque acting on the lifting line can be expressed as a function of known parameters, i.e., the induced velocities and a sectional drag-to-lift ratio $\kappa(r) = C_D(r)/C_L(r)$, and the circulation distribution (Kerwin & Hadler, 2010).

$$Q = \rho Z\left[\int_{r_h}^{R}\left(V - u_a^*\right)\Gamma r dr - \int_{r_h}^{R} \kappa(r)\left(\omega r + u_t^*\right)\Gamma r dr\right], \tag{1}$$

where V is the current velocity; ωr the rotational speed; ρ and Z are the fluid density and the number of turbine blades, respectively. In the inviscid formulation, $\kappa(r)$ is taken to be zero, and thus the second term in the integral vanishes. In order to maximize Q, we take a derivative on Q with respect to Γ to find its extreme value.

$$\frac{\partial Q}{\partial \Gamma} = 0. \quad (2)$$

It gives a system of differential equations in a linearized form (with higher order terms disregarded in the present method) and can be solved for the optimal circulation over the lifting line in a discretized form. The optimal circulation, together with the induced velocities, completes Eqn. (1), determining the optimal torque. The efficiency of the turbine $\left(C_{POW}\right)$ is then obtained following Eqn. (3) where the turbine power coefficient C_{POW} has the same definition as its efficiency η. More details on the lifting line method can be found in Kerwin & Hadler (2010).

$$C_{POW} = \eta = \frac{Q\omega}{\frac{1}{2}\rho\pi V^3 R^2}. \quad (3)$$

Given the optimal circulation and induced velocities, the optimum blade geometry can be generated based on Prandtl's lifting line equation. In the method, the line vortex on the key blade is represented by horseshoe vortices. Then, the total inflow velocity $V^*(r)$ can be decomposed as:

$$V^*(r) = \sqrt{\left[V - u_a^*(r)\right]^2 + \left[\omega r - u_t^*(r)\right]^2}, \quad (4)$$

Thus, the modified wake pitch angle β_i can be written as:

$$\tan tan\, \beta_i(r) = \frac{V - u_a^*(r)}{\omega r - u_t^*(r)}. \quad (5)$$

Once the optimum circulation is given, pitch and camber distributions on the blade can be determined by assuming a chord distribution and the blade section profile (NACA $a = 0.8$ in this paper). The lift coefficient $C_L(r)$ then can be determined from the predicted circulation $\Gamma(r)$ (the Kutta-Joukowski law), total velocity $V^*(r)$, and sectional chord length $c(r)$ as follows:

$$C_L(r) = \frac{L(r)}{0.5\rho\left[V^*(r)\right]^2 c(r)} = \frac{2\Gamma(r)}{V^*(r)c(r)}. \quad (6)$$

Eqns. (5) and (6) determine the sectional pitch angle $\theta(r)$ and maximum camber $f_{max}(r)$, which are in a function of the corresponding lift coefficient (Eqn. (6)):

$$\theta(r) = \beta_i(r) - \alpha_0 C_L(r), \quad (7)$$

$$f_{max}(r) = f_0 C_L(r), \quad (8)$$

where α_0 and f_0 are the ideal angle of attack and ideal maximum camber for a certain section when $C_L = 1.0$, given as 1.54 and 6.79×10^{-2}, respectively in this paper. The modeled geometry can then be analyzed using BEM, which provides the pressures on the blade and radial circulation on the blade. Comparison of this with the original optimal circulation out of the lifting-line method will be presented as a validation of the design and analysis methods.

A nonlinear optimization technique

Unlike the traditional blade definition, where radial distributions of the propeller profile are given, the present nonlinear optimization method uses B-splines defining all points on the blade surface uniquely. One of the advantages of this method is that the blade can be defined with fewer numbers of parameters using B-splines. This is a crucial aspect in the design problem since the number of design parameters directly affects the computing efficiency of the optimization method. The hydrodynamic quantities, such as the turbine forces, are evaluated and updated multiple times as the design process continues. Using cubic B-spline surfaces, the turbine blade is defined as follows:

$$\vec{P}(u,w) = [x(u,w), y(u,w), z(u,w)] = \sum_{i=0}^{N_u-1} \sum_{j=0}^{N_w-1} \vec{d}_{ij} N_{i,4}(u) N_{j,4}(w), \tag{9}$$

where $\vec{d}_{ij}$ are the B-spline control points; $N_{i,4}(u)$ and $N_{j,4}(w)$ the B-spline bases of order 4; u and v the parameters for B-splines; and N_u and N_w are the numbers of B-spline control points in u- and w-directions, respectively. A cubic B-spline surface for a turbine blade by a 4×4 vertex polygon net is shown in Figure 2[1].

[1] Note that, in this figure, hub and tip vertices are moved after adjustments such that they are on the hub and tip surfaces, respectively.

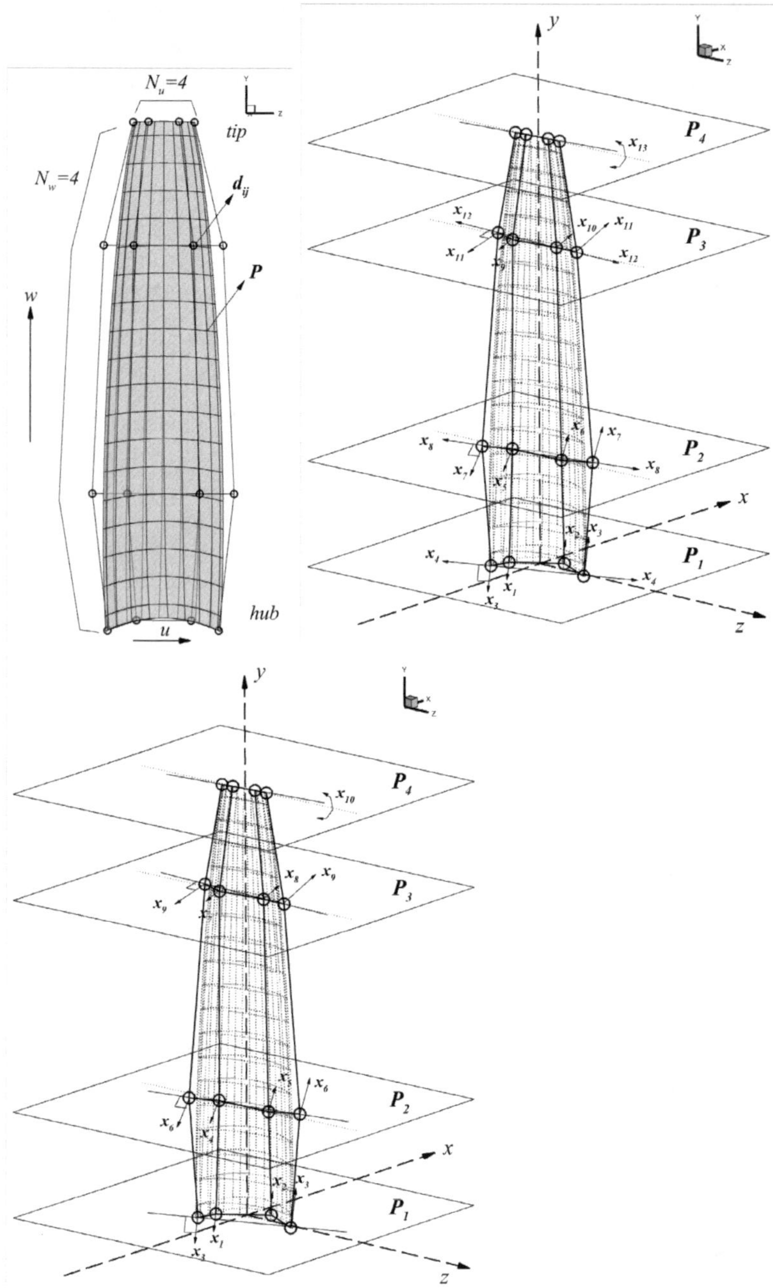

Figure 2 Blade geometry from BEM using cubic B-splines with 4×4 polygon net (left); number of variables (vertices) and their movements when the chord length is determined by the method (middle) or given as a constraint (right)

In the turbine problem, the general nonlinear constrained optimization problem is defined as follows:

$$maximize \ \ f(\vec{x}) \tag{10}$$

$$subject\ to \ \ h_i(\vec{x}) = 0 \quad i = 1, 2, \ldots, l \tag{11}$$

where $f(\vec{x})$ is the *objective function*; $\vec{x}$ the solution vector of $n = 13$ or 10 components in case the chord is determined by the method or given, respectively; and $h_i(\vec{x})$ is the equality constraint. The converged vector $\vec{x} = [x_1, x_2, \dots, x_n]^T$ should satisfy all the constraints to be the solution for a given problem. In the present method, the design objective is to maximize the torque for a given thrust such that the turbine can produce the highest efficiency; therefore, the equality constraint will be set on the thrust coefficient. There are no inequality constraints, such as maximum allowable cavity area, cavity volume velocity, or maximum allowable cavity length in this paper since the turbine is subject to fully-wetted flow.

A flow chart of the present algorithm is shown in Figure 3. The design algorithm starts with an initial turbine geometry represented by a design variable vector $\vec{x}_0$ of n components. The first stage is to obtain linear approximations of the objective and constraint functions. BEM is implemented for the $1 + n$ geometries $\vec{x}_j$ $(j = 0, 1, \dots, n)$ obtained by slightly perturbing the initial geometry $\vec{x}_0$. This is to evaluate the coefficients of the linear functions in terms of the design variables for the next linear optimization stage.

$$f(\vec{x}) = F_f + \vec{x}^T G_f h_1(\vec{x}) = F_{h_1} + \vec{x}^T G_{h_1} h_2(\vec{x}) = F_{h_2} + \vec{x}^T G_{h_2}$$
$$\vdots h_l(\vec{x}) = F_{h_l} + \vec{x}^T G_{h_l} \tag{12}$$

where $\vec{x}$ is the design variable vector; F_f, G_f, F_{h_1}, G_{h_1}, ..., F_{h_l} and G_{h_l} are the coefficients of the linear functions corresponding to the first-order Taylor series expansions around the initial guess $\vec{x}_0$ in case those are taken as the function values at $\vec{x}_0$. The maximization problem (Eqn. (10)) is then solved numerically for the linear objective and constraint functions to find an improved geometry $\vec{x}_k$. The method of multipliers is used to solve the constrained maximized problem. BEM is run for any change in $\vec{x}_k$ to produce/update the corresponding objective and constrained functions. This process is repeated $n(n + 1)/2$ times completing the linear optimization stage.

In the following quadratic optimization process, quadratic approximations of $f(\vec{x})$ and $h_i(\vec{x})$ are defined as:

$$f(\vec{x}) = F_f + \vec{x}^T G_f + \frac{1}{2}\vec{x}^T H_f \vec{x} h_1(\vec{x}) = F_{h_1} + \vec{x}^T G_{h_1} + \frac{1}{2}\vec{x}^T H_{h_1} \vec{x} h_2(\vec{x}) = F_{h_2} + \vec{x}^T G_{h_2} + \frac{1}{2}\vec{x}^T H_{h_2} \vec{x}$$
$$\vdots h_l(\vec{x}) = F_{h_l} + \vec{x}^T G_{h_l} + \frac{1}{2}\vec{x}^T H_{h_l} \vec{x} \tag{13}$$

where F_f, G_f, F_{h_1}, G_{h_1}, ..., F_{h_l} and G_{h_l} are the coefficients of the quadratic functions in terms of the design variables. H_f, H_{h_1}, H_{h_2}, ..., H_{h_l} are the approximations of the Hessian matrices near the optimum solution. The maximization problem is then solved for the quadratic functions, and the new solution vector $\vec{x}_k$ is obtained and saved at each iteration for later use. The iteration is repeated until the change in $\vec{x}$ falls below a prescribed tolerance ϵ. More details on the method in the case of hydrofoil or propeller problems can be found in Mishima & Kinnas (1996) and Mishima & Kinnas (1997).

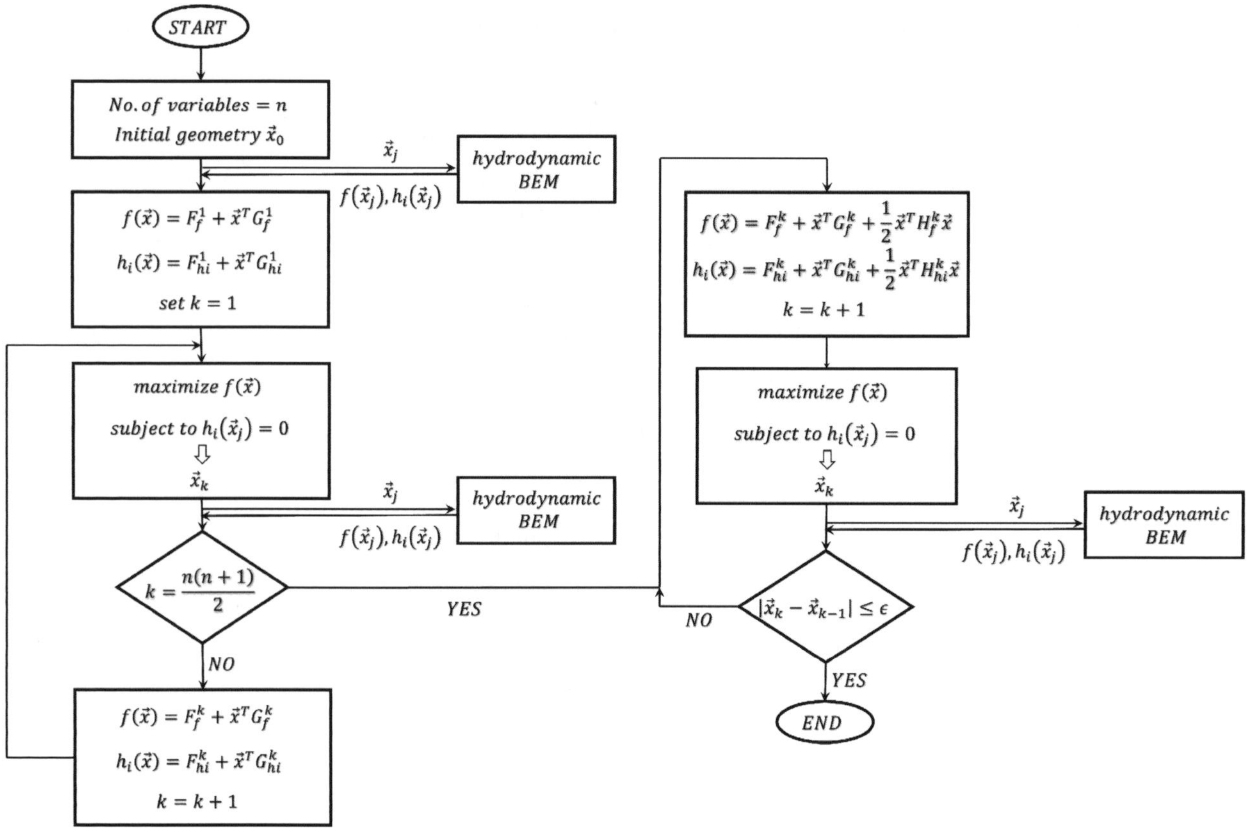

Figure 3 Flow chart of the optimization method coupled with the hydrodynamic BEM.

A boundary element method: evaluation of hydrodynamic quantities

As shown in Figure 3, the analysis of the flow around a turbine is required at each iteration. Since this analysis is called multiple times during the optimization process to update the solution vector $\vec{x}$, the evaluation of hydrodynamic quantities *must be* computationally efficient. BEM has been proven to be adequate for this task, and its brief description is presented in this section.

In the hydrodynamic BEM for turbine problems, the total flow field $\vec{q}$ (Figure 4) can be decomposed into the incoming flow $\vec{U}_{in}$ and perturbation velocity $\vec{u}$ due to the presence of the turbine in inviscid and irrotational flow:

$$\vec{q} = \vec{U}_{in} - \vec{\omega} \times \vec{r} + \vec{u}. \tag{14}$$

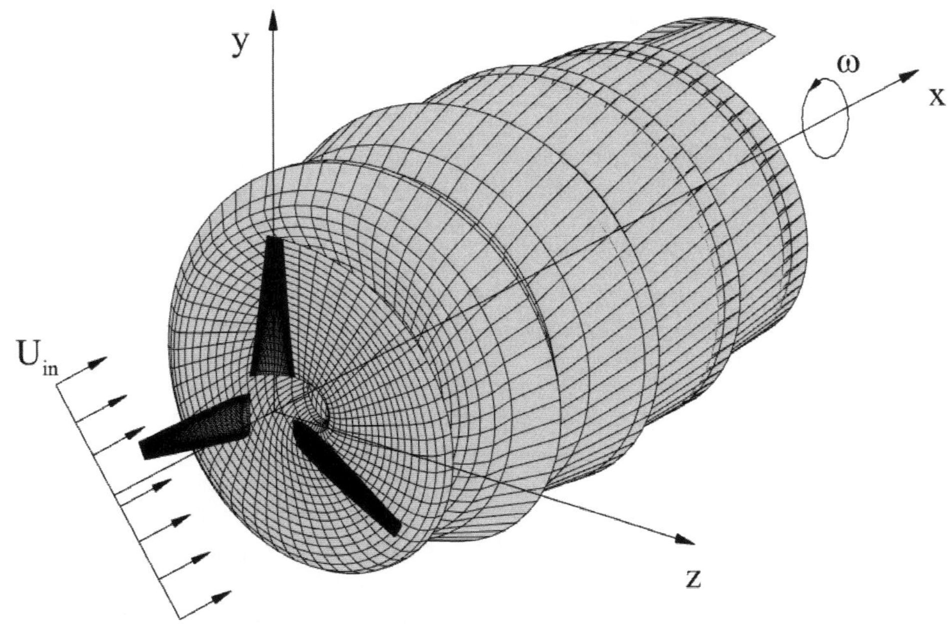

Figure 4 Turbine blades and the key wake subject to general inflow; the x, y, and z axes represent the turbine fixed coordinate system.

The perturbation velocity $\vec{u}$ can be found by taking a gradient on the velocity potential ϕ, which satisfies the Laplace equation:

$$\vec{u} = \nabla\phi, \tag{15}$$

$$\nabla^2\phi = 0. \tag{16}$$

The perturbation potential on the turbine surface S_T can be expressed as a superposition of potentials induced by a continuous source and dipole on S_T, as well as a continuous dipole distribution on the trailing wake S_W. Based on Green's third identity in potential flow, the velocity potential ϕ_p at point p on the turbine surface should satisfy:

$$2\pi\phi_p = \iint_{S_T}\left[\phi_q \frac{\partial G(p;q)}{\partial n_q} - G(p;q)\frac{\partial \phi_q}{\partial n_q}\right]dS + \iint_{S_W} \Delta\phi_W \frac{\partial G(p;q)}{\partial n_q} dS, \tag{17}$$

where $G(p;q) = 1/R(p;q)$ is Green's function; $R(p;q)$ the distance between control point p and variable point q; n_q the unit normal vector pointing into the fluid; and $\Delta\phi_W$ is the wake strength (i.e., the potential jump across the wake surface or the trailing edge of the turbine blade). The kinematic boundary condition provides the unknown source strength on the turbine surface. Once the unique solutions of the velocity potential are obtained, the turbine-induced velocity (Eqn. (15)) can be evaluated anywhere in the flow field by taking a gradient on Eqn. (17). More details on the panel method or viscous RANS applied to turbine problems can be found in Kim et al. (2021), Xu & Kinnas (2010), Baltazar & Falcão de Campos (2009), Salvatore et al. (2018), and Otto et al. (2012).

RESULTS

Hydrodynamic analysis of turbine using BEM in comparison with RANS

In the first section, numerical predictions from the lifting line method are first compared with results from BEM for validations of each method. The model turbine is obtained from the lifting line optimization method with the number of blades $Z = 3$ at $TSR = 6$. Tests are made under the uniform inflow without any yaw. For a given TSR, the power coefficient $\left(C_{POW}\right)$ and thrust coefficient $\left(C_T\right)$ are evaluated based on the predicted thrust (T) and torque (Q) following the equations:

$$Tip\ Speed\ Ratio\ (TSR) = \frac{\omega R}{V_s} = \frac{\pi n D}{V_s} = \frac{\pi}{J_s}, \tag{18}$$

$$Advance\ ratio\ \left(J_s\right) = \frac{V_s}{nD}, \tag{19}$$

$$Thrust\ Coefficient\ \left(C_T\right) = \frac{T}{\frac{1}{2}\rho V_s^2 A}, \tag{20}$$

$$Power\ Coefficient\ \left(C_{POW} = \eta\right) = \frac{Q\omega}{\frac{1}{2}\rho V_s^3 A}, \tag{21}$$

where $A = \pi R^2\ \left[m^2\right]$ and ω is the turbine angular speed $[1/s]$; n the turbine rotational frequency $[1/s]$; $D = 2R$ the turbine diameter $[m]$; and V_s is the inflow speed far upstream $[m/s]$. Figure 5 shows the model turbine represented in a lifting-line or 3D turbine geometry based on Prandtl's lifting line equation from both methods. In BEM, the geometry is discretized using 80 chordwise and 20 spanwise panels using a cosine and a constant spacing, respectively. Although both methods equally adopt the full wake alignment model for their trailing wake, wake panels show some differences with the distance from the turbine mainly because one simplifies the blade into a concentrated chord while the other fully considers 3D variations in the blade geometry.

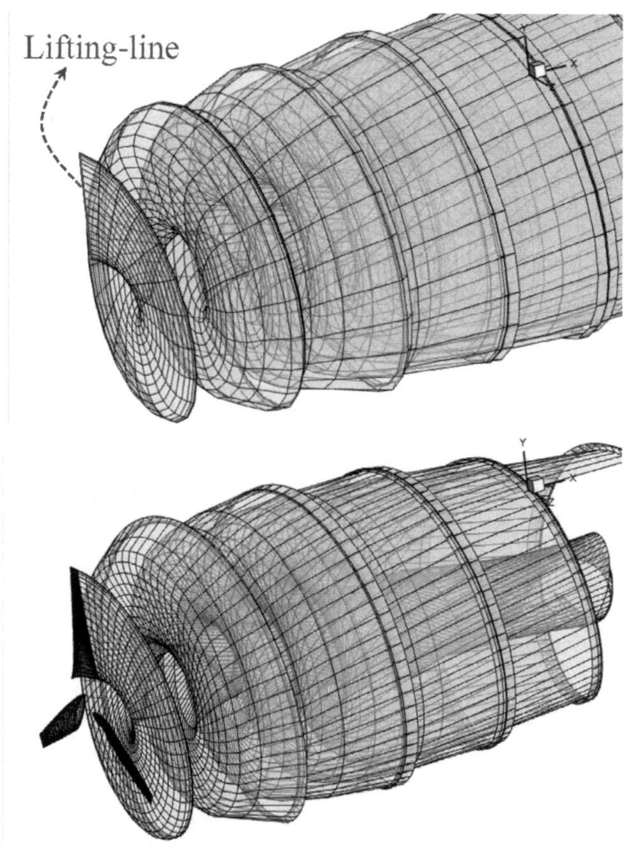

Figure 5 Turbine geometry represented by (a) a lifting-line model and (b) BEM panels and their fully-aligned trailing wake; both approaches use the full wake alignment.

Figure 6 shows the comparison of circulations from the lifting line optimization method and BEM with the original and decreased thickness. Tests are also made with/without the hub model in BEM to see its impact on the blade loading. The circulation in the figure is defined following the equation where Γ is the predicted circulation.

$$G_s = \frac{\Gamma}{2\pi R V_s}, \tag{22}$$

First, the hub has an appreciable impact on the turbine loading close to the blade root, and its impact seems to be quite minor away from the root region $(r/R > 0.4)$. In case the hub is removed from the BEM model, the circulation near the blade tip and root seems quite close to that from the lifting-line optimization method. Differences are captured around the midspan, where the blade thickness mainly causes the difference. BEM further decreases its thickness from the original $(t = 1.0)$ to its half thickness $(t = 0.5)$ as shown in Figure 6(right). In this scenario, the BEM prediction gets much closer to that from the lifting-line method even in the midspan area, which is a good sign that both methods produce a close answer under proper assumptions.

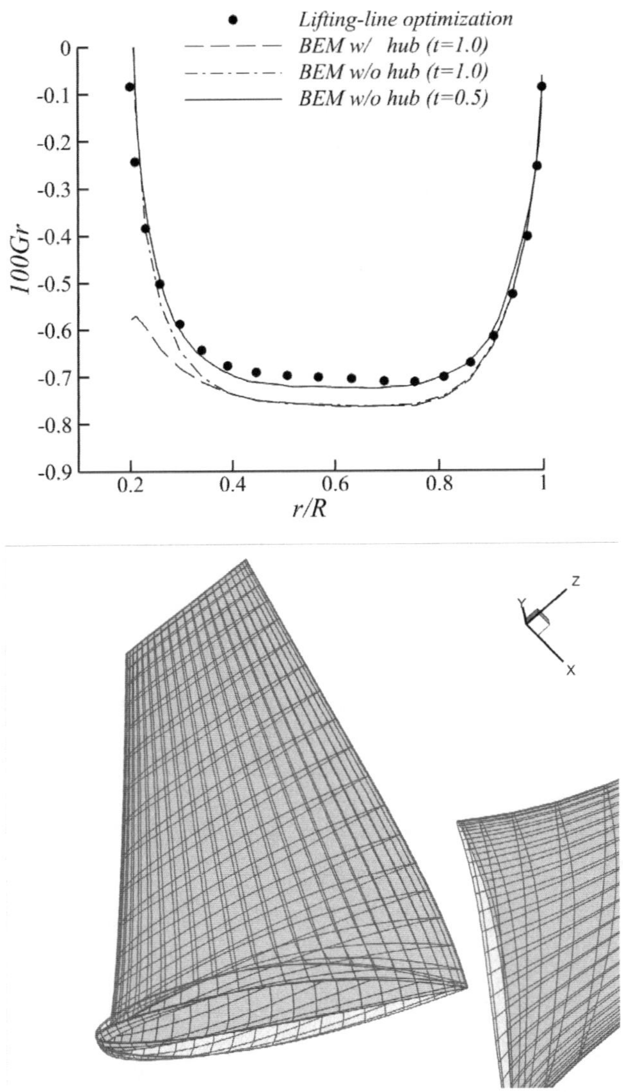

Figure 6 Comparison of circulations from the lifting-line optimization method and BEM with the (red) original and (blue) decreased thickness on the blade.

This blade is further tested using viscous RANS simulations to see how the viscous effects influence the blade pressures, as pointed out in the previous research. Figure 7 shows the computational domain used in RANS[2] with boundary conditions applied on the surrounding boundaries. Given the periodicity of the problem under uniform inflow, the periodic boundary condition is utilized to model only the key-blade and computational domain with approximately 8 million unstructured polyhedral cells. The $k - \omega$ SST turbulent model is adopted with the Reynolds number of $R_e = \frac{V_s D}{\nu} = 1.0\times10^6$. The SIMPLEC and QUICK schemes are used for the pressure-velocity coupling and to solve for the unknown quantities, respectively. The computation is terminated when the convergence residuals for continuity and momentum equations fall below 1.0×10^{-6}. It took about 2.5 hours for this model to reach the converged blade forces on 4 Intel Xeon Platinum 8160 cores. The size of the computational domain is given wide enough to successfully isolate the turbine blade from the tunnel effect from the surrounding boundaries. It has to be pointed out that BEM, for the same geometry as discussed earlier, only took less than 4 minutes on 48 cores; therefore, it is fair to say that BEM is time efficient in the analysis of tidal turbine performance.

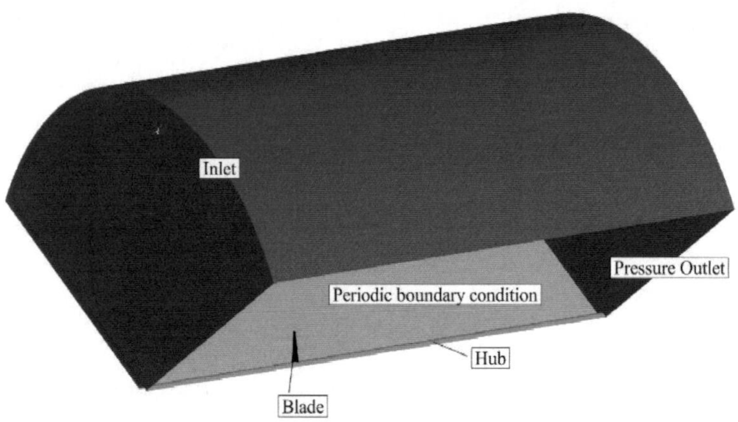

Figure 7 RANS computational domain with boundary conditions applied on its boundaries

The fully aligned wake from BEM is presented in Figure 8 at TSR=6. The wake panels are plotted on the 2D $x - y$ plane through the turbine axis, overlaid to contour plots of the vorticity magnitude from RANS simulations. The locations of the fully aligned wake are in reasonable agreement with those from RANS. It is encouraging that the present wake alignment method places the roll-ups behind the blade tip on close locations as those from RANS results. BEM seems to capture the exact locations of the tip vortices with its panels. The agreement is somewhat mitigated near the blade root mainly because RANS includes an infinite hub model in its calculation while BEM does not. The agreement also seems to be worse downstream. It is because the diffusion effect shown in the viscous simulations cannot be captured by the present inviscid panel method where the shedding vortices are considered concentrated on the wake panel edges. Given that the turbine performance is mainly affected by the panels close to the blade, the good agreement in the near field wake should be sufficient to predict the turbine performance by BEM.

[2] In all RANS calculation, ANSYS/Fluent (ver. 20.1) is utilized.

Figure 8 Fully aligned wake plotted on the $x - y$ plane from BEM with overlaid contour plots of the vorticity magnitude [1/s] from RANS at TSR=6.

Figure 9 shows the predicted pressure coefficients $\left(-C_p = \frac{P_0 - P}{0.5\rho n^2 D^2}\right)$ from BEM and RANS. P_0 and P in the figure represent the pressure far upstream along the turbine axis and the predicted local pressure, respectively. For the tested TSR, the comparisons show quite satisfactory agreement around the blade surface between BEM and RANS, especially toward the leading edge. Given that BEM only uses a standard cosine spacing with much fewer elements on the blade than RANS, this much of agreement between BEM and RANS seems encouraging. Some differences are present near the trailing edge because of the flow separation captured in RANS (Figure 10). To improve the deviation, it might be best that BEM is coupled with a boundary layer solver in the near future to see if that in any situation can improve the correlation.

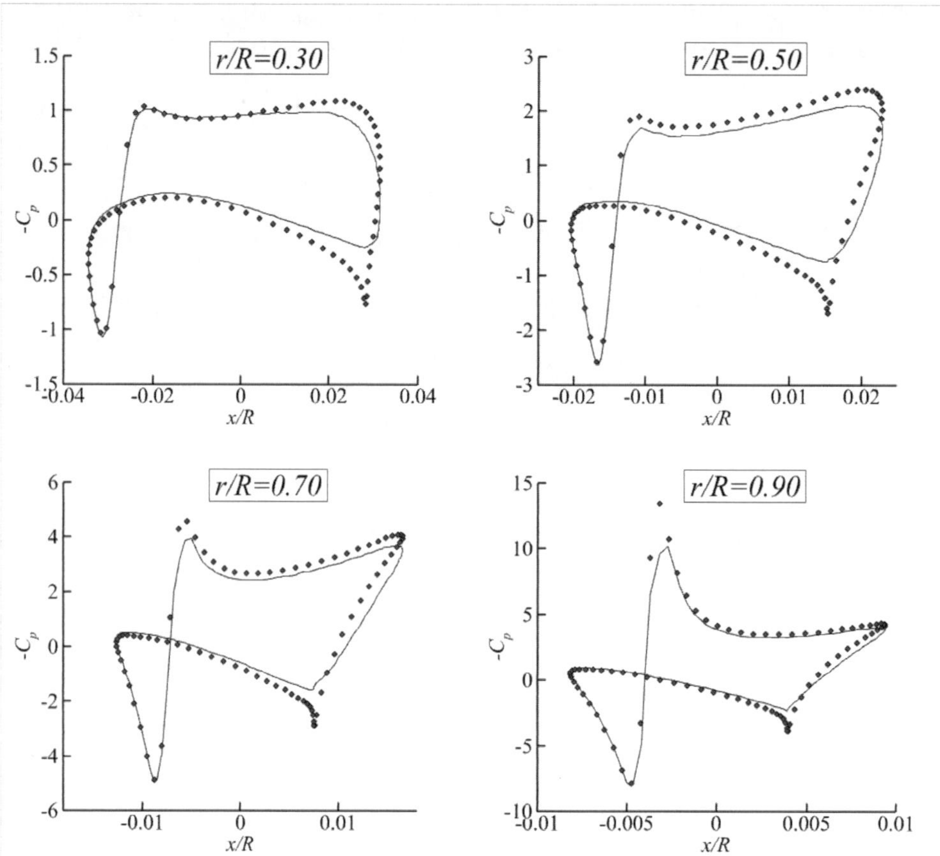

Figure 9 Comparisons of the pressure coefficient ($-C_p$) on the blade between (symbols) BEM and (solid lines) RANS at denoted radial locations; $TSR = 6$.

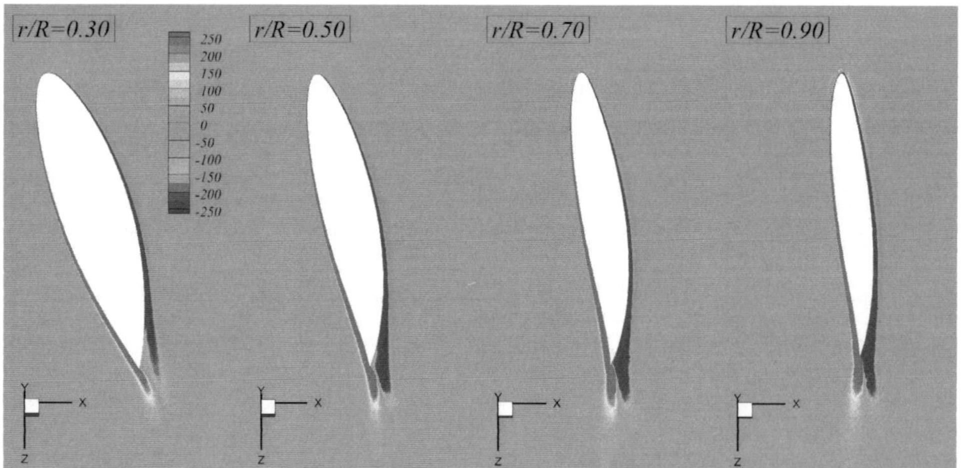

Figure 10 Vorticity magnitude [1/s], plotted on several radial stations predicted by RANS.

Numerical optimization method: statement of the problem

In this section, the design of a 3D turbine blade by the numerical optimization method is presented. The design of a turbine is performed in uniform inflow. The hub geometry is neglected in BEM when evaluating the hydrodynamic quantities. It is to avoid a panel issue on the hub during the iterative optimization process due to the very low pitch of the model blade. The goal is to determine the turbine geometry that produces the maximum torque for a specified thrust. Different from a propeller problem where the torque is usually required to be minimum due to the opposite direction of the blade rotation to torque, the design goal here for a turbine blade is to maximize torque for the best power output $\left(C_{POW}\right)$. The blade mean camber surface is defined by B-splines, as described in the methodology. The effective wake is given as *uniform inflow*, and the thrust coefficient (K_T) and TSR are specified as given design requirements. The design is performed in a fully-wetted flow ($\sigma_n = \infty$) and the gravity effect (hydrostatic effect) is neglected with infinite Froude number ($F_n = \infty$). The design requirements considered in this application are summarized as follow:

- $K_T = \frac{T}{\rho n^2 D^2} = 0.11$
- $TSR = 6$
- $\sigma_n = F_n = \infty$
- $maximize\ C_{POW}$

where T is the predicted propeller thrust from the hydrodynamic analysis. The initial K_T is taken from BEM output, wherein the optimal blade from the lifting line method is used. Thickness and chord distributions and zero skewness are given as the geometrical constraints. In other words, thickness, chord, and skewness are taken from the initial value and *not* designed internally by the method. Those profiles are also taken from the model turbine shown in the previous section.

Ideally, the turbine design has to be free of an equality condition for K_T, meaning the K_T produced by the optimum design can be any different value from its initial condition. In this paper, however, the equality constraint is still given for K_T, based on which the torque is maximized to achieve the highest efficiency. The number of blades (Z) is given 3, and the hub to blade radius ratio is fixed 0.2. Since the chord length is not allowed to change, B-spline vertices are allowed to move only in the direction normal to the chordwise segments in the $y - z$ plane, as shown in Figure 2(right). The optimized blade geometry from the present method is shown in Figure 11, together with its 4×4 B-spline vertices. In BEM, 40×40 panels are used to discretize the blade geometry in chordwise × spanwise directions, respectively.

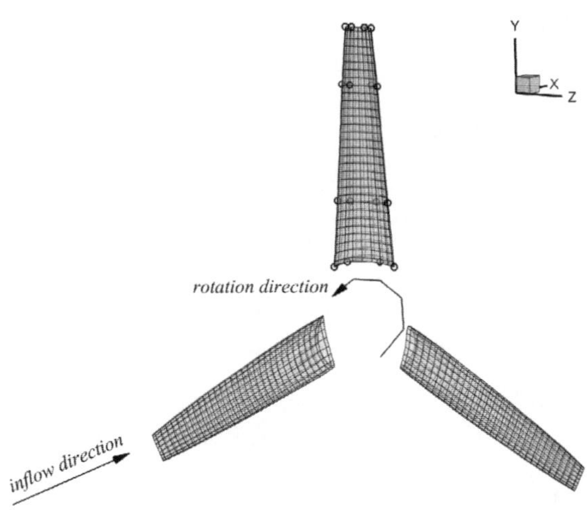

Figure 11 Optimum blade geometry modeled by BEM with 4×4 B-spline vertices

In the test, two different wake alignment models are used to see their impact on the turbine performance and design, i.e., a simplified scheme (called PSF-2 wake alignment model, proposed by Greeley & Kerwin, 1982) and the blade pitch alignment scheme. The FWA scheme cannot be fully incorporated in the current design method because of its relatively longer computing time. It is basically for a steady run and based on an iterative algorithm to produce a fully aligned wake. However, a certain numerical modification can be incorporated in the future such that FWA comes into play in the design process after completing a certain number of iterations, which is expected to expedite the overall computing efficiency even with FWA. The PSF-2 type alignment model has the advantage of fast BEM calculations and been proven to be accurate for propeller problems in a moderately loaded condition. The blade pitch alignment simply aligns the tailing wake based on the blade pitch without any consideration on the induced velocity from the trailing wake. In the following test, the FWA is only applied on the converged blade from the optimization method using the PSF-2 model. In other words, the PSF-2 model is basically used in the BEM part of the design method throughout the process, and FWA is only applied on the converged blade to see how the hydrodynamic performance is affected by different wake alignment models. Incorporating FWA in the design process is currently ongoing research, and we expect the overall performance would be different from what is shown out of the present method using PSF-2. Figure 12 shows the trailing wake aligned using different alignment models mentioned above.

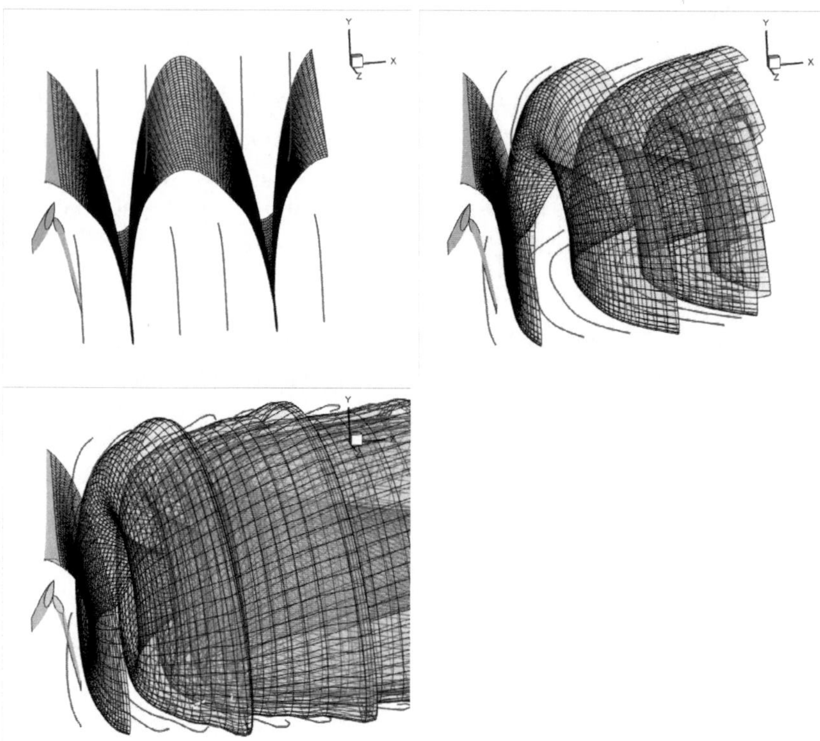

Figure 12 Optimal turbine blade and its wake aligned using different wake alignment models: (left) blade pitch alignment, (middle) PSF-2, and (right) FWA.

Figure 13(a) shows the convergence history of the turbine forces (K_T and $10K_Q$) and power coefficient (C_{POW}) predicted by the present method. The frictional drag coefficient, C_f is given 0.009 constantly over the blade surface in this case. Importantly, the efficiency looks undergoing some variations, increases, and eventually converges to a certain value of about 0.4. The K_T constraint is shown to be successfully satisfied, and the convergence history shows a quite mild and solid trend toward the final solution. Considering that the accuracy of wake approximation depends on the turbine operating conditions and specific blade geometries, simplified techniques, such as the blade pitch alignment, might not work properly at some loading conditions, as shown in Figure 13(b). The converged efficiency using the blade pitch alignment is predicted higher than the previous case using PSF-2, which looks somewhat non-physically high efficiency. The overpredicted result from the blade pitch alignment technique might be ascribed to the ignored details of the flow behind the trailing edge.

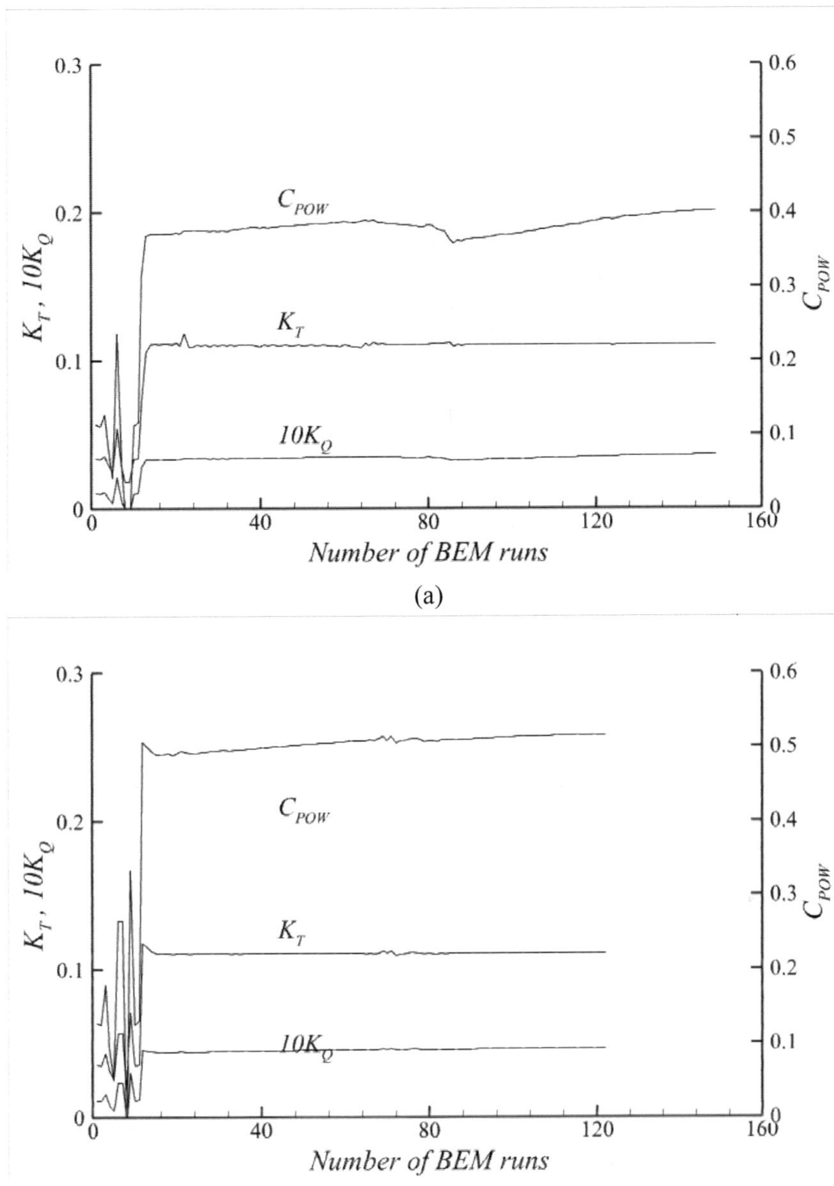

Figure 13 Convergence history of the predicted forces and efficiency out of the optimization method using (a) PSF-2 and (b) blade pitch alignment.

Figure 14 shows the predicted force performance out of the design method using two wake alignment models (i.e., PSF-2 and blade pitch alignment) and from BEM using the designed blade with FWA. Viscous RANS simulations are conducted separately to validate the results from the present method. As the wake alignment model becomes more elaborate, predictions from the present method get closer to those from RANS simulations. Notably, the efficiency from PSF-2 looks much closer to the RANS result compared to that from blade pitch alignment. If the FWA is implemented in BEM, the agreement is more improved, and the present method produces a quite close efficiency as the RANS prediction. This might be because FWA aligns the downstream wake following the local stream, allowing radial expansions of the wake panels as well as better representations of tip vortices in the wake tip region, as shown in Figure 15. For the other two models, on the other hand, the radial expansion of the trailing wake is suppressed below $r/R = 1.0$, and the roll-ups at the wake tip are shown to be neglected. From this, it is evident that the trailing wake significantly influences the turbine loading, which has been proven to be true also for the propeller problem. In conclusion, it is expected that FWA, once it is fully involved in the present design method, would produce the best power performance comparable to RANS results.

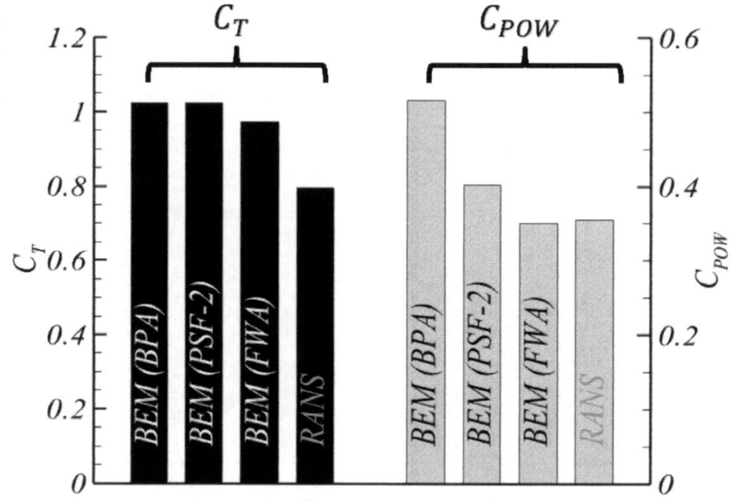

Figure 14 Predicted force performance from the design method, BEM, and RANS; different wake alignment models are tested in the present method.

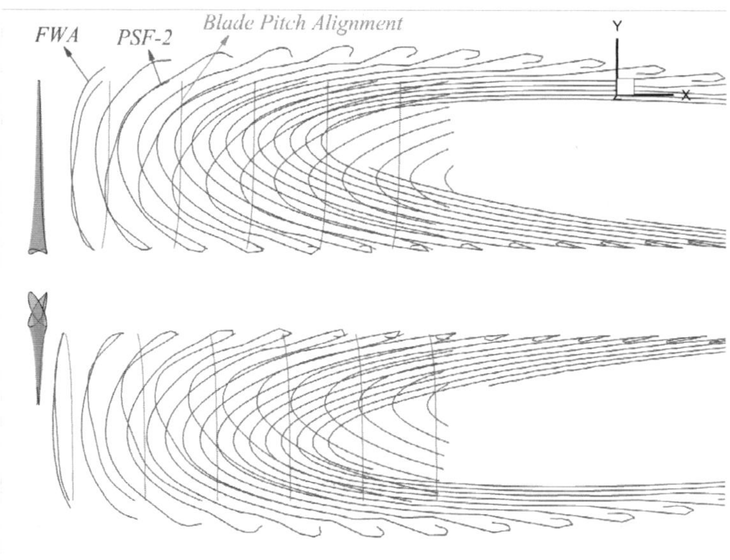

Figure 15 Wake panels aligned with different wake alignment models in BEM; they are plotted on $x - y$ plane.

Lastly, Figure 16 shows the predicted pressure distributions on the blade from the present method using the initial/designed blade using two different wake alignment models. They show somewhat close results near the root and around the midspan. The designed blade puts more pressure toward the trailing edge than the initial one. Importantly, the pressure on the designed blade looks more favorable toward the blade tip since it more suppresses the pressure peak near the leading edge; therefore, is better for cavitating performance.

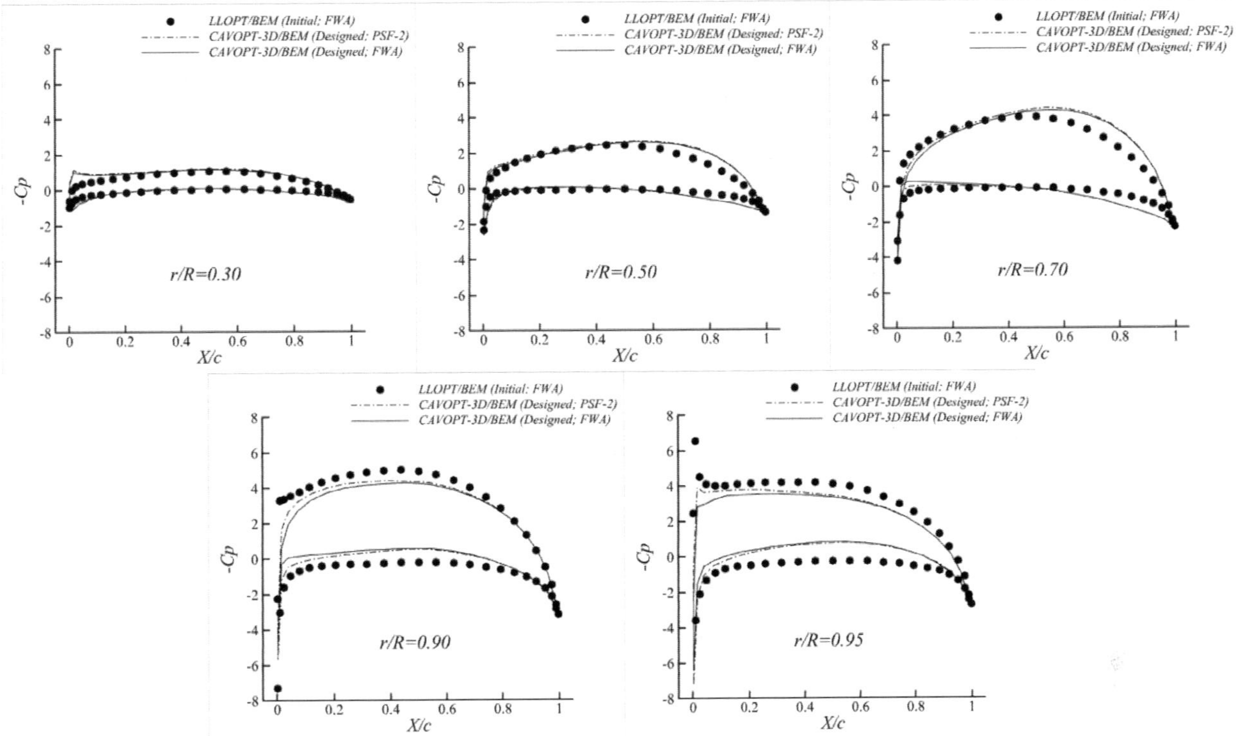

Figure 16 Pressure distributions on the blade at several radial stations (denoted at each figure) from the initial/designed blade using PSF-2 and FWA models

CONCLUSIONS

In this paper, a numerical optimization technique, which has been proven to be successful in marine propeller problems, is coupled with a boundary element method to design a ocean current turbine. Based on the initial blade geometry obtained from the lifting line method, the optimization method is implemented for a given thrust as a design constraint in order to maximize the turbine efficiency. For validation purposes, BEM is first used to solve a single turbine problem using the model geometry from the lifting line optimization method. The BEM predictions are compared with those from viscous RANS simulations, and the agreement in the downstream wake and pressure distributions on the blade is shown to be satisfactory.

In the numerical optimization part, the predicted results show that the design objectives (maximize the torque for a given thrust) and constraints (given chord, thickness, skew distributions) are successfully satisfied. The present method is shown to be able to increase and find the optimal efficiency based on its initial geometry. Considering the importance of the trailing wake on the turbine performance, different wake alignment models are tested in the numerical optimization technique to investigate their impact on the optimal performance. It is shown that excessive simplification on the trailing wake can non-physically increase the turbine efficiency. In case a better wake alignment model (PSF-2 or FWA), which aligns the trailing wake based on the local stream, is implemented, the turbine efficiency falls within a reasonable range and shows improved performance compared to RANS performance.

Future work includes coupling a boundary layer solver with the present method to improve the pressures distributions on the blade and consequently thrust performance. It is also essential to incorporate the most elaborate wake alignment scheme, FWA, into the present optimization method to improve its comparisons with viscous RANS simulations. Besides, the equality constraint on the thrust has to be removed from the present method such that the optimal power performance can be obtained regardless of any change in the turbine thrust during the design process.

Acknowledgements

Support for this research was provided by the U.S. Office of Naval Research (Grant Numbers N00014-18-1-2276 and N00014-21-1-2488; Dr. Ki-Han Kim) and by Phases VIII and IX of the "Consortium on Cavitation Performance of High Speed Propulsors".

References

[1] R. Charlier, Sustainable co-generation from the tides: A review, *Renewable and Sustainable Energy Reviews*, 7 (3), 2003, pp. 187-213.

[2] M. Grabbe, E. Lalander, S. Lundin, and M. Leijon, A review of the tidal current energy resource in Norway, *Renewable and Sustainable Energy Reviews,* 2009, 13 (8), pp. 1898-1909.

[3] S. Allsop, C. Peyrard, P. Thies, E. Boulougouris, and G. Harrison, Hydrodynamic analysis of a ducted, open centre tidal stream turbine using blade element momentum theory, *Ocean Engineering,* 2017, 141, pp. 531-542.

[4] W. Batten, A. Bahaj, A. Molland, and J. Chaplin, Experimentally validated numerical method for the hydrodynamic design of horizontal axis tidal turbines, *Ocean Engineering,* 2007, 34 (7), pp. 1013-1020.

[5] C. Belloni, R. Willden, and G. Houlsby, An investigation of ducted and open-centre tidal turbines employing CFD-embedded BEM, *Renewable Energy,* 2017, 108, pp. 622-634.

[6] R. Malki, A. Williams, T. Croft, M. Togneri, and I. Masters, A coupled blade element momentum computational fluid dynamics model for evaluating tidal stream turbine performance, *Applied Mathematical Modeling,* 2013, 37 (5), pp. 3006-3020.

[7] S. Turnock, A. Phillips, J. Banks, and R. Nicholls-Lee, Modeling tidal current turbine wakes using a coupled RANS-BEMT approach as a tool for analysing power capture of arrays of turbines, *Ocean Engineering,* 2011, 38 (11-12), pp. 1300-1307.

[8] S. Kim, Y. Su, and S. A. Kinnas. A BEM/RANS Interactive Method Applied to an Axial Tidal Turbine Farm. *Journal of Ship Research*, 65 (4), 2021, pp. 320-345.

[9] W. Xu, and S. A. Kinnas. Performance prediction and design of marine current turbines, *Transaction of Society of Naval Architects and Marine Engineers*, 118, B31-B38.

[10] J. Baltazar, and J.A.C. Falcao de Campos, Unsteady Analysis of a Horizontal Axis Marine Current Turbine in Yawed Inflow Conditions With a Panel Method, 1st International Symposium on Marine Propulsors (SMP'09), June 2009, Trondheim, Norway.

[11] F. Salvatore, and Z. Sarichloo, and D. Calcagni, Marine Turbine Hydrodynamics by a Boundary Element Method with Viscous Flow Correction, *Journal of Marine Science and Engineering*, 6 (53), 2018, pp. 1-30.

[12] W. Otto, D. Rijpkema, and G. Vaz, Viscous-flow calculations on an axial marine current turbine, In the proceedings of 31st International Conference on Ocean, Offshore and Arctic Engineering (OMAE2012), July 1-6, Rio de Janeiro, Brazil.

[13] J. E. Kerwin, and J. B. Hadler, Principles of Naval Architecture Series: Propulsion, *The Society of Naval Architects and Marine Engineers (SNAME)*, 2010.

[14] S. Mishima, and S. A. Kinnas, A numerical optimization technique applied to the design of two-dimensional cavitating hydrofoil sections, *Journal of Ship Research*, 40 (1), 1996, pp. 28-38.

[15] S. Mishima, and S. A. Kinnas, Application of a numerical optimization technique to the design of cavitating propellers in non-uniform flow, *Journal of Ship Research*, 41 (2), 1997, pp. 93-107.

[16] P. E. Griffin, and S. A. Kinnas, A Design Method for High-Speed Propulsor Blades, *Journal of Fluid Engineering,* 1998, 120 (3), pp. 556-562.

[17] D. H. Menéndez Arán, and S. A. Kinnas, On Full Aligned Lifting Line Model for Propellers: An Assessment of Betz Condition, *Journal of Ship Research*, 58 (3), 2014, pp. 130-145.

[18] D. S. Greeley, and J. E. Kerwin, Numerical Methods for Propeller Design and Analysis in Steady Flow, Society of Naval Architects and Marine Engineers: Jersey City, NJ, 90, 1982, pp. 415-453.

[19] S. Kim, S. A. Kinnas, and W. Du, Panel method for ducted propellers with sharp trailing edge duct with fully aligned wake on blade and duct, *Journal of Marine Science and Engineering,* 2018, 6 (3), pp. 1-22.

[20] S. Kim, and S. A. Kinnas, Prediction of cavitating performance of a tip loaded propeller and its induced hull pressures, *Ocean Engineering,* 2021, 229 (108961), pp. 1-17.

[21] S. A. Kinnas, K. Cha, and S. Kim, Comprehensive Design Method for Open or Ducted Propellers for Underwater Vehicles, In the proceedings of the SNAME Maritime Convention, Providence, Rhode Island, US, October 2021.

[22] S. A. Kinnas, W. Xu, Y.-H. Yu, and L. He, Computational Methods for the Design and Prediction of Performance of Tidal Turbines, *Journal of Offshore Mechanics and Arctic Engineering*, 134 (1), 2012, pp. 1-10.

[23] D. H. M. Arán, Y. Tian, and S. A. Kinnas, Effect of Wake Alignment on Turbine Blade Loading Distribution and Power Coefficient, *Journal of Offshore Mechanics and Arctic Engineering*, 141 (4), 2019, pp. 1-11.

SNAME
THE INTERNATIONAL COMMUNITY
FOR MARITIME AND OCEAN PROFESSIONALS

COUPLED CONTROL FOR MOTION COMPENSATED OFFSHORE OPERATIONS

BRUNO ROSSIN
MARIN – THE NETHERLANDS

ATHEENDRA SREENIVASAN
MARIN – USA

BAS DE KRUIF
MARIN – THE NETHERLANDS

Proceedings of the 27th Offshore Symposium, February 22nd 2022, Houston, Texas
Texas Section of the Society of Naval Architects and Marine Engineers

ABSTRACT

The use of motion compensated equipment has become more frequent in modern DP operations, especially the ones related to the installation, maintenance and decommissioning of offshore wind turbines. In these operations, while low-frequency motions are compensated by DP systems, wave-frequency motions are compensated by gripper frames, heave compensators, hexapods, gangways and other type of motion compensated equipment. A study was performed to investigate if the coupling of a dynamic positioning system with a motion compensation device could improve the accuracy and efficiency of offshore motion compensated operations. The reference case consisted of the positioning of a payload, hanging on a ship-mounted crane, by means of a single horizontal active tugger line and the ship's DP system. Four controllers – consisting of two levels of awareness and two tuning conditions – were designed and implemented in a numerical model. The four strategies were then evaluated in terms of position offsets and actuator usage under a number of sea-states. Results demonstrated that coupled controllers are able to decrease the position deviations of the motion compensated object and decrease the usage of the motion compensation device. The paper describes the control strategies, numerical models and results of the simulations.

Keywords: dynamic positioning, motion compensation, coupled control.

INTRODUCTION

The increased use of motion compensated devices in offshore wind operations – such as gangways, pile grippers, tugger lines and heave compensators – where a large variety of wind turbine components need to be installed, can create interactions with the vessel's dynamic positioning (DP) system, or vice versa. As indicated by contractors, these operations are commonly limited to calm seas, leading to costly downtime periods. If the control accuracy during these operations could be increased such that operational windows could be extended to rougher weather conditions, it would increase the operability of the vessels and decrease the overall operational costs. Aiming at understanding these interactions, this study investigates if coupling of control systems could improve the accuracy of DP-assisted motion compensated operations, when compared to the traditional stand-alone control strategies followed by the industry.

Typically, a DP system is responsible for counteracting the low-frequency motions of a combined vessel-motion compensated system, while wave-frequency motions are attenuated by the motion compensated device. Final positioning is then influenced by two control systems, which may act in favor or against each other. The hypothesis to be verified was that, by integrating the DP and a motion compensated system, better use of the available actuators could be achieved by increased awareness between the two systems.

The envisaged operation used as reference for this study consisted of the transfer of wind-turbine components from a floating barge to a DP vessel, using the vessel's crane, which was equipped with a payload compensation device (PCD). The two control systems to be integrated were the vessel's DP system and the payload compensation device, which, for this study, was modelled as a single horizontal active tugger line connecting the payload compensation device to the vessel's crane. The crane was oriented perpendicular to the vessel, in such a way that the payload compensation device acted on the vessel's sway direction only. The simulation study made use of a full 6 degrees-of-freedom model. The DP and the payload compensation device act simultaneously to position the payload onto a earth-fixed horizontal position.

The study was split into different phases. First, a detailed non-linear model, including the vessel and crane, was built. The barge was not modelled. Second, this model was linearized for design of four controllers, consisting of two levels of awareness – named independent and integrated – and two tuning conditions. In the independent strategy, the DP and tugger line controllers were designed without knowledge of each other. In the integrated strategy, a single MIMO (multi-input multi-output) controller was designed. The different tuning conditions were used to verify the robustness of coupled controllers in sight of practical implementation. The tuning was done by minimizing a system norm, trying to minimize the subjective influence of a designer. Finally, the four designed controllers were implemented in the original non-linear model, which was used for final verification and evaluation under a number of wave conditions.

The comparison of the different control strategies focuses on the horizontal offsets of the payload. In addition, operational efficiency is also analyzed by means of actuator usage. The design procedure, results and conclusions obtained from this study provide useful insights that can be useful in similar coupled control problems involving a DP system and a motion compensation device.

MODELLING AND CONTROL DESIGN

The design goal of the control system is to minimize the variance on the payload, while respecting the actuator limits. A model based approach, H2-design [3], is employed to minimize the influence of the designer. The design approach is furthermore well suited to work in the frequency domain, allowing one to work with wave spectra that affect the motions. In order to use a model based approach, the first step is to linearize the available detailed numerical model around its working point. The detailed numerical model is elaborated upon in the next section and the linearization was automatically obtained by means of a finite differences method.

The H2-design approach minimizes the (summed) variance of selected outputs for the disturbances acting on the system. The wave spectrum that induces a ship and payload motion are used in the control design as the disturbances. The motion spectrum is calculated for two different wave periods: $T_P = 6s$ and $T_P = 12s$, and a significant height of Hs=1.88m. These conditions were defined based on operational aspects of the operation. The resulting motion spectrum is included in the control design [4].

In the independent control design the interaction between the two systems is willfully ignored and the output variance of the ship position is minimized for the DP controller, and the output variance of the payload is minimized for the tugger line controller. The trade-off between actuator usage and accuracy is tuned such that the actuators did not saturate for more than 3% of the time. This approach gives a controller for the DP system that can be rewritten as a Kalman state estimator with state feedback [5]. The DP controller for the wave spectrum with a Tp=6s could be tuned much stiffer than the DP controller for the wave spectrum with Tp=12s.

In the integrated control design a Multi-Input Multi-Output controller is designed in a single design step. The output we are interested in is the pay load motion only in this situation. Again, the accuracy is increased until the actuators saturated for approximately 3% of the time. As the effect of the payload on the ship motion is marginal in comparison to the effect the other way around, the ship measurements did contribute to the control signal of the tugger line controller, but not the other way around. This simplified the implementation: the original DP controller could be used. The tugger line controller did use the ship's position and roll. Again, it was found that for shorter wave periods the DP could be tuned stiffer.

Table 1: Control strategies designed for optimum accuracy.

Control strategy	PCD awareness level	Tuning condition	DP gain
Ind 6	Independent	HS = 1.88m and TP = 6.0s	7
Ind 12	Independent	HS = 1.88m and TP = 12.0s	5
Int 6	Integrated	HS = 1.88m and TP = 6.0s	7
Int 12	Integrated	HS = 1.88m and TP = 12.0s	5

TIME-DOMAIN MODEL

The time-domain model was built using MARIN's software aNySIM [1]. The software has reached large development momentum since 2020, when a dedicated team within MARIN was set for its development, documentation, maintenance and support. aNySIM consists of a versatile and modular framework that grants the user great modelling flexibility. The framework is especially useful to support the incremental design phases of a project, allowing design variations, static calculations, frequency-domain analysis and time-domain simulations. In addition to its stand-alone use, aNySIM can accommodate interface to third-party systems, as for digital-twin modelling. Furthermore, it can also be deployed on MARIN's clusters, basins and real-time simulators, assuming a central role on the integration of MAIRN's technologies.

The numerical model used for modelling of the system and implementation of the controllers is presented in Figure 1. Three objects have been modelled in the study – the vessel, the payload compensation device (PCD) and the payload. The vessel has an associated hydrodynamic file which contains the frequency dependent added mass, damping and the first and second order wave forces. These were calculated with MARIN's potential code DIFFRAC [2]. The vessel was also assigned with linear hydrostatics, wind and current coefficients and a small amount of roll damping to account for viscous effects. The crane was assumed to be rigidly attached to the vessel. Thus, only the crane tip was modelled as a line connection point from which hang the PCD and payload. A small amount of damping was applied to the crane wires to avoid instabilities. The PCD and payload were modelled as two rigid bodies with mass and rotational inertia and had wind load coefficients defined as well. For better verification of the linear modelling described in the previous sections, the DP actuators and tugger line actuator were modelled as force vectors. Physical limitations, such as latency and saturation, were accounted for in the DP and PCD controllers.

Figure 1: Vessel, payload compensation device (smallest block) and payload (largest block).

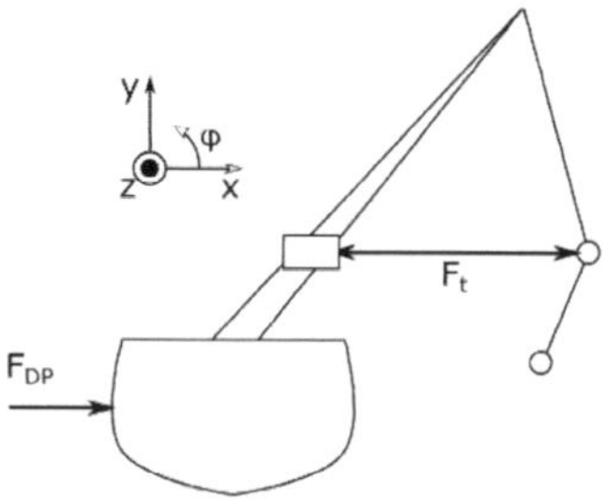

Figure 2: Indication of DP actuator forces (F_{DP}) and PCD actuator force (F_T).

The DP system is included in numerical model following the same topology of a DP system installed onboard a real vessel, uncoupled from the (numerical model of the) vessel. The same principle was used for inclusion of the controller of the payload compensation device in the numerical model. Such approach enables easy replacement of controllers while the plant model remains untouched. As observed from Figure 3 and Figure 4, the state of the vessel is known by the integrated controller of the payload compensation device. The DP and PCD controller implementations used in the simulations follow from the combinations defined by the four control strategies described in Table 1.

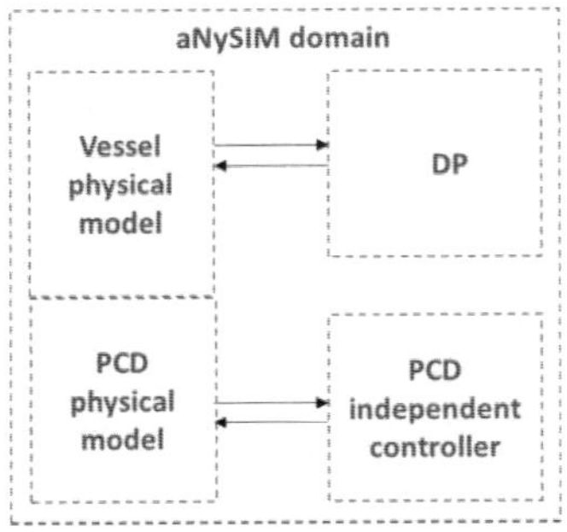

Figure 3: Schematic overview of independent controller in the numerical model.

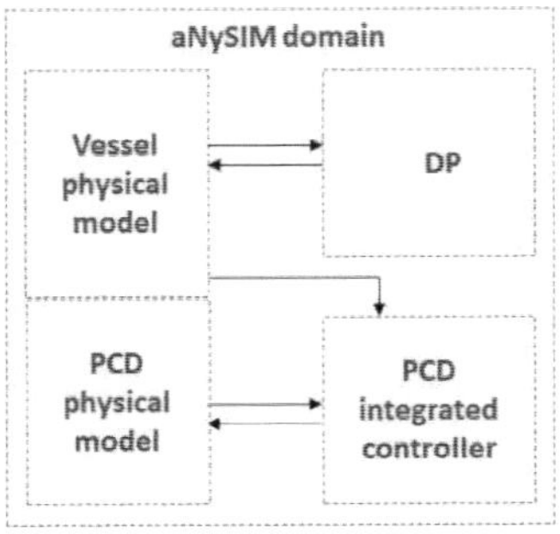

Figure 4: Schematic overview of integrated controller in the numerical model..

SIMULATION AND ANALYSIS

A set of 16 time-domain simulations was run. Each of the four control strategies was evaluated against four sea-states. The sea-states were modelled as a Jonswap wave spectrum with a shape factor of 3.3, had a significant wave height (H_S) of 1.88m and peak periods (T_P) of 6s, 8s, 10s and 12s. The sea-states were run from a incidence of 270 degrees – beam-on seas, as shown in Figure 5. The beam-on waves are aligned with the direction of actuation of the tugger line of the payload compensation device. Each simulation was run for 3,5 hours, where only the last 3,0 hours were accounted for statistics.

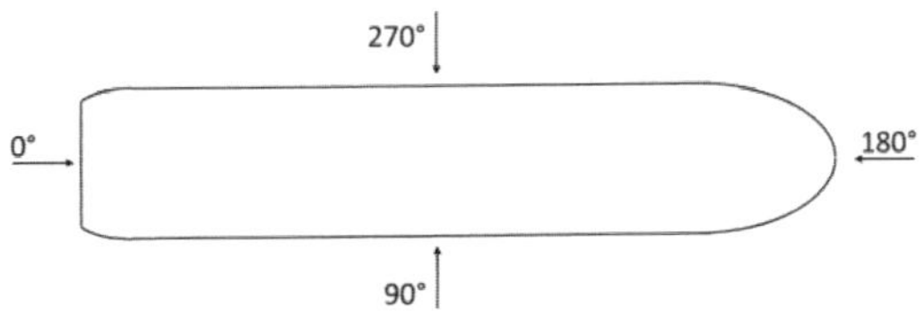

Figure 5: Definitions applied to wave incidence angles.

The results of the simulations were analyzed with regards the actuator usage and offset of the payload. The actuator usage was defined as the average of the squared actuation forces and offset was defined as the horizontal distance of a body (vessel/PCD/payload) to its initial position, which is also its earth-fixed position setpoint. Since the same trend in positioning is observed for the payload and payload compensation device, the discussions in the next paragraphs refer to the payload compensation device only.

The thrust and offset values obtained from the simulations are a result of the actuator's physical properties defined for the simulations, which were estimated based on the available information. These values can differ for different vessels and objects being lifted. Therefore the relative values (differences in thrust/offset between the different strategies) are mostly interesting for the purpose of this study.

RESULTS

Figure 6 to Figure 11 present the offset variance and maximum offset of the vessel, payload compensation device and payload along the vessel's sway direction.

The vessel positioning is generally worse with controllers tuned for 12s waves, especially for peak periods of 6s and 8s. From the analysis of the coupled system, it is known that the payload compensation device has little influence on the vessel positioning. Thus the worse vessel positioning at these sea-states is a consequence of the lower DP gain level – equal to 5 – used with the controllers tuned for 12s waves. At shorter wave periods (around 1rad/s), the second order wave forces become considerably larger (Figure 12 shows the sway drift force), resulting in larger vessel excursions when the DP is tuned for 12s (0.52rad/s) waves.. For the longer 12s waves, the different DP gains provide similar position accuracy.

For positioning of the payload compensation device, it is again seen that the controllers tuned for 12s waves are worse at the shorter period waves. This is partly influenced by the bad positioning of the vessel. Note, however, that the control strategy applied has also a large influence on the motions of the payload compensation device. This is observed, for instance, by comparing the numbers obtained from the 12s waves in Figure 6 and Figure 8. Despite the same position accuracy of the vessel, the position accuracy of the payload compensation device is totally changed by the control strategy applied. Looking at the controllers tuned for the 6s waves, the integrated system performs better at the lower periods but is slightly worse at the 12s period compared to the independent system.

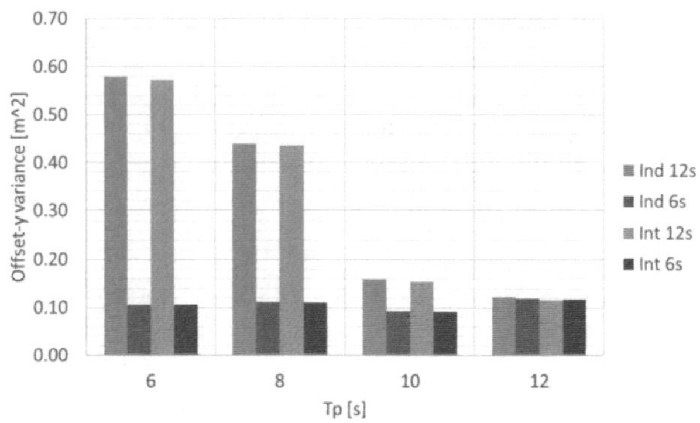

Figure 6: Vessel offset-y variance. Beam-on waves.

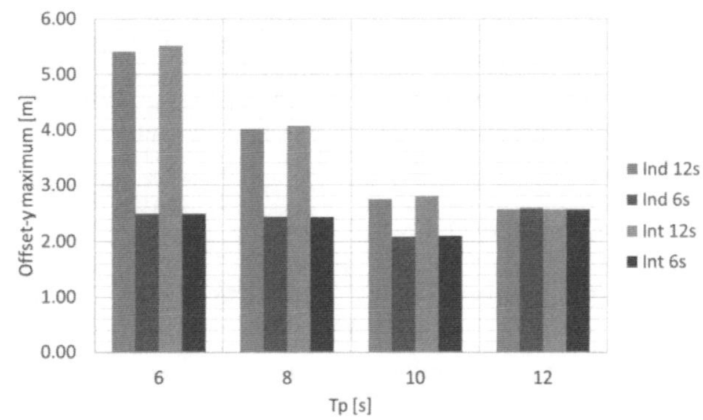

Figure 7: Vessel maximum offset-y. Beam-on waves.

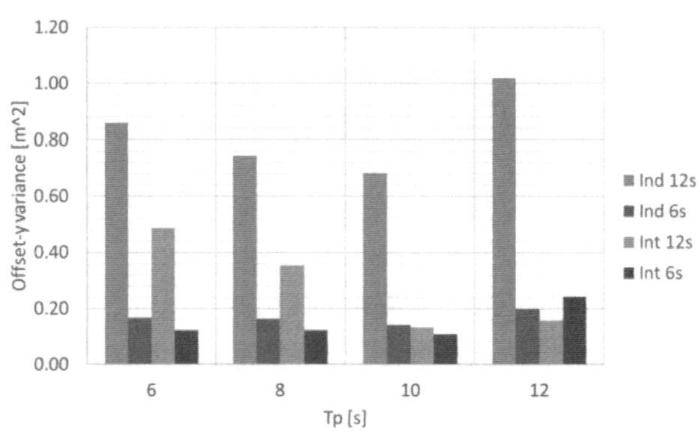

Figure 8: PCD offset-y variance. Beam-on waves.

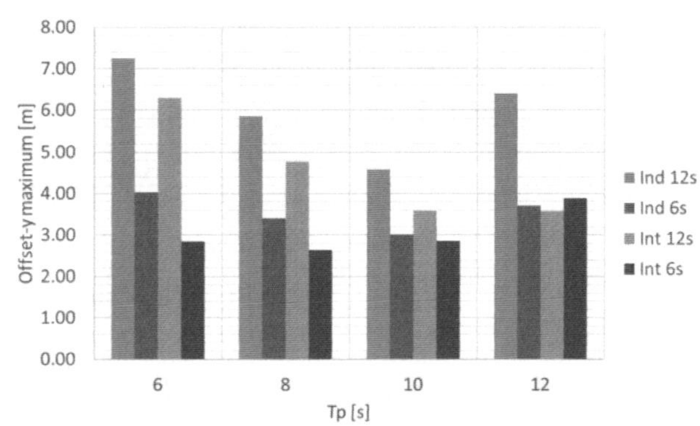

Figure 9: PCD maximum offset-y. Beam-on waves.

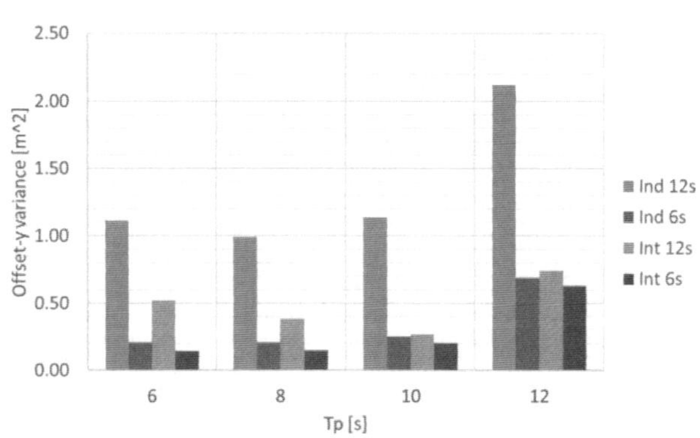

Figure 10: Payload offset-y variance. Beam-on waves.

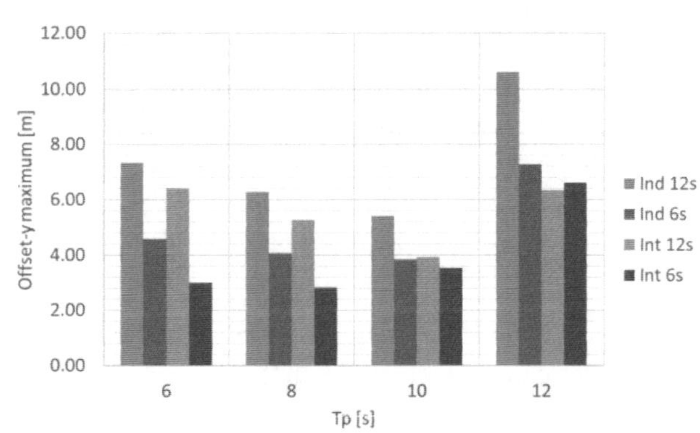

Figure 11: Payload maximum offset-y. Beam-on waves.

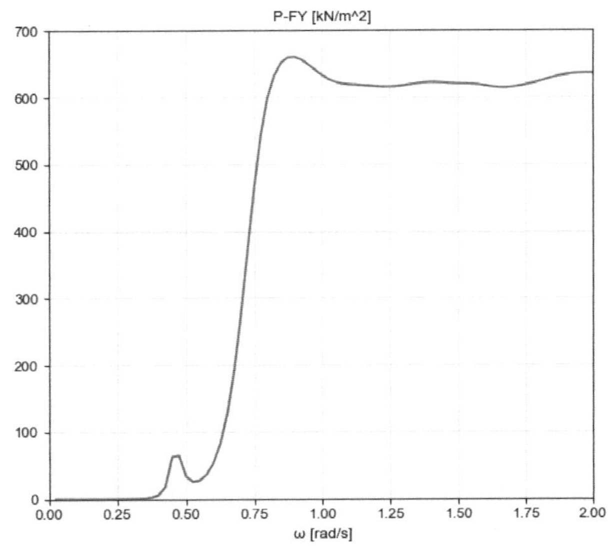

Figure 12: Vessel mean wave drift force on sway for beam-on waves.

The numbers from Figure 9 are summarized in Table 2, together with the relative difference in positioning of the payload compensation device achieved with integrated controllers. Improved positioning of the payload compensation device is observed in all cases, except with Int 6s at 12s waves. Nevertheless, it was only 5% higher than the independent system. The average offset reduction obtained with integrated controllers tuned for 6s waves was 13%, in comparison to 24% obtained with integrated controllers tuned for 12s waves. At the same time, the controllers tuned for 6s presented the lowest absolute average maximum offset.

The worst positioning of the payload compensation device happens with the independent controller tuned for a peak period of 12s, at all sea-states. The best positioning is generally observed with integrated controllers tuned for a peak period of 6s. It is also observed that, among all controllers, the largest minimum offset of the payload compensation device happens at 12s waves. This follows from the higher excitation, at this peak period, of the counter[1] motion mode of the system.

Table 2: Payload compensating device maximum y-offset and relative difference due to integrated controllers.

				T_P [s]				Average	
				6	8	10	12		
Controller	6s	Ind	[m]	4.0	3.4	3.0	3.7	3.5	3.8
		Int	[m]	2.8	2.6	2.9	3.9	3.1	
		Diff	[%]	70	78	95	105	87	
	12s	Ind	[m]	7.2	5.8	4.6	6.4	6.0	5.3
		Int	[m]	6.3	4.8	3.6	3.6	4.6	
		Diff	[%]	87	82	78	56	76	

Figure 13 and Figure 14 show the spectral difference of the motions of the payload compensation device, for waves of 12s and 6s and the four control strategies evaluated. The peaks at 0.25rad/s and 0.5rad/s correspond to the similar[2] and counter motion mode of the system, respectively. It is observed that the coupled controllers mitigate the motions of the payload compensation device mostly around 0.25rad/s, which is the bandwidth for which the controllers were designed.

[1] Ship moves opposite to the PCD and payload, with a natural period of 11.2s (0.56rad/s).

[2] Ship, PCD and payload move at the same direction, with a natural period of 26.2s (0.24rad/s).

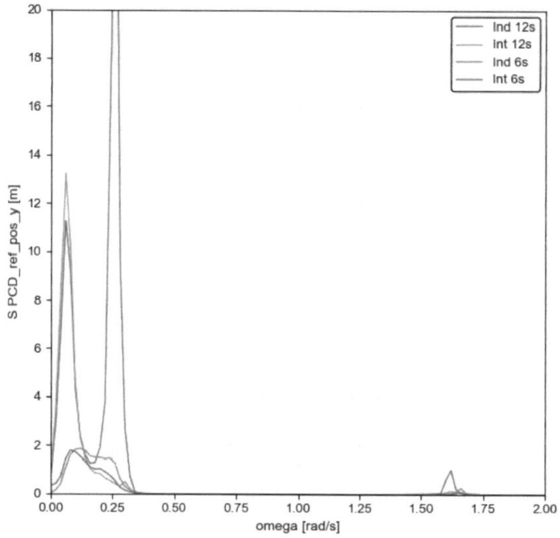

Figure 13: Spectrum of PCD motion at 6s (1.0 rad/s) waves.

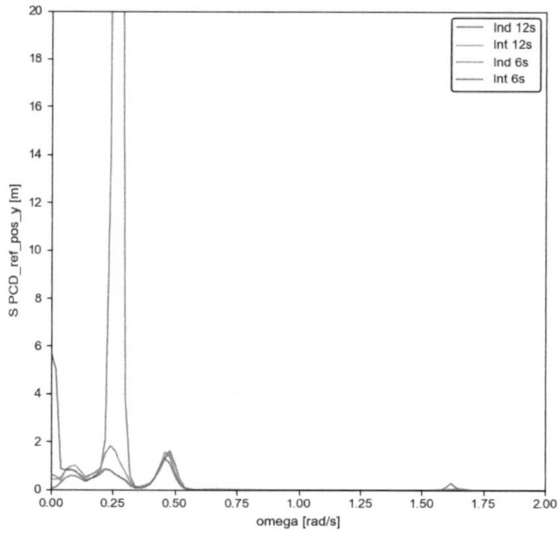

Figure 14: Spectrum of PCD motion at 12s (0.5 rad/s) waves.

The actuator usage of the DP system, payload compensation device and combined systems are summarized, respectively, from Figure 15 to Figure 17, for each sea-state and control strategy applied. As expected, due to the larger inertia of the ship and higher power capacity of the DP actuators, the combined actuator usage is dominated by the DP forces. The controllers tuned for 6s – which make use of a DP gain of 7 – present slightly higher demand, especially for wave peak periods of 10s and 12s.

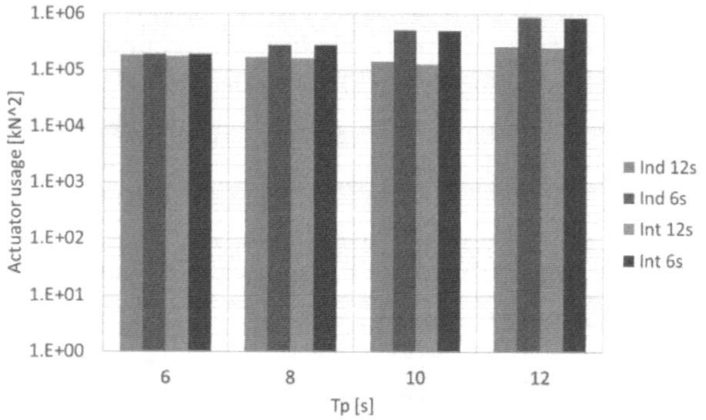

Figure 15: DP force-y usage. Beam-on waves.

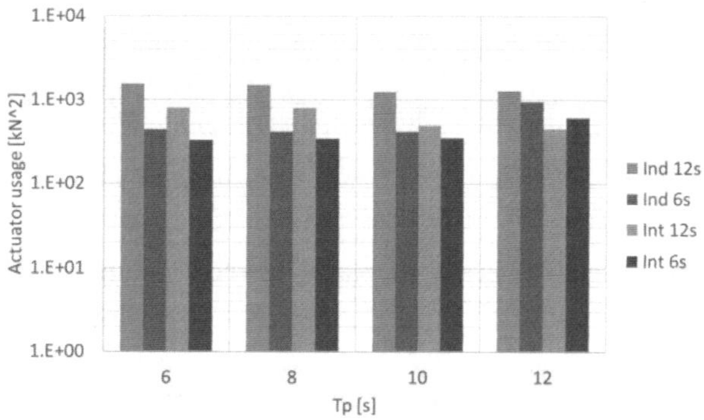

Figure 16: PCD force-y usage. Beam-on waves.

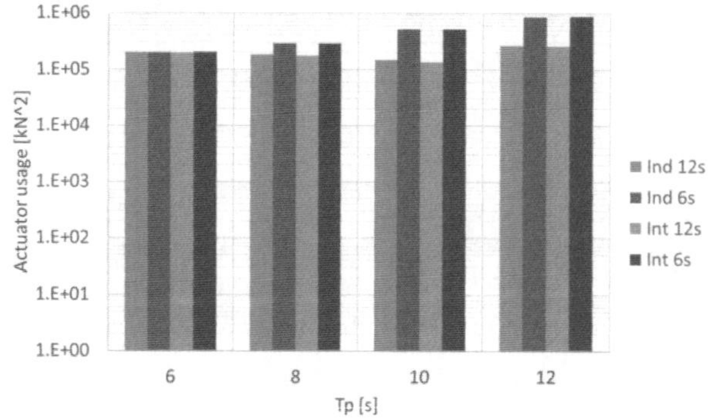

Figure 17: Combined force-y usage. Beam-on waves.

The numbers related to actuator usage of the payload compensation device, presented in Figure 16, are also summarized in Table 3. The average reduction in actuator usage presented by integrated controllers tuned for 6s waves was of 24%, in comparison to 55% obtained with integrated controllers tuned for 12s waves. At the same time, the controllers tuned for 6s presented the lowest absolute average actuator usage.

Table 3: Payload compensating device actuator usage and relative difference due to integrated controllers.

				T_P [s]				Average	
				6	8	10	12		
Controller	6s	Ind	[kN^2]	439	417	416	960	558	784
		Int	[kN^2]	330	345	352	608	409	
		Diff	[%]	75	83	84	63	76	
	12s	Ind	[kN^2]	1548	1516	1237	1280	1395	1015
		Int	[kN^2]	806	798	490	445	635	
		Diff	[%]	52	53	40	35	45	

The independent controller tuned for 12s waves presented the highest demand, regardless of the sea-state. Despite its higher actuator usage, this controller also led to the worst positioning of the payload compensation device, as shown in Table 2. Integrated controllers led to lower actuator usage of the payload compensation device under all combinations of sea-states and tuning conditions, with the controller tuned for 6s waves being the most efficient one. In addition to its lowest actuator usage, this controller also provided the best position accuracy of the payload compensation device, as can be seen by comparing Figure 9 and Figure 16.

CONCLUSION

Integrated controllers led to improved position accuracy of the payload compensation device in seven out of the eight wave conditions evaluated. In the case where worse position accuracy was achieved with integrated controllers – Int 6s at 12s waves – it was only 5% higher than the independent controller. The average offset reduction obtained with integrated controllers tuned for 6s waves was 13%, in comparison to 24% obtained with integrated controllers tuned for 12s waves. At the same time, the controllers tuned for 6s presented the lowest absolute average maximum offset and therefore can be considered the most accurate strategy of this study.

The use of integrated controllers decreased the usage of the motion compensation device in all wave conditions evaluated. The average reduction in usage of the motion compensation device by integrated controllers tuned for 6s waves was 24%, in comparison to 55% obtained with integrated controllers tuned for 12s waves. At the same time, the controllers tuned for 6s presented the lowest absolute average actuator usage.

Due to the larger inertia of the ship, the total actuator usage is dominated by the DP forces. Decreasing the usage of the motion compensation device has neglectable influence on the instantaneous fuel consumption. This can, however, reduce their wear and maintenance costs. In addition, the right control strategy can improve accuracy and widen operational window. Consequently, overall lower fuel consumption and costs can be achieved by the use of coupled controllers.

In future studies, the use the same DP settings for all controllers might facilitate the evaluation of coupled controllers. This will decrease the influence of the vessel positioning on the positioning of the payload and enable better comparison of independent and integrated control strategies.

ACKNOWLEDGEMENTS

The authors would like to Van Oord and Seaqualize the cooperation and insights to produce this paper. The work was founded by the Dutch government, as part of the TKI Wind of Zee program.

REFERENCES

[1] MARIN (2021). aNySIM-XMF; Time domain simulations for ships and offshore vessels. URL https://www. https://www.marin.nl/facilities-and-tools/software/anysim.

[2] MARIN (2021). DIFFRAC; Calculation of wave loads and motions using 3-D potential theory including wave diffraction. URL https://www. https://www.marin.nl/facilities-and-tools/software/diffrac.

[3] S. Skogestad and I. Postlethwaite, Multivariable feedback control: analysis and design, vol. 2. Wiley New York, 2007.

[4] B. J. de Kruif and B. Rossin, "Pendulation control for dynamical positioning capable ship ; considerations on actuator usage," in 13th IFAC Conference on Control Applications in Marine Systems, Robotics, and Vehicles, 2021.

[5] O. Cadet, "Introduction to Kalman Filter and its Use in Dynamic Positioning Systems," in Dynamic Positioning Conference, 2003.

SNAME
THE INTERNATIONAL COMMUNITY
FOR MARITIME AND OCEAN PROFESSIONALS

Virtual Telepresence for the Future of ROV Teleoperations: Opportunities and Challenges

Pengxiang Xia
PhD student, University of Florida

Kevin McSweeney
Manager, American Bureau of Shipping (ABS)

Feng Wen
Coordinator, American Bureau of Shipping (ABS)

Zhuoyuan Song
Assistant Professor, University of Hawaii, Manoa

Michael Krieg
Assistant Professor, University of Hawaii, Manoa

Shuai Li
Assistant Professor, The University of Tennessee, Knoxville

Xiao Yu
Assistant Professor, University of Florida

Kent Crippen
Professor, University of Florida

Jonathan Adams
Assistant Professor, University of Florida

Eric Jing Du
Associate Professor, University of Florida (Corresponding author)

Proceedings of the 27th Offshore Symposium, February 22nd, 2022, Houston, Texas
Texas Section of the Society of Naval Architects and Marine Engineers

ABSTRACT

Underwater robots, including Remote Operating Vehicles (ROV) and Autonomous Underwater Vehicles (AUV), are currently used to support underwater missions that are either impossible or too risky to be performed by manned systems. In recent years the academia and robotic industry have paved paths for tackling technical challenges for ROV/AUV operations. The level of intelligence of ROV/AUV has increased dramatically because of the recent advances in low-power-consumption embedded computing devices and machine intelligence (e.g., AI). Nonetheless, operating precisely underwater is still extremely challenging to minimize human intervention due to the inherent challenges and uncertainties associated with the underwater environments. Proximity operations, especially those requiring precise manipulation, are still carried out by ROV systems that are fully controlled by a human pilot. A workplace-ready and worker-friendly ROV interface that properly simplifies operator control and increases remote operation confidence is the central challenge for the wide adaptation of ROVs.

This paper examines the recent advances of virtual telepresence technologies as a solution for lowering the barriers to the human-in-the-loop ROV teleoperation. Virtual telepresence refers to Virtual Reality (VR) related technologies that help a user to feel that they were in a hazardous situation without being present at the actual location. We present a pilot system of using a VR-based sensory simulator to convert ROV sensor data into human-perceivable sensations (e.g., haptics). Building on a cloud server for real-time rendering in VR, a less trained operator could possibly operate a remote ROV thousand miles away without losing the minimum situational awareness. The system is expected to enable an intensive human engagement on ROV teleoperation, augmenting abilities for maneuvering and navigating ROV in unknown and less explored subsea regions and works. This paper also discusses the opportunities and challenges of this technology for ad hoc training, workforce preparation, and safety in the future maritime industry. We expect that lessons learned from our work can help democratize human presence in future subsea engineering works, by accommodating human needs and limitations to lower the entrance barrier.

Keywords: ROV, Virtual Reality, Telepresence

INTRODUCTION

Augmenting human abilities in subsea engineering, such as offshore construction and inspection (e.g., floating cities and offshore wind farms) and subsea exploration and operations (e.g., offshore mining, subsea cables, and energy harvesting), provides a historic opportunity for the new economic growth and scientific discoveries [1]. Underwater infrastructure is also critical to tackling the unprecedented challenge of the climate crisis - various disaster models project a 45-87% increase of severe hurricanes in the next three decades in the US coastal areas [2], threatening more than 30% of the US population [3]. There is an urgent need to transform the future workforce partnering with new subsea robot technologies to meet the societal, environmental, and engineering needs of grand significance.

Subsea engineering remains a highly specialized area with a high barrier for broader participation. Most subsea jobs require a strict professional preparation (science, engineering, and diving knowledge) that takes many years of training. The existing subsea workforce consists of marine engineers, professional divers, and robot service providers, and only a small portion holds licenses for certain tasks. But the current subsea professionals may not possess all the needed knowledge from other domains such as structural engineering knowledge. This high barrier of subsea works has substantially limited the workforce for an aggressive human ocean exploration roadmap. The future subsea engineering should be participatory, meaning that the majority of the interested populations should not experience serious difficulties in entering a subsea career.

The realization of this goal depends heavily on the efficient utilization of underwater robots such as Remote Operating Vehicles (ROV) and Autonomous Underwater Vehicles (AUV) as remote surrogates for future subsea workers. However, the harsh subsea environment and difficult robot controls are major obstacles for the future subsea workforce to utilize or team up with underwater robots. Underwater robot operations are particularly challenged by the uncertainties of the subsea environments, including the low visibility, chaotic ocean currents, interference from subsea ecosystems, and less understood subsea properties of materials and systems [4]. Existing Human-Robot Collaboration (HRC) methods for land applications are insufficient and not directly applicable to meet the unique challenges of subsea engineering. There is also little knowledge about the implications of subsea engineering operations in human perception, training, education and health, such as work-related stress, sensory loads and fear of being underwater [5].

To fill the gap in ROV operations, our team proposes a new Virtual Reality (VR) based teleoperation approach called "Human-Robot Sensory Transfer", which transfers and simulates robot sensor data of subsea workplaces as human-perceivable multimodal sensory feedback, including visual, tactile, haptics, and force feedback to enable a shared perception in complex subsea workplaces. In the remainder of this paper, we will discuss the basics of the proposed method, and its potential challenges and opportunities.

METHODOLOGY

1. Problem Related to ROV Operations

Underwater robots currently are used to support underwater missions that are either impossible or too risky to be performed by manned systems. The original design philosophy of ROV was driven by automation for replacing humans in repetitive or dangerous subsea works. In recent years the academia and robotic industry have paved paths for tackling technical challenges for autonomous ROV such as underwater manipulation [6], power supplies [7] and data telecommunication networks [8]. The level of intelligence of ROV has increased dramatically as a result of the recent advances in low-power-consumption embedded computing devices and machine intelligence [9]. Nonetheless, operating precisely underwater is still extremely challenging to achieve for a fully autonomous system due to the inherent challenges associated with the underwater environments [10]. As a result, close-range operations, especially those requiring manipulation, are still carried out by ROV systems that are fully controlled by a human pilot. A workplace-ready and worker-friendly ROV interface that properly simplifies control and increases remote operation confidence is the central challenge for the wide adaptation of ROVs.

Compared to the ROV automation solutions, the technical specifications for human-in-the-loop challenges are less thought out. Established algorithms, methods, and systems for automation are not effective for supporting tasks with an intensive human engagement, such as overwhelming information for underwater environmental understanding; difficulty of remote robot controls, the missing sensations of human operators underwater (e.g., inability to directly sense water flow), and maneuver and navigation in unknown and less explored subsea regions and works. To democratize human presence in the future subsea engineering works, a new design paradigm for ROV that accommodates human needs and limitations are needed to lower the entrance barrier. The technical innovation should also meet the requirements for new subsea jobs and technology certification.

Our team is funded by National Science Foundation (NSF) to design and explore *Human-Robot Sensory Transfer*, a new Human-Underwater Robot Interaction method based on the virtual telepresence and sensory augmentation technologies, to significantly lower the barrier for underwater robot operations. We envision that the human-robot sensory transfer method to be *virtual*, in which the solutions are based on a Cloud infrastructure where human agents located at different locations can access multiple underwater robots remotely. The multimodal sensory augmentation in an immersive virtual environment will enable an "idiot-proof", intuitive teleoperation experience with minimum training needs. We also expect the method to be *ad hoc*, meaning that the resources (human and robots) can be easily matched and rematched, and the work schedule is adaptive, etc. The following section describes our approach.

2. The Proposal Human-Robot Sensory Transfer Method for ROV Teleoperation

Fig.1 illustrates the architecture of the proposed human-robot sensory transfer system for simplifying ROV teleoperation. The proposed system consists of five modules, including *underwater robot (ROV)*, *subsea sensing*, *workplace model*, *robotic simulation*, and *human operator*. The details of each module are described as follows.

- **Subsea Sensing Module**: Subsea workplaces will be substantially different from our established workplaces, and thus the human operator may not perceive the environmental data in a desired way. As a result, we expect the sensing system to capture the key characteristics of a dynamic underwater environment for sensory augmentation. Subsea sensing module builds on a multi-level sensor network to collect real-time subsea environmental data pertaining to the hydrodynamic features and temperature changes. The sensor network captures three levels of sensor data to meet the sensing needs, including *far field hydrodynamic status* based on an acoustic Doppler current profiler (ADCP) [11] to collect underwater wave profiles in the 3 meter-20 meter proximity; and *near field* (0.1 meter – 3 meter) and *micro field* (<0.1 meter) turbulence based on *in-situ* sensors equipped on the ROV. The far field sensing aims to identify the sudden change of underwater wave in a bigger scale, or the so-called "seamount" (Fig.1). The existence of seamounts often poses significant risks to the teleoperation of ROV as they usually intensify the tidal flow and water stratification, interrupting the on-gong ROV operation profiles. As for the in-situ sensor, we leverage an *artificial skin* – an innovative distributed sensor system that allows a flexible add-on sensor package. The artificial skin sensing system enables fast disturbance detection and rejection to improve vehicle control accuracy. The design of the in-situ sensor is based on our pervious works [12-14] on lateral-line sensory mechanism within fish which consists of specialized "hair-cells" throughout the body surface capable of detecting pressure gradients and shear stress. Sensing signals include temperature, pressure gradient, and shear stress sensors distributed throughout a custom shell designed to fit to the surface of the ROV.

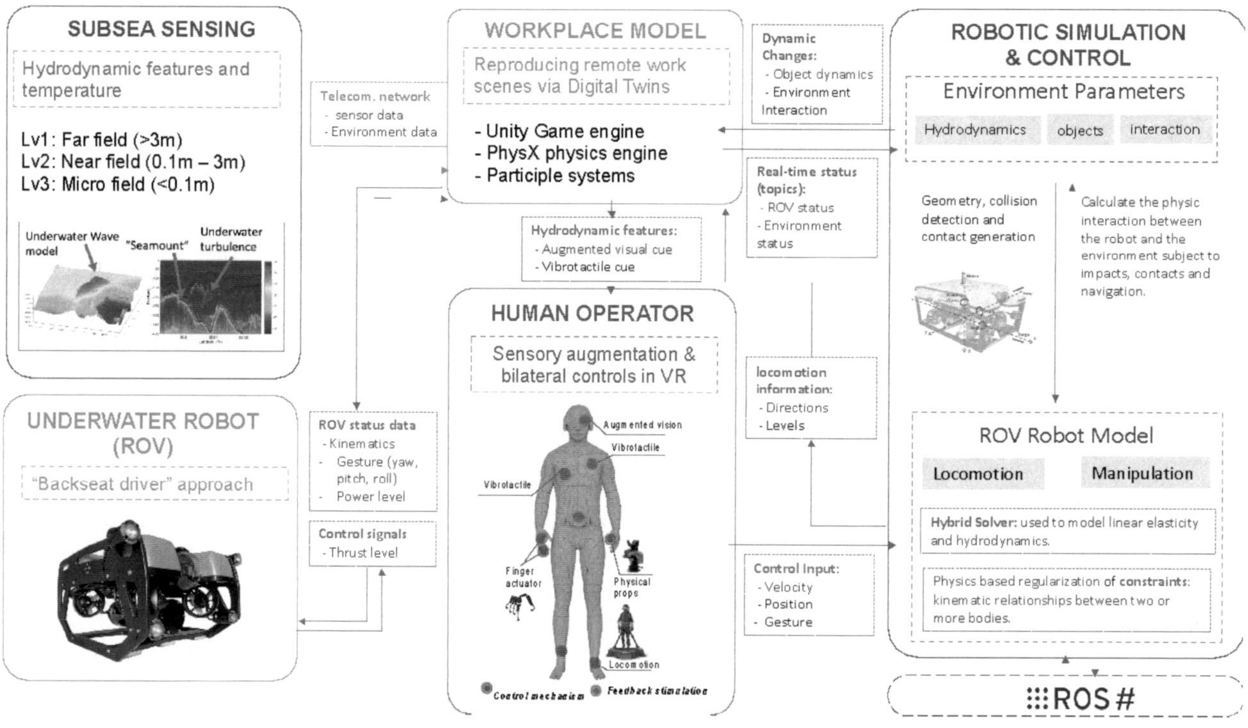

Fig.1: Proposed underwater human-robot interaction for enhanced ROV teleoperation.

- **Workplace Model Module:** The real-time sensor data is then used to model spatiotemporal dynamics of a subsea zone in the vicinity of the robot. To generate an immersive visualization of the subsea workplace, a game engine Unity v2020.1 [15] is used. Unity can model the far field sensor data as vectors and render the entire space as Virtual Reality displays. Another key feature of the proposed system is to convert the hydrodynamic features into human-perceivable sensations, i.e., vibrotactile cues. To realize this function, a physics game engine NIVIDIA PhysX is used (version 9.19) to simulate underwater [16]. Specially, the smoothed-particle hydrodynamics (SPH) particle method [17] of PhysX is used to simulate the hydrodynamic changes based on the sensor data. The raw data is used to determine the initial conditions of the particle emitters. Then a collision detection mechanism is used to examine the collision events between each particle and the virtual ROV model. The collisions frequency and magnitude will be used to generate haptics of different levels (see the next section)
- **Human Operator Module:** Although in ROV operations many of the environmental parameters can be collected and displayed via monitor screens, it requires a significant training effort to comprehend and react in a timely manner. To mitigate the perceptual-motor malfunction and reduce training needs, we use a virtual telepresence environment based on VR and a sensory augmentation simulator, or the human operator module. Two functions are built for human operator module including the sensory augmentation and the bilateral control. Below we introduce the method for sensory augmentation. Bilateral control will be described in the next section.

For sensory augmentation, hydrodynamic features are simulated as the visualized vectors via a virtual reality (VR) headset, and via and vibrotactile via a haptic suit. For the far field data, clusters of vectors are rendered to indicate the overall hydrodynamic flows necessary for the operator's navigation decision-making, including fluid directions, speed, and gradient extensions. Colored vectors are rendered in the simulation to indicate temperature changes. These clusters of vectors flash along with the direction the same as flow data, with amplitude and frequency indicating the flow speed. A more frequent swing and larger amplitude indicate faster and stronger water flows. Compared to traditional camera view feedback, VR provides more enriched spatial information with immersive and interactive visual feedback. Such a hierarchical system design can simplify the overall piloting effort during routine operations and reduce operation inaccuracy due to human errors. For the near field waterbody surrounding the ROV, a SPH particle system is applied to simulate the physical interactions with the ROV in a realistic way. In this study, Obi Fluid [18, 19] is selected as the core near field particle simulation method. The activated particle number is set to 650 for balancing the simulation fidelity and CPU cost. To simulate physical interactions between the waterbody and the ROV, seven particle emitters are set around the ROV model, as demonstrated in Fig. 2a. The initial parameters of these emitters, such as particle initial direction and speed, are based on the received hydrodynamic data from ROV pressure sensors. Whatever actions the human operator initiates would trigger the status changes to the emitters. For example, an emitter in front of the ROV model will generate a particle flow if operators perform a forward operation. A total of 40 sensors are deployed in the virtual ROV model (a digital replica of the real ROV) to collect the collision events between the particles and the ROV model (Fig2b). We use a dynamic collision detector, *PxParticleFlag::eCOLLISION_WITH_DYNAMIC*, to examine whether a particle collided with the dynamic rigid body (i.e., the virtual ROV model) during the last simulation step. Then two methods from PhysX are used to read position and velocity information, including *PxParticleReadDataFlag::ePOSITION_BUFFER* and *PxParticleReadDataFlag::eFLAGS_BUFFER*. The collected parameters are used to calculate the vibration magnitude of a haptic suit with the same number of vibrators (Fig.2c) using the following formula:

$$m_i = \frac{(e^{0.4F_{sensor}}-1)}{(e^{0.4F_{sensor}}+1)} + \frac{(h-h_{min})}{2(h_{max}-h_{min})} \qquad \text{...Eq(1)}$$

Where m_i refers to the magnitude of the ith vibrator on the haptic suit, F_{sensor} represents the flow intensity sent by the sensors, h is the subsea depth of the ROV, h_{min} is the minimum subsea depth of workplace, and h_{max} is the maximum subsea depth of workplace. The subsea hydro pressure is converted into a range of 0 to 0.5, which ensures the final haptic suit vibration intensity would not be larger than 1.5cm/s^2. As such, the human operator can "see" and "feel" the real-time fluid changes analogous to what most fish are capable of.

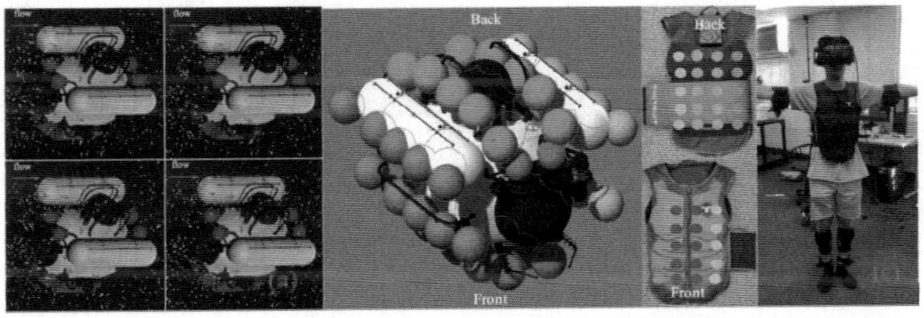

Fig.2: Simulating hydrodynamic features on human operator via a haptic suit

In addition to the reality reconstruction of the near water, we will also apply MS MRTK [20] (an augmented reality toolkit) to enhance color perception [21], such as highlighting the object boundaries, shapes and volumes that will help human operator concentrate on the work zone, and visualizing occluded objects (see-through effects). Although other subsea environmental features may also be relevant (such as salinity and turbidity), we will focus on the above three categories of information for a better system reliability. We will also examine sensory stimulation strategies for user adaptation: *Sensory Prioritization* to decide what feedback is most needed for a given situation; *Sensory Filtering* to determine what sensory feedback shall be blocked to avoid cognitive overload; and *Sensory Manipulation* to simulate one sense (e.g., force) in another form of sense (e.g., vision). Remote robots will be a natural extension of human senses.

- **Robotic Simulation and Control Module.** Physics engine simulation data and sensor data from the remote ROV need to be transferred to Robot Operating System (ROS) seamlessly to enable ROV simulation and controls. Building on our previous work [22], we will examine a data synchronization system for VR and robotic systems.

The system features two functions: converting environmental parameters extracted from the workplace model (hydrodynamics, objects, and interactions) to ROS to rebuild the 3D scene in ROS Gazabo for robot simulation, and to enable the control commends for the ROV. Rosbridge is used to provide a JSON API for transferring data between ROS and Unity [23]. Rosbridge also provides a WebSocket server for web browsers to interact with, serving as a connecting between ROS and the network [23]. ROS server converts ROV dynamics data into JSON messages via rosbridge and publishes it to the website or receives JSON message from Internet and converts it to ROS message [23, 24]. On the Unity side, we use ROS#, a set of open-source software libraries in C#, for communicating with ROS from .NET applications, in particular, Unity [25]. ROS# establishes a WebSocket in Unity so that Unity can connect to a computer with a specific IP address through the network and transfer data. It also helps build nodes that publish and subscribe topics from ROS in Unity. ROS# converts data into JSON and publishes it or converts the received data into the original format. We grant ROS server and Unity's WebSocket the same IP address, so that ROS server can publish the processed topics to the ROS platform, and the Unity can subscribe to all topics on ROS platforms through ROS#.

The robotic simulation and control module also supports an intuitive control of the remote ROV via natural body motions. As shown Fig.3. Human control input parameters, including local rotation of HTC trackers, body postures and a secondary auxiliary controller, are designed to match ROV control parameters such as rotation, moving and some specific control functions. The local rotation of human body is sent to ROV for pitch, roll and yaw control, which ensures ROV's orientation consistent with human's body motion. Human body postures are designed to control ROV moving in the subsea environment. For example, ROV pitch down when human operator leans forward. A secondary auxiliary control method, HTC VIVE controller is introduced for vertical up and down operations and function control. Specifically, the y axis input value of touchpad was used to control up and down for ROV, with the x axis value for speed control. The Boolean value of trigger button was designed to control the system on and off.

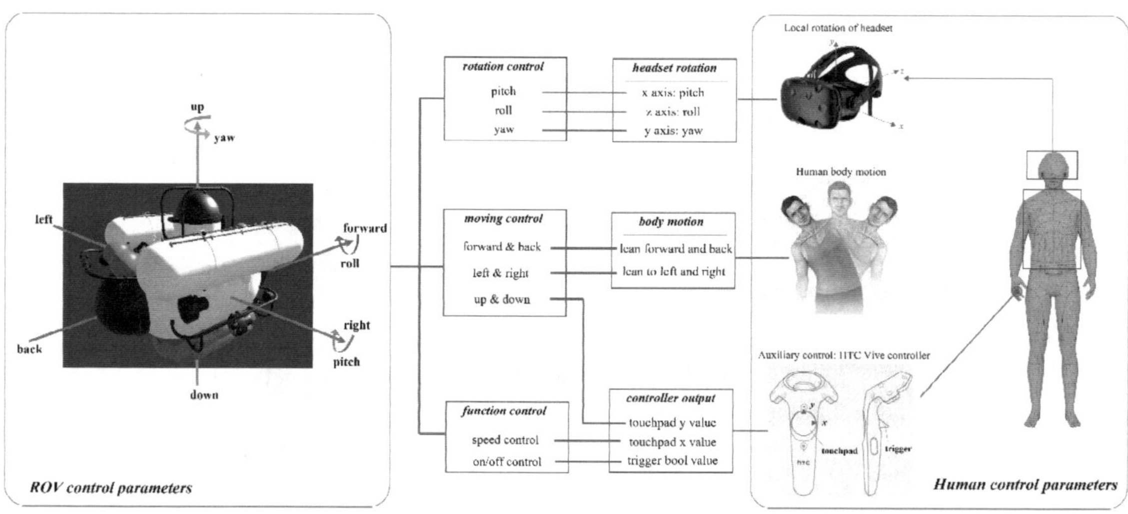

Fig.3: Natural body-motion control

Another need for seamless ROV teleoperation is to render the kinematics features of the remote ROV in VR (e.g., speed, gesture etc.). This is because the locomotion control signals from the human operator are not always realized on the remote ROV due to the dynamic subsea environment. For instance, a human operator may lean forward by 10 degrees to command the corresponding 10-degree negative pitch of the ROV. Nonetheless the real ROV may only demonstrate a 5-degree pitch due to the liquid viscosity underwater. As such, the reactions of the ROV kinematics must be regenerated despite what controls are given by the human operator. In our system, we don't rely on the real ROV kinematics data (collected from the onboard sensors) because of the possible tracking errors or telecommunication latencies. Instead, we rely on the real-time ROS Gazebo simulation to recover the predicted ROV kinematics status. The challenge would be to reproduce the robotic dynamics in ROS Gazebo in a precise and accurate manner. We will use a hybrid solver that solves both the linear elasticity and hydrodynamic changes of the simulated ROV in Gazebo, such as [26].

- **Subsea Robot (ROV) Module.** The last module of the proposed system is the remote ROV. The vehicle used in this project is based on a BlueROV2 platform with a heavy configuration. The BlueROV2 is an open-source underwater vehicle equipped with six thrusters in a vectored configuration. With the heavy retrofit kit, four additional vertical thrusters are added to the vehicle, providing the vehicle with all six degrees of freedom. The based vehicle is powered by an onboard 18Ah battery, giving 2-3 hours battery life with a single charge. On top of the base platform, our vehicle is upgraded with a Jetson Xavier NX backseat computer to perform high-level sensor fusion and closed-loop control autonomy. The vehicle is outfitted with a bottom-facing single-beam echosounder that measures distances with respect to the seafloor for up to 50 meters, a 360-degree scanning imaging sonar for underwater perception. In addition, the vehicle is equipped with a Nortek Doppler velocity log with current profiling capability that allows the vehicle to measure the relative velocity with respect to the seafloor as well as the far-field flow velocity for augmenting the operator's situational awareness in the digital-twin environment. We developed a "backseat driver" computing method to realize the open loop control needs. As discussed earlier, there may be a disconnection between the control commends issued by the human operator and the actual reaction of the ROV due to the changing hydrodynamic conditions in the subsea workplace. As a result, a resolver is used to generate the correct rendering of ROV kinematics in VR. The same applies to the controlling of the real ROV, as the real system also needs to match the control commends and mirror the behaviors in the VR environment. The same hybrid solver will be applied on the backseat driver computer.

Demonstration Case

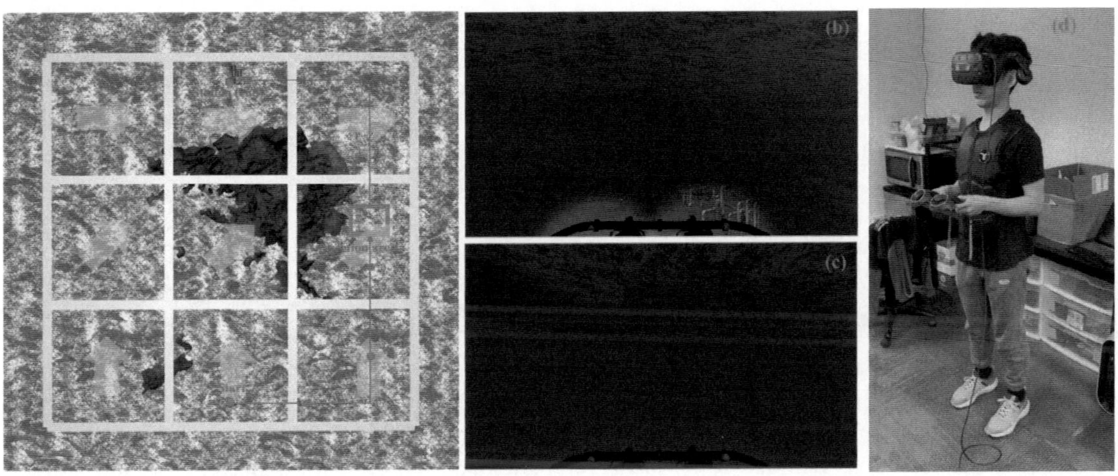

Fig.4: Demonstration case. (a) Task map; (b) Inspection area; (c) view of station; (d) Human operator with control devices.

We performed a test case to prove the concept. As shown in **Fig.4**, the operator needed to complete a subsea inspection in the inspection area then navigated to the station to end the task. The task route and ocean flow conditions are shown in **Fig. 4a**. In this task, operators could finish a series of basic operations including moving forward, up, down and turning around. With the combination of different flow conditions and ROV operations, the haptic suit would generate significantly different vibration patterns on the body of the operator. In total, five different representative cases were selected to show the varying haptic patterns, including right flow with forward operation (case 1), back flow with forward operation (case 2), 45-degree back flow with forward operation (case 3), forward flow with forward operation (case 4), and forward flow with up operation (case 5).

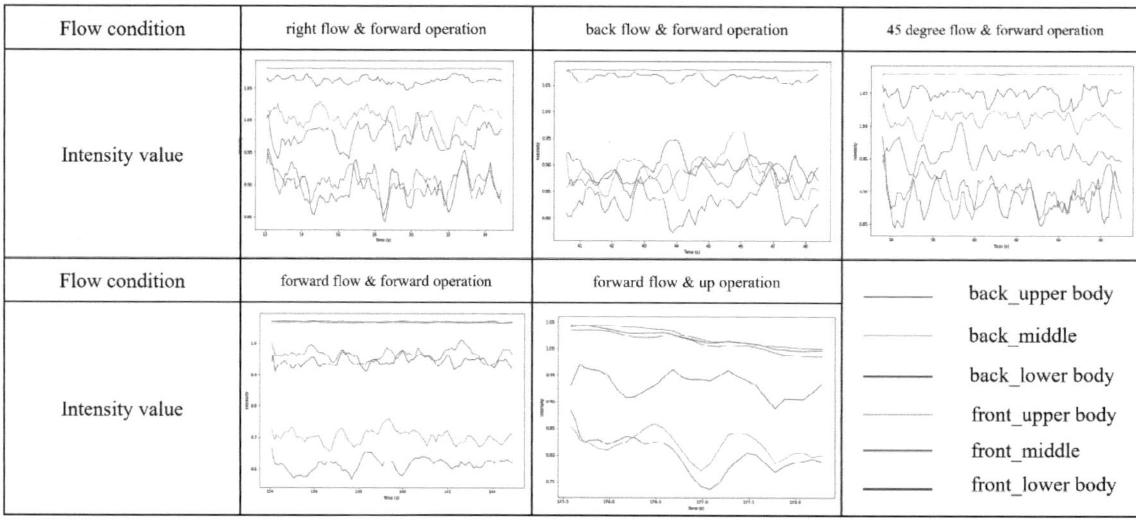

Fig.5: Haptic intensity pattern for five cases

Fig. 5 demonstrates the different haptic patterns of average intensity value in each case. For different combinations of flow conditions and operations, significant difference could be observed in the haptic feedback. For example, for the back flow with forward operation, a higher haptic intensity was observed in the front part. The reason might be that moving operation influence more than the fluid speed we set. For forward flow with forward operation, a similar higher haptic intensity patten was observed in the front part, but the difference between front and back is much greater. For forward flow with up operation, a general higher intensity was observed in the front and upper channels. The result showed that there was a significant difference in vibration patterns for different operations and flow conditions. By transferring underwater hydrodynamic conditions to human body sensation, participants could identify different ROV positions and locomotion conditions effectively.

Discussion

1. Opportunities

We expect this new method of ROV teleoperations to advance a booming subsea engineering industry that requires a strong integration between human intelligence and robots to tackle environmental complexity (e.g., volatile water flow) and task dynamics (e.g., navigation with low visibility). At present, the needs for subsea robots are diverse, from inspection ROVs (smaller scale and single-purposed) to workhorse ROVs (bigger scale and versatile) for shallow water and deep-sea activities. Without losing the generalizability, this method is expected to enable a much closer human-ROV collaboration for subsea inspection and survey, i.e., the maneuver and navigation controls of remote ROVs for sensor data collection and scanning of vessels and subsea structure inspection in offshore zones. It can make the key tasks easier, including *navigation* (localization and state estimation), *control* (path planning and maneuvering through complex environments) and *perception* (for robot position control and the inspection task). Subsea inspection and survey are among the most common subsea robotic operations. The relevant tasks include but are not limited to subsea infrastructure inspection, geological survey, marine habitat monitoring, pollution assessments, ship-hull inspections, unexploded ordnance (UXO) surveys, contraband detection, aquaculture monitoring facility inspection, search and rescue, and archaeology exploration and surveys. Applications in other industries where the same technologies apply include hydro power-dam wall inspections, blockage detection at penstock intake, bridge and pier structure monitoring, nuclear plant inspection and operations, and survey and monitoring in lake environments. Subsea robotic inspection is urgently needed and represents unique challenges in both workplaces (diverse underwater environments) and works (dynamic and instant decision making, path planning, operational and decisional uncertainties).

This research is strongly positioned for better accessibly and inclusion because it aims to lower the career barrier for a traditionally highly professional area. The proposed underwater human-robot interaction approach will greatly simplify the requirement on engineering, science, and robotics knowledge for subsea engineering and underwater robot operation jobs. The marine and offshore sectors are dominated by male workers. The new job opportunities will help ease the inequality in gender, with that more women will feel more welcomed because of the safer work environment created by the virtual telepresence technology. The current subsea engineering also poses strict requirements on age, because of the mental and physical load in certain tasks. The sensory augmentation method for robotic control will mitigate the age requirement, promoting career longevity. The new technology will also help salvage the careers of experienced workers who have suffered from career injuries, such as diving diseases. Furthermore, because our investigation will focus on reskilling and upskilling the workforce that is threatened by automation, lessons learned in this project will help workforce transformation and secure job opportunities for US workforces in various industries. It includes construction and mining industries that have been dominant by underrepresented groups of workers.

We also anticipate potential economic impacts of our technology. We predict that the new technology will reduce costs through four channels: (i) Reduced training requirements. Current ROV operation and commercial diving are skill-intensive, with higher barriers to entry and expensive training[27, 28]. We expect the new technology to reduce these barriers. This reduces costs by reducing training time, expenses, and salary requirements for high skilled workers. (ii) Increased capital productivity. Currently, sub-sea inspections and maintenance are performed by the personnel onboard (POB), which carries high logistical, transportation, safety, and compensation costs – if operators can direct an ROV from a VR suit in an office park, rather than helicoptering out to a ship or platform, cost savings may be significant[29]. (iii) Increased capital utilization. Industry feedback (pilot interviews with three ROV service providers) indicates that ROVs may be used on only 50% of days because of the limited availability of high-skill, trained operators – this is unusually low[30]. Reducing operator training and reducing the costs of matching an operator with the robot can allow each piece of equipment to sit unused less often, lowering costs. (iv) Investment incentives. Reallocating human-performed tasks to HRC is especially incentivized in the low interest rate environment that long-run yields predict will characterize the next several decades, because capital costs for robots are amortized over years. Our plan for estimating industry growth is described as follows.

2. Challenges

There are still many challenges for us to resolve to make this technology viable. The first challenge relates to the technological maturity. Realizing the proposed system would require a significant change to the current ROV designs. It includes equipping the ROV systems with new sensors that can collect high-fidelity underwater environment data, such as pressure sensors on the surface of ROV and Doppler sensors for far field hydrodynamic sensing. The initial cost may become a burden for many businesses. A new data and telecommunication infrastructure is also needed for transferring the potentially big amount of data to support the human-robot sensory transfer. Without concrete evidence on the benefits of the proposed method, it would be extremely difficult for industry and academia to make the aggressive move in transforming the current systems.

In addition, we recognize that the technology innovation alone would not be sufficient for attracting the current and future maritime workforce to enter the new subsea engineering robot teleoperation sector. The lack of trust issue has been seen in many new technology adoptions. The workforce technology transformation cannot be capitalized without appropriate training and commitment. The training program needs to be integrated into the current training protocols to gradually adapt and change the transformative workforce. But there are many unknowns. First, we don't know how and to what extent a new ROV teleoperation experience affects productivity and safety for subsea robot operations. We expect that the introduction of the new human-robot interaction will inevitably result in behavioral and psychological changes of future underwater robot operators. Evidence must be collected to verify the benefits of the proposed technology, and more importantly, the possible side effects. Previous HRC literature has observed undesired side effects of introducing shared autonomy in workplaces, such as bully behaviors [31-33], distrust [34-37], loss of self-confidence in tasks [38-40], and loss of situational awareness [41, 42]. Traditional subsea inspection is among one of the most dangerous occupations [43, 44]. We don't know if the proposed HRI technology can enforce the required safety protocols in subsea inspection. Second, even given the productivity and training benefits attributed to the proposed new HRI technology, we still don't know if it is enough to increase the willingness for workforces from other declining industries to enter a subsea career. Considerations for changing career path can be complex. According to the Rhodes and Doering Model [45], factors include traits, demographic factors, human capital (e.g., education and occupational tenure), job satisfaction, motivation, organizational factors, and self-identification. We need to measure the career change motivation in addition to work performance. Third, we don't know how the proposed technology affects the training and education for a transformative workforce. Successful adult learning involves a complex process of transformation in perspective that is deeply tied to context and influenced by a catalyzing dilemma [46]. Findings from prior research indicates the importance of individual agency, shared experience, critical reflection, and the need for time [47]. However, little attention has been given to understanding how the unique attributes of a more applied professional setting might serve as an affordance [48]. Few studies have explored how a general transformative learning principle can be informed for the unique attributes of participants from diverse backgrounds, cultures, etc., or how emerging technologies might influence the processes and outcomes [49]. Especially, later stage adults are understudied, yet this segment is growing more rapidly than any other age group. There is a critical need for in-depth analysis of multiple theoretical perspectives of transformative learning, particularly involving comparative analysis. Applying various theoretical orientations for transformative learning to a context involving VR and HRI as an emerging technology would offer the opportunity to better understand both the nature and practice of a workforce transformation.

Conclusion

Subsea engineering operations heavily rely on ROV, and the performance depends on the seamless interaction between ROV and the human operator. Due to the dynamics of subsea environments such as the uncertainty of turbulence, affected visibility, and the interference with subsea ecosystems, control of subsea ROV is challenging for human operators who have never exposed to such an environment. Even with the tremendous amount of practice, the human operator can easily misjudge the situation and make wrong decisions in subsea operations. This paper introduces the basic information of a proposed human-robot collaboration method based on a VR and haptic simulation system. To explore the potentials of utilizing VR to accomplish ROV control, a haptic suit is combined with VR goggles to bring an immersive subsea environment to the operator. A comprehensive physics model is first transferred to VR equipment, depicting a detailed underwater environment including turbulence and interaction with ambient objects. Then the haptic suit executes the corresponding pressure signals to the operator's body in real-time. The feedback signals from haptic suit provide the human operator with realistic sensation of subsea situations peripheral to the robot. This immersive virtual environment ensures ROV operators of real-time awareness of the proximity conditions and prediction of changes. As a result, a less-trained human operator can pilot the ROV based on his/her intuition and experience to maximize the performance and avoid potential mistakes. A variety of opportunities have been identified, including supporting the easier underwater inspection tasks, improve the access and inclusion of subsea engineering, and the long-term economic benefits. The challenges related to technological transformation and training/adoption issues have also been identified.

Acknowledgements

This material is supported by National Science Foundation (NSF) under award 2128895.

References

[1] Shimizu, "OCEAN SPIRAL, Deep Sea Future City Concept." [Online]. Available: https://www.shimz.co.jp/en/topics/dream/content01/.

[2] T. R. Knutson *et al.*, "Dynamical downscaling projections of twenty-first-century Atlantic hurricane activity: CMIP3 and CMIP5 model-based scenarios," *Journal of Climate,* vol. 26, no. 17, pp. 6591-6617, 2013.

[3] U. C. B. Census. "About 60.2M Live in Areas Most Vulnerable to Hurricanes." https://www.census.gov/library/stories/2019/07/millions-of-americans-live-coastline-regions.html (accessed.

[4] M. Sahi, "Challenges for Underwater Robots," 2015. [Online]. Available: https://www.tractica.com/automation-robotics/challenges-for-underwater-robots/.

[5] N. Anderson, "Thalassophobia: Fear Of The Sea," *Spazio Humano,* no. 2, 1990.

[6] S. Sivčev, J. Coleman, E. Omerdić, G. Dooly, and D. Toal, "Underwater manipulators: A review," *Ocean Engineering,* vol. 163, pp. 431-450, 2018.

[7] R. S. McEwen, B. W. Hobson, L. McBride, and J. G. Bellingham, "Docking control system for a 54-cm-diameter (21-in) AUV," *IEEE Journal of Oceanic Engineering,* vol. 33, no. 4, pp. 550-562, 2008.

[8] S. Sendra, J. Lloret, J. M. Jimenez, and L. Parra, "Underwater acoustic modems," *IEEE Sensors Journal,* vol. 16, no. 11, pp. 4063-4071, 2015.

[9] M. L. Seto, *Marine robot autonomy*. Springer Science & Business Media, 2012.

[10] G. Antonelli and G. Antonelli, *Underwater robots*. Springer, 2014.

[11] R. Kostaschuk, J. Best, P. Villard, J. Peakall, and M. Franklin, "Measuring flow velocity and sediment transport with an acoustic Doppler current profiler," *Geomorphology,* vol. 68, no. 1-2, pp. 25-37, 2005.

[12] M. Krieg, K. Nelson, and K. Mohseni, "Distributed sensing for fluid disturbance compensation and motion control of intelligent robots," *Nature machine intelligence,* vol. 1, no. 5, pp. 216-224, 2019.

[13] M. Krieg, I. Sledge, and K. Mohseni, "Design considerations for an underwater soft-robot inspired from marine invertebrates," *Bioinspiration & biomimetics,* vol. 10, no. 6, p. 065004, 2015.

[14] M. Krieg, P. Klein, R. Hodgkinson, and K. Mohseni, "A hybrid class underwater vehicle: Bioinspired propulsion, embedded system, and acoustic communication and localization system," *Marine technology society journal,* vol. 45, no. 4, pp. 153-164, 2011.

[15] Unity. "Official Website." https://unity.com (accessed.

[16] N. PhysX. "PhysX Engine Dscription." https://www.nvidia.com/en-us/drivers/physx/physx-9-19-0218-driver/ (accessed.

[17] J. J. Monaghan, "Smoothed particle hydrodynamics," *Annual review of astronomy and astrophysics,* vol. 30, no. 1, pp. 543-574, 1992.

[18] VirtualMethod. "Obi Fluid." https://assetstore.unity.com/packages/tools/physics/obi-fluid-63067 (accessed.

[19] Obi. "Obi: Unified particle physics for Unity." http://obi.virtualmethodstudio.com/ (accessed.

[20] Microsoft. "MRTK: Mixed Reality Toolkit." https://github.com/microsoft/MixedRealityToolkit-Unity (accessed March 5, 2021).

[21] J. L. Gabbard, M. Smith, Coleman Merenda, G. Burnett, and D. R. Large, " Perceptual Color-Matching Method for Examining Color Blending in Augmented Reality Head-Up Display Graphics," *IEEE Transactions of Visualization & Computer Graphics (In review),* 2020.

[22] T. Zhou, Zhu, Q., Dixon, W., Du, J. , "Intuitive Telerobotic Controls based on Real-Time Digital Twins Simulation," *Automation in Construction (under review),* 2020.

[23] C. Crick, G. Jay, S. Osentoski, B. Pitzer, and O. C. Jenkins, "Rosbridge: Ros for non-ros users," in *Robotics Research*: Springer, 2017, pp. 493-504.

[24] M. Quigley *et al.*, "ROS: an open-source Robot Operating System," in *ICRA workshop on open source software*, 2009, vol. 3, no. 3.2: Kobe, Japan, p. 5.

[25] GitHub. "ROS# GitHub page." https://github.com/siemens/ros-sharp (accessed.

[26] S. Chitta, I. Sucan, and S. Cousins, "Moveit![ros topics]," *IEEE Robotics & Automation Magazine,* vol. 19, no. 1, pp. 18-19, 2012.

[27] B. o. L. S. BLS, "Occupational Employment Statistics, Marine Engineers and Naval Architects," 2021. [Online]. Available: https://www.bls.gov/oes/current/oes172121.html

[28] B. o. L. S. BLS, "Occupational Employment Statistics, Commercial Divers," 2021. [Online]. Available: https://www.bls.gov/oes/2018/may/oes499092.html

[29] D. McLeod, "Emerging capabilities for autonomous inspection repair and maintenance," in *OCEANS 2010 MTS/IEEE SEATTLE*, 2010: IEEE, pp. 1-4.

[30] J. J. Beaulieu and J. Mattey, "The workweek of capital and capital utilization in manufacturing," *Journal of Productivity Analysis,* vol. 10, no. 2, pp. 199-223, 1998.

[31] B. J. I. w. C. Whitby, "Sometimes it's hard to be a robot: A call for action on the ethics of abusing artificial agents," vol. 20, no. 3, pp. 326-333, 2008.

[32] L. D. Riek, T.-C. Rabinowitch, B. Chakrabarti, and P. Robinson, "How anthropomorphism affects empathy toward robots," in *Proceedings of the 4th ACM/IEEE international conference on Human robot interaction*, 2009: ACM, pp. 245-246.

[33] C. Bartneck and J. J. I. S. Hu, "Exploring the abuse of robots," vol. 9, no. 3, pp. 415-433, 2008.

[34] P. A. Hancock, D. R. Billings, K. E. Schaefer, J. Y. Chen, E. J. De Visser, and R. J. H. F. Parasuraman, "A meta-analysis of factors affecting trust in human-robot interaction," vol. 53, no. 5, pp. 517-527, 2011.

[35] A. Freedy, E. DeVisser, G. Weltman, and N. Coeyman, "Measurement of trust in human-robot collaboration," in *Collaborative Technologies and Systems, 2007. CTS 2007. International Symposium on*, 2007: IEEE, pp. 106-114.

[36] T. Sanders, K. E. Oleson, D. R. Billings, J. Y. Chen, and P. A. Hancock, "A model of human-robot trust: Theoretical model development," in *Proceedings of the human factors and ergonomics society annual meeting*, 2011, vol. 55, no. 1: SAGE Publications Sage CA: Los Angeles, CA, pp. 1432-1436.

[37] P. A. Hancock, D. R. Billings, and K. E. J. E. i. D. Schaefer, "Can you trust your robot?," vol. 19, no. 3, pp. 24-29, 2011.

[38] J. L. Burke, R. R. Murphy, M. D. Coovert, and D. L. J. H. C. I. Riddle, "Moonlight in Miami: Field study of human-robot interaction in the context of an urban search and rescue disaster response training exercise," vol. 19, no. 1-2, pp. 85-116, 2004.

[39] J. D. Lee and N. J. I. j. o. h.-c. s. Moray, "Trust, self-confidence, and operators' adaptation to automation," vol. 40, no. 1, pp. 153-184, 1994.

[40] K. Goddard, A. Roudsari, and J. C. J. J. o. t. A. M. I. A. Wyatt, "Automation bias: a systematic review of frequency, effect mediators, and mitigators," vol. 19, no. 1, pp. 121-127, 2011.

[41] C.-M. Huang and A. L. Thomaz, "Effects of responding to, initiating and ensuring joint attention in human-robot interaction," in *RO-MAN, 2011 IEEE*, 2011: IEEE, pp. 65-71.

[42] V. Micelli, K. Strabala, and S. Srinivasa, "Perception and control challenges for effective human-robot handoffs," in *Robotics: Science and systems workshop on rgb-d cameras*, 2011.

[43] OSHA, "Dysbarisms and Other Selected Health Effects, Commercial Diving - Hazards and Solutions," 2020. [Online]. Available: https://www.eagle.org/ABSEaglePrograms/es/es-search.jsp

[44] HSE. "HSE and Diving, U.K. Health and Safety Excutive." https://www.hse.gov.uk/diving/how.html (accessed.

[45] S. R. Rhodes and M. Doering, "An Integrated Model of Career Change," *Academy of Management Review,* vol. 8, no. 4, pp. 631-639, 1983.

[46] E. W. Taylor, "An update of transformative learning theory: A critical review of the empirical research (1999–2005)," *International journal of lifelong education,* vol. 26, no. 2, pp. 173-191, 2007.

[47] E. Taylor, "Fostering Mezirow's transformative learning theory in the adult education classroom: A critical review," *Canadian Journal for the Study of Adult Education,* vol. 14, no. 2, pp. 1-28, 2000.

[48] E. W. Taylor and M. J. Snyder, "A critical review of research on transformative learning theory, 2006-2010," *The handbook of transformative learning: Theory, research, and practice,* pp. 37-55, 2012.

[49] V. Wang, G. Torrisi-Steele, and E. Reinsfield, "Transformative learning, epistemology and technology in adult education," *Journal of Adult and Continuing Education,* p. 1477971420918602, 2020.

SNAME

THE INTERNATIONAL COMMUNITY
FOR MARITIME AND OCEAN PROFESSIONALS

MODELING ASPECTS OF RUNNING DUCTILE FRACTURE PROPAGATION IN HIGH PRESSURE LINE PIPES

MARCELO PAREDES
DEPARTMENT OF OCEAN ENGINEERING, TEXAS A&M UNIVERSITY, GALVESTON, TX 77554

Proceedings of the 27th Offshore Symposium, February 22nd 2022, Houston, Texas
Texas Section of the Society of Naval Architects and Marine Engineers

ABSTRACT

Multi-physics simulations are becoming increasingly important in the time where full-scale test facilities and materials are turning out costly. Moreover, the new transition trend of energy production from fossil fuels to renewable sources like solar, wind and hydrogen is pushing forward the development of new technologies for harvesting, storage and transportation. For the latter, the existing pipeline network is intended to be utilized for moving a mixture of hydrogen with natural gases from clean hydrogen production centers to consumers. Understanding the coupling of fluid-structure interaction with thermodynamics of the mixtures is imperative to assure more reliable and fail-safe operations of the infrastructure for production and transportation. In this short survey, a phenomenological fracture model along with coupled Eulerian-Lagrangian/Fluid-Structure interaction model is reviewed. It is shown that the obtained simulation results are in good agreement with experimental tests at laboratory scale and full-scale settings.

Keywords: pipeline failure, fluid-structure interaction, CH_4 mixtures

INTRODUCTION

In the U.S., the pipeline network of natural gas is a highly integrated network that transports this commodity throughout the country. This pipeline system is comprised of more than 3 million miles of mainline and other pipelines that link natural gas production areas and storage facilities with consumers. Its delivery rate in 2019 was approximately 30 trillion cubic feet (Tcf) of natural gas to more than 77 million customers [1]. A recent survey shows that in a 9-year period (2010-2018), the millions of miles of pipelines have resulted in the following: more than 5,500 total incidents that include about 600 injuries, more than 125 fatalities, more than 800 fires, almost 300 explosions, more than $4 billion in damages, and almost 30,000 people who had been evacuated [2]. This suggests a more exhaustive control on operations and regulations are necessary being placed in order to guarantee safety and reliability. Moreover, new challenges are coming to the already installed pipeline infrastructure. The Department of Energy (DoE) recently launched the HyBlend initiative that pursues the utilization of such network to move mixtures of hydrogen and natural gas from clean hydrogen production facilities to consumers including industrial, commercial and residential end users [3]. This project aims to address technical barriers to blending hydrogen in natural gas pipelines as materials compatibility and utilization along with high-fidelity computational models development. The latter becomes increasingly important in an age where full-scale burst tests are becoming expensive to be carried out on daily basis. In this manuscript, therefore, a brief review on computational modeling of propagating ductile fracture in line pipes by mean of a multi-physics phenomenological approach is done. As highlight, the coupled Eulerian-Lagrangian (CEL) framework, built-in capability in Abaqus/Explicit [4], is exploited along with material user-subroutine to define the thermodynamics of the escaping gas, mechanical response of pipe material and backfill soil and fluid-structure interaction. All defined within a single computer-aided engineering environment to simulate running ductile fracture in pressurized pipelines, typically, employed for transporting gases in onshore installations. The review begins with a description of the phenomenological model used to define the existing constitutive relations between stresses and strains under dynamic loading scenarios. Then, it is followed by the methods employed to describe the thermodynamics of gases and the backfill soil effects in conjunction with their numerical implementation. Finally, some published results are shown and discussed within this framework.

PHENOMENOLOGICAL MODEL

Rate dependent plasticity model

The mechanical behavior of a typical pipeline steels can be described by a quadratic yield function in the form

$$f(\sigma, k) = \bar{\sigma} - k = \sqrt{(\mathbf{P}\boldsymbol{\sigma}) \cdot \boldsymbol{\sigma}} - k = 0 \tag{1}$$

where $\boldsymbol{\sigma}$ and k represents the symmetric Cauchy stress tensor and deformation resistance, respectively. $\boldsymbol{P}$ is the semi-positive definite matrix of the material stress ratios measured at different loading directions conveniently given in the following form[5].

$$\mathbf{P} = \begin{bmatrix} P_{11} & P_{12} & P_{31} & 0 & 0 & 0 \\ P_{12} & P_{22} & P_{23} & 0 & 0 & 0 \\ P_{31} & P_{23} & P_{33} & 0 & 0 & 0 \\ 0 & 0 & 0 & P_{44} & 0 & 0 \\ 0 & 0 & 0 & 0 & P_{55} & 0 \\ 0 & 0 & 0 & 0 & 0 & P_{66} \end{bmatrix} \tag{2}$$

The coefficients of $\boldsymbol{P}$ matrix are related to Hill's anisotropy coefficients as [6]

$$\begin{bmatrix} P_{11} \\ P_{22} \\ P_{33} \\ 2P_{12} \\ 2P_{23} \\ 2P_{31} \\ P_{44} \\ P_{55} \\ P_{66} \end{bmatrix} = \begin{bmatrix} G+H \\ F+H \\ F+G \\ -2H \\ -2F \\ -2G \\ 2N \\ 2M \\ 2L \end{bmatrix} \tag{3}$$

The expansion of the product in Eq. (1) will be set equal to Hill's equivalent stress

$$\begin{aligned} \bar{\sigma}^2 = {} & (G+H)\sigma_{11}^2 + (F+H)\sigma_{22}^2 + (F+G)\sigma_{33}^2 - 2H\sigma_{11}\sigma_{22} - 2F\sigma_{22}\sigma_{33} - 2G\sigma_{11}\sigma_{33} \\ & + 2N\sigma_{12}^2 + 2M\sigma_{31}^2 + 2L\sigma_{23}^2 \end{aligned} \tag{4}$$

rearranging Eq. (4) by assuming that the anisotropy is negligible in the out-of-plane shear components and the selected reference material orientation is along the longitudinal axis (e_1), the $\boldsymbol{P}$ matrix can be simplified significantly into three independent parameters, $\boldsymbol{P}_{12}$, $\boldsymbol{P}_{22}$ and $\boldsymbol{P}_{44}$. For the special case of the von Mises isotropic yield function is retrieved by setting $\boldsymbol{P}_{12} = -0.5$, $\boldsymbol{P}_{22} = 1$ and $\boldsymbol{P}_{44} = 3$.

$$\mathbf{P} = \begin{bmatrix} 1 & P_{12} & -(1+P_{12}) & 0 & 0 & 0 \\ P_{12} & P_{22} & -(P_{22}+P_{12}) & 0 & 0 & 0 \\ -(1+P_{12}) & -(P_{22}+P_{12}) & 1+2P_{12}+P_{22} & 0 & 0 & 0 \\ 0 & 0 & 0 & P_{44} & 0 & 0 \\ 0 & 0 & 0 & 0 & 3 & 0 \\ 0 & 0 & 0 & 0 & 0 & 3 \end{bmatrix} \tag{5}$$

Flow Rule

The flow rule defines the amplitude of the incremental plastic strain tensor in the appropriate stress space. Classical plasticity theory requires that the plastic strain increment lie normal to the yield surface as the stress derivative of function f (Eq. 1). This associate flow condition does not always hold true for all cases when anisotropy is present in [7]. Instead, a plastic potential function $\boldsymbol{g}$ is used to account for directional effect on plastic strain evolution in the form

$$d\varepsilon_p = d\lambda \frac{\partial \mathrm{g}}{\partial \boldsymbol{\sigma}} \tag{6}$$

where $d\lambda$ is nonnegative scalar multiplier while $\partial \mathbf{g}/\partial \boldsymbol{\sigma}$ is the gradient of the plastic potential from which plastic strain vector is computed. In the same fashion as before $\boldsymbol{g}$ function is defined as a quadratic function in stress space

$$\mathrm{g}(\boldsymbol{\sigma}) = \sqrt{(\mathbf{G}\boldsymbol{\sigma}) \cdot \boldsymbol{\sigma}} \tag{7}$$

where $\boldsymbol{G}$ is a positive definite matrix written in the following form

$$\mathbf{G} = \begin{bmatrix} 1 & G_{12} & -(1+G_{12}) & 0 & 0 & 0 \\ G_{12} & G_{22} & -(G_{22}+G_{12}) & 0 & 0 & 0 \\ -(1+G_{12}) & -(G_{22}+G_{12}) & 1+2G_{12}+G_{22} & 0 & 0 & 0 \\ 0 & 0 & 0 & G_{44} & 0 & 0 \\ 0 & 0 & 0 & 0 & 3 & 0 \\ 0 & 0 & 0 & 0 & 0 & 3 \end{bmatrix} \tag{8}$$

Note that in the special case of $\boldsymbol{P} = \boldsymbol{G}$ the associate flow rule is retrieved.

Rate dependent isotropic hardening

The functional variable k represents the multiplicative decomposition of the hardening response to include the effects of strain rate and temperature softening on the plastic flow of the material in the form [8]

$$k(\bar{\varepsilon}_p, \dot{\bar{\varepsilon}}_p, T) = k_\varepsilon(\bar{\varepsilon}_p) k_{\dot{\varepsilon}}(\dot{\bar{\varepsilon}}_p) k_T(T) \tag{9}$$

where the first strain hardening k_ε term corresponds to the rate independent deformation resistance given as the combined Swift-Voce isotropic strain hardening law that describes the plastic flow behavior in the post-necking region [9], [10]

$$k(\bar{\varepsilon}_p) = \alpha A(\varepsilon_0 + \bar{\varepsilon}_p)^n + (1-\alpha)\left(k_0 + Q\left(1 - e^{-\beta\bar{\varepsilon}_p}\right)\right) \tag{10}$$

where A, ε_0 and n are material parameters of Swift model and k_0, Q and β are coefficients of Voce equation. Through the α parameter, the contribution of each hardening model is weighted to determine the influence of the plastic work done (δw_p) on the deformed material during each increment of plastic strain $(d\bar{\varepsilon}_p)$. Moreover, the strain rate hardening function $k_{\dot{\varepsilon}}$ and the temperature softening function k_T correspond to classical Johnson-Cook form [8]

$$k_{\dot{\varepsilon}}\left(\dot{\bar{\varepsilon}}_p\right) = \begin{cases} 1 & \dot{\bar{\varepsilon}}_p < \dot{\varepsilon}_0 \\ 1 + C\ln\left[\dfrac{\dot{\bar{\varepsilon}}_p}{\dot{\varepsilon}_0}\right] & \dot{\bar{\varepsilon}}_p \geq \dot{\varepsilon}_0 \end{cases} \tag{11}$$

$$k_T\,(T) = \begin{cases} 1 & T < T_r \\ 1 + \left(\dfrac{T - T_r}{T_m - T_r}\right)^{\mu} & T_r \leq T \leq T_m \end{cases} \tag{12}$$

where C is the strain rate sensitivity, $\dot{\varepsilon}_0$ is the reference strain rate, μ is the temperature softening exponent, T_r is the reference temperature and T_m is the melting temperature. On the other hand, the temperature evolution is accounted for as purely mechanical expression, which temperature changes as plastic works increases in the form [11]

$$\rho C_p dT = \omega\left(\dot{\bar{\varepsilon}}_p\right)\eta_k\,\bar{\sigma}\cdot d\bar{\varepsilon}_p \tag{13}$$

herein, ρ, C_p and η_k are density, heat capacity and Taylor-Quinney coefficient, respectively. The functional plastic strain parameter $\omega\left(\dot{\bar{\varepsilon}}_p\right)$ defines the condition where the plastic work turns into heat (adiabatic) or simply dissipates into the air (isothermal) as follows

$$\omega\left(\dot{\bar{\varepsilon}}_p\right) = \begin{cases} 1 & \dot{\bar{\varepsilon}}_p \geq \dot{\varepsilon}_\alpha \\ \dfrac{\left(\dot{\bar{\varepsilon}}_p - \dot{\varepsilon}_{it}\right)^2\left(3\dot{\varepsilon}_\alpha - 2\dot{\bar{\varepsilon}}_p - \dot{\varepsilon}_{it}\right)}{(\dot{\varepsilon}_\alpha - \dot{\varepsilon}_{it})^3} & \dot{\varepsilon}_{it} < \dot{\bar{\varepsilon}}_p \leq \dot{\varepsilon}_\alpha \\ 0 & \dot{\varepsilon}_{it} > \dot{\bar{\varepsilon}}_p \end{cases} \tag{14}$$

where $\dot{\varepsilon}_{it}$ and $\dot{\varepsilon}_a$ are the isothermal and adiabatic limit strain rates, respectively. For simplicity, $\dot{\varepsilon}_{it} = \dot{\varepsilon}_0$. The adjustment for rate dependent parameters is done by getting the fitting parameters of the strain rate experiments detailed in next section. Moreover, thermal softening constants were obtained from the open literature for typical ferritic steels since it is believed that small variance might exist but not so seriously to affect the results. On the other hand, the normality condition is adopted between strain and stress components increment via the work-conjugate principle of the equivalent plastic strain in the form

$$\delta w_p \equiv \sigma : d\varepsilon_p = \bar{\sigma} d\bar{\varepsilon}_p \tag{15}$$

Damage model

A damage indicator to predict fracture initiation under dynamic loads is proposed. An integral form of the damage indicator (D) serves to describe fracture initiation when it reaches a critical value, typically equal to unity in the following form

$$D\left(\bar{\varepsilon}_p,\ \dot{\bar{\varepsilon}}_p\right) = \int_0^{\bar{\varepsilon}_f} \frac{d\bar{\varepsilon}_p}{\bar{\varepsilon}_f(\eta, \bar{\theta}, \dot{\bar{\varepsilon}}_p)} \tag{16}$$

where $d\bar{\varepsilon}_p$ is the equivalent plastic strain increment, while $\bar{\varepsilon}_f$ is a functional relation that defines fracture strain as a function of stress triaxiality (η), Lode angle ($\bar{\theta}$) and strain rate ($\dot{\bar{\varepsilon}}_p$). The rate independent Modified Mohr-Coulomb model [12] to describe stress state dependence fracture initiation along with Johnson-Cook failure model for strain rate effect are employed as follows [10]

$$\bar{\varepsilon}_f(\eta, \bar{\theta}, \dot{\bar{\varepsilon}}_p) = \phi(\dot{\bar{\varepsilon}}_p) \cdot \bar{\varepsilon}_f(\eta, \bar{\theta}) \tag{17}$$

where

$$\phi(\dot{\bar{\varepsilon}}_p) = \begin{cases} 1 & \dot{\bar{\varepsilon}}_p < \dot{\varepsilon}_0 \\ 1 + \kappa \ln \frac{\dot{\bar{\varepsilon}}_p}{\dot{\varepsilon}_0} & \dot{\bar{\varepsilon}}_p \geq \dot{\varepsilon}_0 \end{cases} \tag{18}$$

and

$$\bar{\varepsilon}_f(\eta, \bar{\theta}) = \left\{ \frac{A}{C_2} \left[C_3 + \frac{\sqrt{3}}{2-\sqrt{3}} (1 - C_3) \left(\sec\left(\frac{\bar{\theta}\pi}{6}\right) - 1 \right) \right] \left[\sqrt{\frac{1+{C_1}^2}{3}} \cos\left(\frac{\bar{\theta}\pi}{6}\right) + C_1 \left(\eta + \frac{1}{3} \sin\left(\frac{\bar{\theta}\pi}{6}\right) \right) \right] \right\}^{-\frac{1}{n}} \tag{19}$$

the κ and $\dot{\varepsilon}_0$ are parameters related to dynamic fracture initiation, while A and n are the same material parameters established by Eq. 3 and the three free-parameters c_1, c_2 and c_3 correspond to the entries of the MMC fracture criterion. Altogether will define the shape and magnitude of the fracture locus at each strain rate level herein considered. Simulated crack propagation in thick components usually requires the use of post-initiation softening to account for progressively local damage at the integration point in the form of a power law [9], [13]

$$\beta = \left(\frac{D_C - D}{D_C - D_0}\right)^h \tag{20}$$

where D_c, D_0 and h are damage parameters, while D is the current magnitude of the damage indicator. Note that Eq. 13 operates only when $D > 1$ and ceases when $D = D_C$. As a result, β will describe a decreasing power law scalar that scale down the undamaged flow stress $\bar{\sigma}$ to the actual damage flow stress $\tilde{\sigma}$ until the critical element is deleted

$$\tilde{\sigma} = \beta \cdot \bar{\sigma} \tag{21}$$

PIPELINE STEEL MATERIALS

In the Oil and Gas industry the most common materials for pipe manufacturing, especially for high-pressure transmission commodities, are steel alloys, which comes in different grade levels. According to the American Petroleum Institute (API 5L), pipeline steels are classified by their minimum yield strength in ksi preceded by letter "X". Among the most popular pipeline steels, X65 and X100 are the prominent for offshore and onshore operations, respectively. In this study, the two above steels were mechanically characterized to determine their mechanical properties (yield and ultimate strengths, ductility, etc.) by means of very careful selected custom-made samples to cover the spectrum of all possible failure modes as function of loading conditions, strain rates and temperature effects. For plasticity features, such as strain hardening, yielding and tensile points, standard uniaxial tensile tests are carried out. This set of information is deemed as basic and is important to define post-necking behavior, whose response is commonly determined iteratively by means of hybrid/numerical approach [14]. However, uniaxial tensile tests not only provide information regarding plasticity, but also information about plastic anisotropy as a result of manufacturing processes and heat treatment, see Figures. 1a and 1b for X65 and X100, respectively. Typically, only the stress-strain information is not always reliable for directional dependency determination and an additional parameter would be required, such as the Lankford ratio. The Lankford ratio is defined as the ratio of the plastic width strain to the plastic thickness strain measures of the gauge section of a prismatic cross-section (rectangular). By means of incompressibility assumption the through-thickness strain is computed and used for describing directional dependency of plastic flow. Figures 1c and 1d show the Lankford coefficients along with stress ratios for X65 and X100, respectively. The colored symbols correspond to experimental measurements while the solid lines were obtained from the phenomenological model outlined in previous section. To summarize uniaxial tension response at different directions, the stress ratios are used, where the yield stress at each loading direction is normalized with respect to the reference yield stress (normally taken in longitudinal pipe direction, $\alpha = 0°$). As it can be seen, both considered materials exhibit a weak directional dependency in terms of strength (blue line), but strong in terms of

Lankford coefficients (orange line). As reference, for isotropic materials the stress ratio and Lankford coefficients are equal unity and for anisotropic ones they are not. For instance, the X65 exhibits a round-house yield stress transition with a slightly shifting in transverse flow curve (red line in Figure 1a and blue triangle at 90^0in Figure. 1c) in about 8% yield stress increases with respect to the reference stress; while the X100 seemingly displays alike behavior, at least, for 0^0 and 45^0, its transverse alternative, on the other hand, exhibits a flat yield plateau, see Figure 1b. Typically, this is attribute to the manufacturing processes and heat treatment that pipe products commonly undergo, in which microstructure characteristics as well as texture configurations are introduced. It is interesting to note that despite the apparent small differences in plastic flow curves (stress ratios), the plastic strains (Lankford coefficients) show strong dependence on loading direction. For example, the pipe metal X65 is more prone to thinning in transverse direction rather than in longitudinal direction as oppose to its alternative X100, in which thinning develops rapidly in longitudinal direction. This information is important for pipe design, especially for high-pressure gas transportation since the hoop (transverse) stresses are the dominant stress state in biaxially-loaded pipes. It is clear from the above that strains evolve differently than stresses as they are strongly influenced by loading direction. Under this scenario, non-associated flow rule fits perfectly since it is based on the principle that the direction of increment of the plastic potential evolution is distinct from the direction of stresses increment. That is, the plastic flow and the stress are not codirectional.

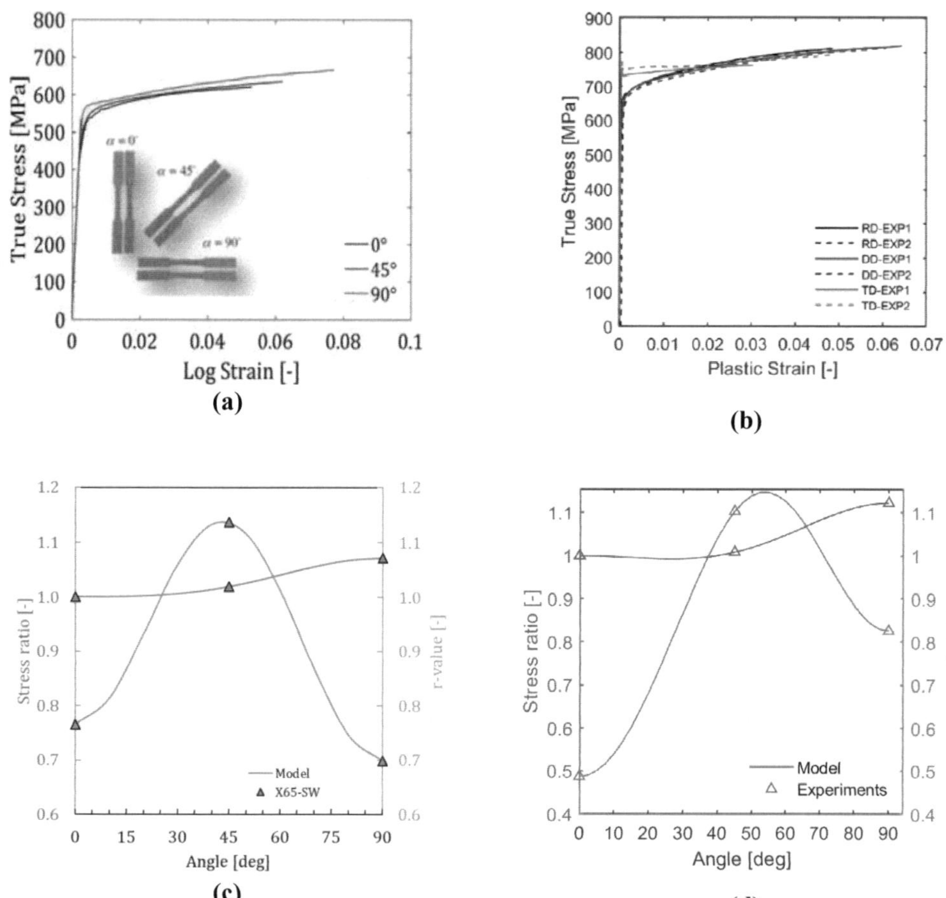

Figure 1: True stress-log. Strain curves and Lankford coefficients for X65 (left column) and X100 (right column) pipeline steels.

In addition to plastic behavior, ductile fracture initiation in metals is commonly preceded by plastic instability or post-necking behavior. The poses question is when ductile failure overcome plastic dislocation by cavitating particles typically associated to second phase particles, carbides and impurities [15]. The rate at which these cavities grow and deform depends strongly upon the stress triaxiality (η) and the Lode angle parameter ($\bar{\theta}$) [16]. A pictorial representation can be drawn when the local equivalent plastic strain is represented as a function of the two above parameters, which cover a wide range of all possible loading modes and stress states, e.g., equibiaxial tension ($\eta \approx 2/3;\ \bar{\theta} \approx -1$), uniaxial tension ($\eta \approx 1/3;\ \bar{\theta} \approx 1$), plane strain tension ($\eta \approx 1/2;\ \bar{\theta} \approx 0$), generalized shearing (($\eta \approx 0;\ \bar{\theta} \approx 0$) and uniaxial compression ($\eta \approx -1/3;\ \bar{\theta} \approx -1$). Any intermediate loading mode will fall within this generalization and will be fitted by a fracture space represented by the triad $\bar{\varepsilon}_p$, η and $\bar{\theta}$ [14]. In anisotropic configurations, fracture initiation is better to be represented by a polar representation as shown in Figure 2. The loading path for each configuration is extracted from the material point that undergoes the highest equivalent plastic strain (location where fracture initiate). The orthogonal axes display the projections of the strain paths with respect to the normalized in-plane stress components. In these polar plots, the anisotropy effect can easily be seen as the strain paths that depart from the bisector lines $\sigma 11 = \sigma 22$ in the in-plane axes [10], [17]. The selected samples employed for fracture initiation characterization are samples mostly loaded by tension such as flat (2 mm thickness), prismatic cross section (rectangular) specimen geometries with certain variations such as samples with central hole (CH); double circular cutouts at the edges (NTs) with various radii as 20, 10 and 6.67 mm; and notch round bars (NRBs) with three distinct radii as 2, 1 and 0.5 mm, see details in refs. [10]. These samples cover all possible stress states to which a pressurized pipe might be subjected to. In general, the model predictions (open color triangles) for all cases are in good agreement with the mapped fracture initiation points (end of the color solid lines) and the impact of anisotropy on strain path is revealed.

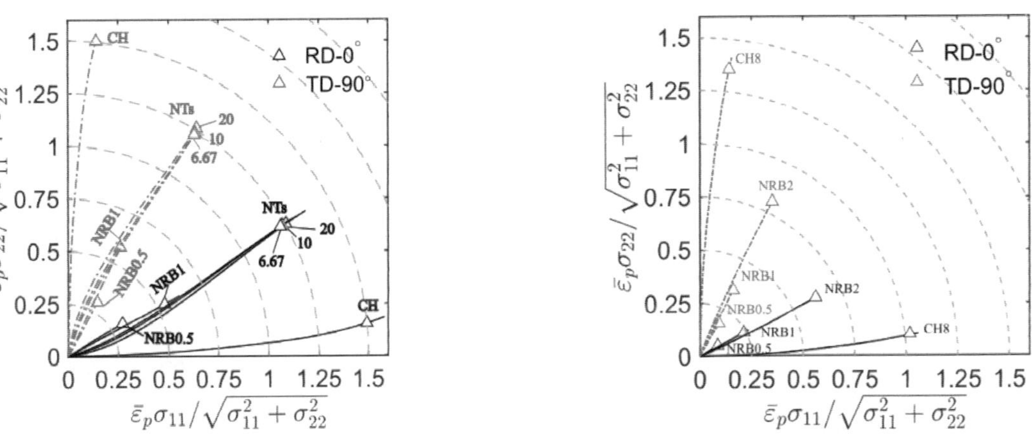

(e) (f)

Figure 2: Polar plots of ductile fracture initiation for a) X65 and b) X100

These properties inherent to mechanical properties of the materials reflect their behavior in quasi-static loading conditions. However, despite the seemingly differences with dynamic conditions and its influence on fracture initiation, these parameters are directly transferable by taking account only strain rate effects on material strength since their influences on fracture initiation is almost negligible, especially for X65 pipeline steel [18].

THERMODYNAMICS OF ESCAPING MIXTURES AND BACKFILL EFFECTS

To model the thermodynamics of gases during depressurization in computational settings, an accurate description of thermodynamic properties of natural gas systems is desired, especially in the oil and gas industry. For this application, non-cubic equations of state (EoSs) are advantageous due to their better density and compressibility description and the GERG-2008 is a new wide-range EoS (Equation of State) for natural gases and other mixtures of 21 natural gas components. It is considered as a standard reference equation suitable for natural gas applications where highly accurate thermodynamic properties are required [19]. A special numerical technique is employed for interpolation and storage of internal variables required to compute the thermodynamics of gases during depressurization. Information of the volumetric behavior of arbitrary fluids is described based on the definition of the pressure as a function of the specific volume v and internal energy U per unit mass $p = f(v, U)$. For given input values of T and v, the pressure p, the partial derivative of the pressure to the internal energy $\partial p/\partial U$ and to the density $\partial p/\partial \rho$ as well as the speed of sound c need to be provided to the user-defined subroutine to resolve for continuity, motion and energy equations within the Euler framework for fluids [20], [21]. This implementation becomes straightforward by means of Look-Up-Table (LUT) technique that allows for interpolation and storage of gas variables during iterations. Figure 3a shows the phase envelope used in this review for methane gas under insentropic conditions (red line path) as a function of its thermodynamics parameters: temperature, T and specific volume, v. Further, Figure 3b displays the decompression curves used by Keim et al. [21] to represent methane and hydrogen mixtures.

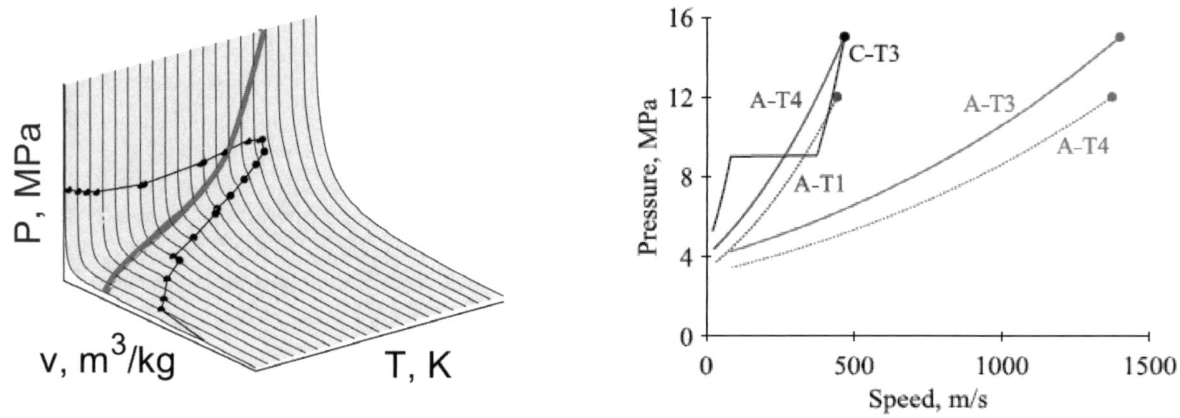

(a) (b)

Figure 3: (a) Phase envelope and isentropic decompression path obtained from GERG 2008 [22]; (b) fluid decompression curves of mixtures studied in ref.[21], whose results pertaining to A-T1 and A-T4 are recasted.

Modeling aspects of the backfill soil is key to improve accuracy in numerical simulations. The description of the backfill material and its mechanical behavior pose a challenging and important task in numerical modeling, since they will strongly influence the pipe opening and inertia during the dynamic ductile crack propagation. Within CEL framework, a Lagrange domain was used to model the soil material in the present fluid-structure interaction model. In the following section, further details about the model setup, mesh arrangement and the domain-coupling algorithm are given. A suitable phenomenological model is employed to predict mechanical response of the soil, such as the Mohr-Coulomb model. This model describes the shear strength of soils and rocks at different effective stresses in accurate form, see details about its formulation and implementation in [21].

NUMERICAL PROCEDURE AND FINITE ELEMENT MODEL

Detailed 3D finite element models were built to adjust material parameters as well as to verify model reliability as shown in Figure 4. For calibration of dynamic fracture model parameters, the Drop Weight Tear Test (DWTT) sample was employed along with its numerical model indicated in Figure 4a. A representative numerical model of 1/4 DWTT test according to the experimental setup. Symmetry boundary conditions (SBCs) in the crack propagation plane 22 (y-axis) and the WT direction 33 (z-axis) is employed for the quarter one, while SBC in WT direction is only applied for the half model. This reduce largely computational effort without losing accuracy on the description of key features. Rigid bodies discretized the hammer and the bearings, which interacted with the solid model of the specimen by surface-to-surface contact definition. Therefore, the Coulomb friction coefficient was assumed by $\mu_0 = 0.1$. The FE-mesh distribution was uniform in the ligament with cubic linear brick elements of reduced integration (ABAQUS element type C3D8R). For the initial verification of the identified model parameters, the ligament elements had edge sizes of $l_e = 0.2$ mm. With respect to the component level simulations, a coarser mesh of $l_e \approx 0.8$ mm was also employed. The FE-mesh density decreased in the remote areas to lower the computation times for the DWTT and later pipe models. For validation/verification, on the other hand, the CEL model required multi-physical interactions between the pipe material, the compressed and the ambient fluid as well as the soil backfill to be modeled. It is used the fluid-structure interaction (FSI) tool provided in Abaqus/Explicit for the description of large scale FSI problems. For CEL models, one Euler domain can be coupled with various Lagrange domains. Further, there are no restrictions regarding the number of materials in each domain or the application of user-defined subroutines to implement constitutive equations for the material description. The experiments carried out by Aihara et al. [23] were taken as reference for this validation test. The full scale burst test is comprised by a 6 mm wall-thickness partially buried pipe with outer diameter, OD = 267 mm filled with methane. The test temperature was 288.15 K at two pressure levels of 12 and 15 MPa. By taking advantage of the geometry symmetry, similar to DWTT, only one-quarter of the pipe model is constructed along with the other interacting mediums such as ambient air, backfill soil and the mixture (in this case methane, CH_4). The appropriate boundary conditions along with the following domains were accurately discretized: the pipe and backsoil models were defined within the Lagrangian domain, while the mixture and ambient air were defined in the Euler domain as displayed in Figure 4b. In total the model was comprised by ≈ 600,000 3D brick elements among special Euler-type (EC3D8R) and Lagrangian-type (C3D8R) elements.

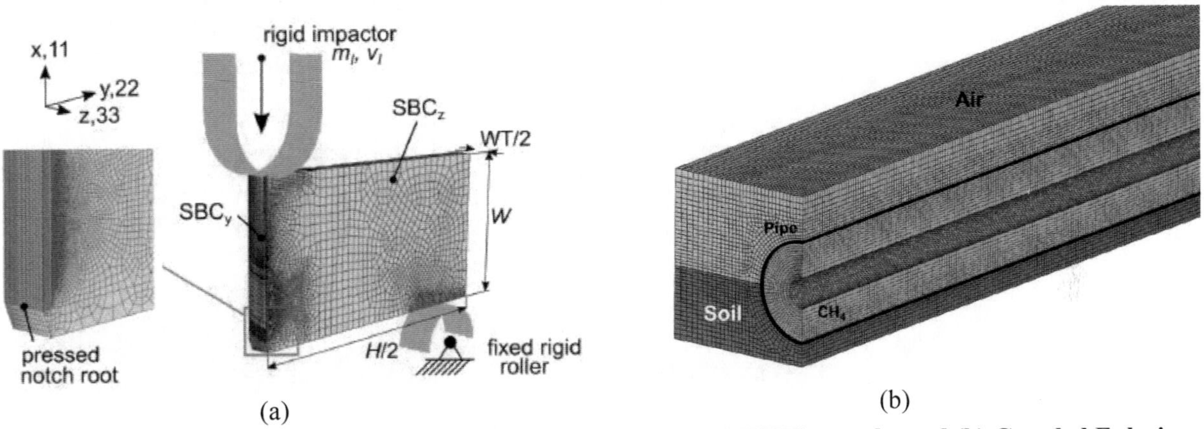

Figure 4: Three dimensional models of (a) Drop Weight Tear Test (DWTT) sample, and (b) Coupled Eulerian-Lagrangian (CEL) model for full scale burst test [10], [21].

Lab-scale simulations

Lab scale simulations are needed, especially for material model calibration under dynamic loadings. Material parameters related to strain rate features are commonly adjusted from experiments performed in well-controlled laboratory space. Here, instrumented drop weight tear tests (DWTT) are usually conducted to provide with the necessary information for the model to be calibrated. One of the salient features of this experiment is its resemblance to the stress states developed at the crack-tip of a full-scale cracked pipe with a running crack at high speed. This is attained by letting drop a heavy hammer from a prescribed height to impact the crack-free specimen's edge to produce dynamic crack initiation and subsequent propagation. Figure 5 shows the experimental and numerical DWTT load–displacement plot. The oscillation in the experimental trace (blue) could be due to bouncing of the hammer after the initial contact. Subsequently, the load increases until fracture initiation occurs. The load sharply decreases in the fracture propagation stage. Since the FE simulation did not show the initial bouncing of the hammer, the numerical load–displacement traces were shifted right by 1.9 mm. As discussed previously, to reduce computation time in simulation is recommendable to exploit geometric symmetry and smaller models as far as one-quarter can be easily built. However, this reduction comes to a certain cost when mesh density is also varied, e.g., the 1/4 model predicts the hardening stage quite well with a load deviation of less than 1% for both the fine and coarse discretization. Moreover, the displacement at maximum load is slightly underestimated in about 3 mm. Similar accuracy, albeit to a lesser extent, is visible in the unloading stage. The model, in general, captures the slope of the experimental load–displacement well for the fine discretization, but fails to capture the remaining of the experimental curve (Disp. > 30 mm). Matching between coarse and fine meshes can be achieved by adjusting one of the fracture parameters given in Equation 19 (parameter c_2). This strategy aims to reduce computational time in large finite element model simulations, mainly for long pipeline models. Because of variation of c_2 only affect the amount of the fracture initiation strain, the stress triaxiality and the Lode angle parameter relation remains unchangeable [12]. Additionally, other simulations were conducted assuming isotropic plasticity and ductile fracture initiation in order to study the anisotropy effect under dynamic loading conditions, see dashed black graph in Figure 5a. As it can be seen, the isotropic model shows higher stresses in the hardening phase, which correspond to the experimental values at maximum load. The anisotropic model slightly underestimates fracture initiation with the load drop corresponding to the experiments in the subsequent fracture propagation phase. Despite the distinctly overestimated stresses in the ♦-range for the isotropic model, the slope of the load–displacement-trace is slightly lower when compared to the anisotropic model. This is an indication that considering the anisotropic effects on plastic response, the model yields result that are in good agreement with measured data. One disadvantage of using reduced models is that features such as chevron marks and tunneling remain hidden, even though the load-displacement response almost coincides with its half-model alternative as indicated in Figure 5. Other alternatives for calibration dynamic fracture propagation is by means of instrumented Charpy V notch test, which consists in breaking small prismatic edge cracked samples in a pendulum. The hammer is dropped freely from a certain height, whose impact energy is associated with fracture toughness. Interested readers are referred to ref. [24] for more details. Nonetheless, for new-generation materials this type of test might not be suitable due to the inability of the small cross-sectional area to capture essential characteristics of the fracture process during impact.

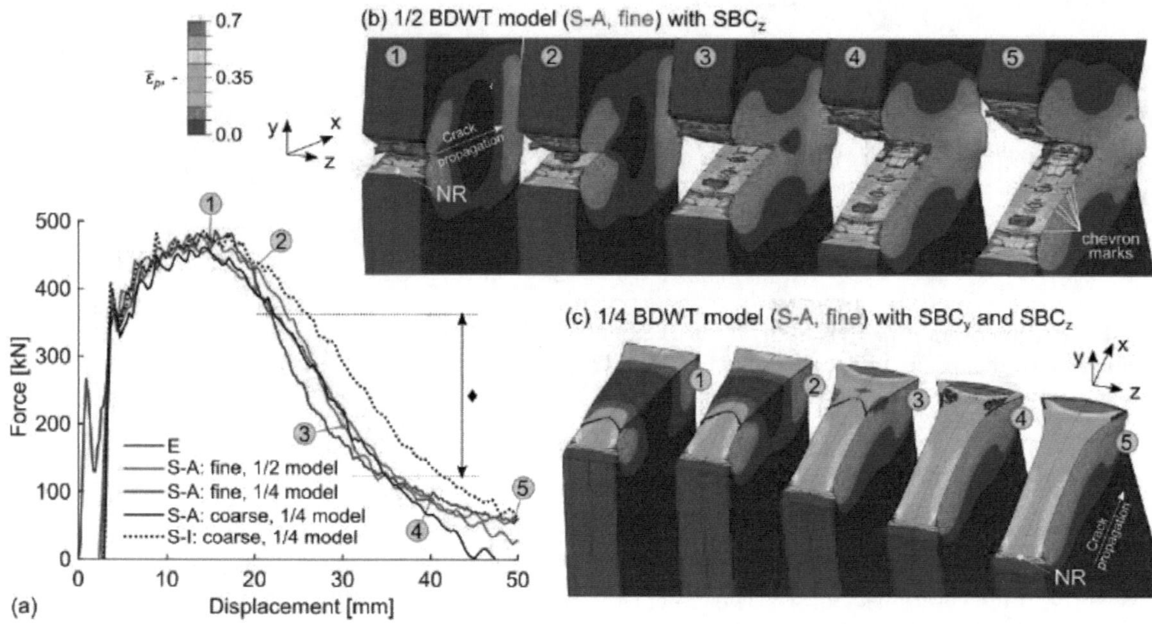

Figure 5: Calibration of the computational model with measured data of DWTT test [10].

FULL-SCALE BURST TEST SIMULATIONS

Once the calibration process has been completed, it is important to test the capability of the proposed phenomenological model for reliability and accuracy. Here, the full-scale burst test results, carried out by Aihara et al. [23], is presented along with numerical simulations performed by Keim et al. [21]. The adjusted material model parameters from lab-scale tests are put to the test with the experimental measurements of fracture speed with respect to the crack advance and depressurization as a function of time as indicated in Figure 6. The full-scale burst tests were performed using a partially buried X65 pipeline steel with outer diameter, OD = 267 mm and wall-thickness, WT = 6 mm. Among the tested gases, the CH_4 gas results are recasted whose depressurization evolution is shown in Figure 3b (blue solid and dash lines), experiments A-T1 and A-T4. To initial conditions were used, $p_{0,i}$ in which the corresponding burst pressure (critical pressure at which a crack starts to propagate) was 12 and 15 MPa. The thermodynamic of the gas, given by the GERG 2008's LUTs in Figure 3a, is coupled with CEL/FSI framework in Abaqus/Explict to simulate running ductile fracture propagation. The model accurately captures this process, especially during crack initiation ($\Delta a \approx 150$ mm), steady crack propagation ($150 < \Delta a < 750$ mm) and crack arrest ($\Delta a > 750$ mm). These three stages of crack extensions are characterized by severe changes in the stress states and loading modes, especially in the arresting stage as the crack driving forces generated by the escaping gas ceases to be significant as the material's fracture toughness overcomes. This is also reflected in the depressurization curve in Figure 6c. As it can be seen, percentage differences are below 5% for all considered cases and the agreement of simulations with experiments is acceptable.

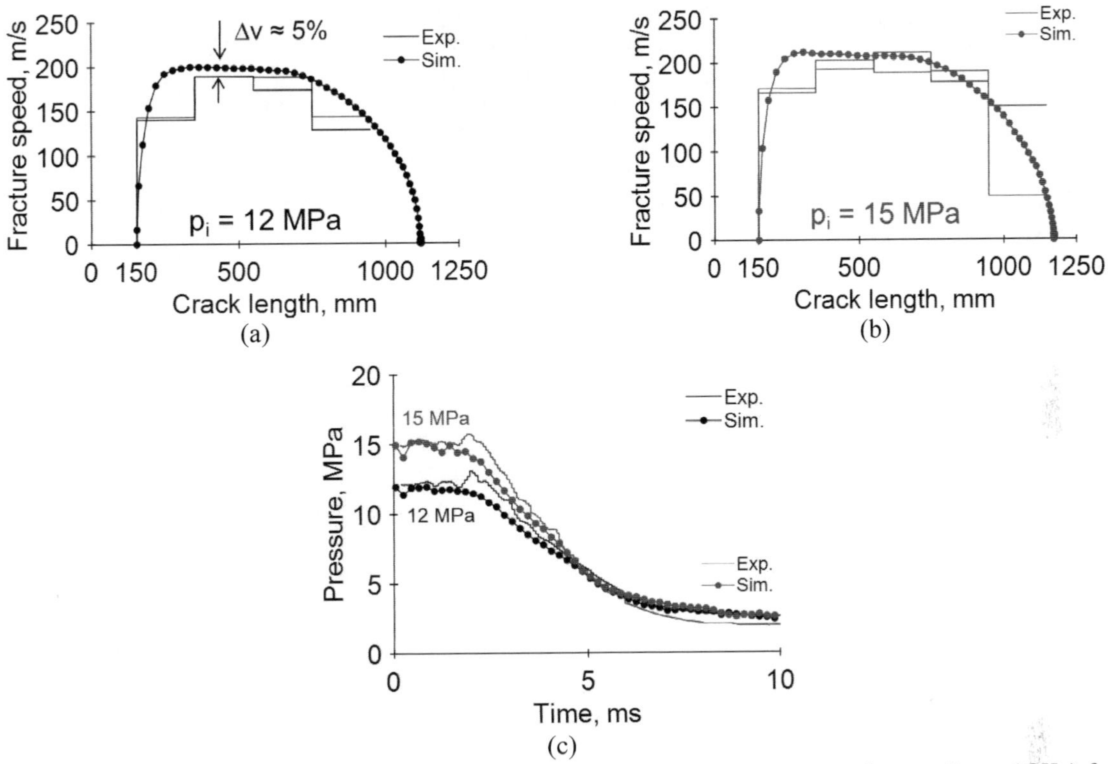

Figure 6: Fracture speed vs. crack length curves of full scale burst tests transporting methane (CH_4) for both initial conditions: (a) $p_{0,i} = 12$ MPa, and (b) $p_{0,i} = 15$ MPa; (c) depressurization curve as a function of time. Simulations and experimental outcomes are displayed as colored dot and solid lines, respectively [21].

A visual comparison proves the effectiveness of the proposed methodology as shown in Figure 7. In contrast, similar results obtained from Nordhagen et al. [25] is also taken for comparison. The resemblance between these two is poor since such a model does not account for backfill effects and the pipe flaps ended up acquiring a sinusoidal shape, which was not observed in experiments. On the other hand, the present model approach not only describe with high-fidelity the deformed pipe shape, but also the backfill soil deformation.

(a) (b) (c)

Figure 7: Visual comparison of the break-open pipe after bursting with numerical simulations performed by (a) Nordhagen et al. [25]; **and (c) Keim et al. [21].**

CONCLUSION

The inclusion of multi-physics features in the proposed model leads to a more realistic finite element simulations with high fidelity to capture even small details. The CEL/FSI framework is a powerful tool provided by Abaqus to couple not only interaction of fluid-structure but also soil-structure interaction by means of defined user-material subroutine. Most of the comparisons displayed in this article speak by themselves in terms of accuracy, high-fidelity and reliability.

REFERENCES

[1] eia - U.S. Energy Information Administration, "Natural gas pipelines," 2021. https://www.eia.gov/energyexplained/natural-gas/natural-gas-pipelines.php#:~:text=Also in Hydrogen explained&text=The U.S. natural gas pipeline,and storage facilities with consumers.

[2] M. Kelso, "Pipelines continue to catch fire and explode," *Fractracker Alliance*, 2020. https://www.fractracker.org/2020/02/pipelines-continue-to-catch-fire-and-explode/.

[3] D.o.E., "HyBlend: Opportunities for Hydrogen Blending in Natural Gas Pipelines," *Hydrogen and Fuel Cell Technologies Office*, 2020. https://www.energy.gov/eere/fuelcells/hyblend-opportunities-hydrogen-blending-natural-gas-pipelines.

[4] A. S. User, "Abaqus 6.14," *Dassault Systèmes Simulia Corp., Provid. RI, USA*, vol. Dassault S, 2014, doi: 10.1017/CBO9781107415324.004.

[5] D. Mohr, M. Dunand, and K. H. Kim, "Evaluation of associated and non-associated quadratic plasticity models for advanced high strength steel sheets under multi-axial loading," *Int. J. Plast.*, vol. 26, no. 7, pp. 939–956, 2010, doi: 10.1016/j.ijplas.2009.11.006.

[6] R. Hill, "The Theory of Plane Plastic Strain for Anisotropic Metals," *Proc. R. Soc. Lond. A. Math. Phys. Sci.*, vol. 198, no. 1054, pp. 428–437, 1949, doi: 10.1098/rspa.1949.0110.

[7] T. B. Stoughton, "A non-associated flow rule for sheet metal forming," in *International Journal of Plasticity*, 2002, vol. 18, no. 5–6, pp. 687–714, doi: 10.1016/S0749-6419(01)00053-5.

[8] G. R. Johnson and W. H. Cook, "A constitutive model and data for metals subjected to large strains, high strain rates and high temperatures. The 7th International Symposium on Ballistics," *Proceedings of the 7th International Symposium on Ballistics*. 1983.

[9] M. Paredes, T. Wierzbicki, and P. Zelenak, "Prediction of crack initiation and propagation in X70 pipeline steels," *Eng. Fract. Mech.*, vol. 168, 2016, doi: 10.1016/j.engfracmech.2016.10.006.

[10] L. Xue, V. Keim, M. Paredes, A. Nonn, and T. Wierzbicki, "Anisotropic effects on crack propagation in pressurized line pipes under running ductile fracture scenarios," *Eng. Fract. Mech.*, vol. 249, 2021, doi: 10.1016/j.engfracmech.2021.107748.

[11] C. C. Roth and D. Mohr, "Effect of strain rate on ductile fracture initiation in advanced high strength steel sheets: Experiments and modeling," *Int. J. Plast.*, vol. 56, pp. 19–44, 2014, doi: 10.1016/j.ijplas.2014.01.003.

[12] Y. Bai and T. Wierzbicki, "Application of extended Mohr-Coulomb criterion to ductile fracture," *Int. J. Fract.*, vol. 161, no. 1, pp. 1–20, 2010, doi: 10.1007/s10704-009-9422-8.

[13] M. Paredes, D. F. B. Sarzosa, R. Savioli, T. Wierzbicki, D. Y. Jeong, and D. C. Tyrell, "Ductile tearing analysis of TC128 tank car steel under mode I loading condition," *Theor. Appl. Fract. Mech.*, vol. 96, 2018, doi: 10.1016/j.tafmec.2017.10.006.

[14] M. Paredes, T. Wierzbicki, and P. Zelenak, "Prediction of crack initiation and propagation in X70 pipeline steels," *Eng. Fract. Mech.*, vol. 168, 2016, doi: 10.1016/j.engfracmech.2016.10.006.

[15] W. M. Garrison and N. R. Moody, "Ductile fracture," *J. Phys. Chem. Solids*, vol. 48, no. 11, pp. 1035–1074, 1987, doi: 10.1016/0022-3697(87)90118-1.

[16] M. Dunand and D. Mohr, "Effect of Lode parameter on plastic flow localization after proportional loading at low stress triaxialities," *J. Mech. Phys. Solids*, 2014, doi: 10.1016/j.jmps.2014.01.008.

[17] M. Paredes, J. Lian, T. Wierzbicki, M. E. Cristea, S. Münstermann, and P. Darcis, "Modeling of plasticity and fracture behavior of X65 steels: seam weld and seamless pipes," *Int. J. Fract.*, vol. 213, no. 1, 2018, doi: 10.1007/s10704-018-0303-x.

[18] P. Fudlailah and M. Paredes, "Fracture Behavior of X65 Q&T Seamless Pipeline Steel Under Different Strain Rates and Stress States," 2020, doi: 10.1115/omae2020-18032.

[19] O. Kunz and W. Wagner, "The GERG-2008 wide-range equation of state for natural gases and other mixtures: An expansion of GERG-2004," *J. Chem. Eng. Data*, 2012, doi: 10.1021/je300655b.

[20] V. Keim, A. Nonn, and S. Münstermann, "Application of the modified Bai-Wierzbicki model for the prediction of ductile fracture in pipelines," *Int. J. Press. Vessel. Pip.*, vol. 171, no. February, pp. 104–116, Mar. 2019, doi: 10.1016/j.ijpvp.2019.02.010.

[21] V. Keim, M. Paredes, A. Nonn, and S. Münstermann, "FSI-Simulation of Ductile Fracture Propagation and Arrest in Pipelines: Comparison with Existing Data of Full-Scale Burst Tests," *Int. J. Press. Vessel. Pip.*, vol. 182, no. November 2019, p. 104067, Feb. 2020, doi: 10.1016/j.ijpvp.2020.104067.

[22] V. Keim, P. Marx, A. Nonn, and S. Münstermann, "Fluid-structure-interaction modeling of dynamic fracture propagation in pipelines transporting natural gases and CO2-mixtures," *Int. J. Press. Vessel. Pip.*, 2019, doi: 10.1016/j.ijpvp.2019.103934.

[23] S. Aihara, E. Østby, H. I. Lange, K. Misawa, Y. Imai, and C. Thaulow, "Burst tests for high-pressure hydrogen gas line pipes," 2009, doi: 10.1115/IPC2008-64166.

[24] A. Cosham, D. G. Jones, K. Armstrong, D. Allason, and J. Barnett, "Analysis of a dense phase carbon dioxide full-scale fracture propagation test in 24 inch diameter pipe," 2016, doi: 10.1115/IPC201664456.

[25] H. O. Nordhagen, S. Kragset, T. Berstad, A. Morin, C. Dørum, and S. T. Munkejord, "A new coupled fluid-structure modeling methodology for running ductile fracture," *Comput. Struct.*, 2012, doi: 10.1016/j.compstruc.2012.01.004.

SNAME
THE INTERNATIONAL COMMUNITY
FOR MARITIME AND OCEAN PROFESSIONALS

FATIGUE LIFE ESTIMATION OF NOTCHED COMPONENTS USING FREQUENCY DOMAIN APPROACH

DIEGO F.B SARZOSA
NAVAL AND OCEAN DEPARTMENT, ENGINEERING SCHOOL, UNIVERSITY OF SÃO PAULO, BRAZIL

VAGNER PASCUALINOTTO JR
NAVAL AND OCEAN DEPARTMENT, ENGINEERING SCHOOL, UNIVERSITY OF SÃO PAULO, BRAZIL

Proceedings of the 27th Offshore Symposium, February 22nd, 2022, Houston, Texas
Texas Section of the Society of Naval Architects and Marine Engineers

ABSTRACT

Engineering structures are constructed to withstand a variety of in-service loading specific to their application. Random vibration excitation is observed in most structural components in the offshore, aerospace, and automotive industries. Likewise, fatigue life estimation for structural elements is fundamental for verifying the design and assurance of structural integrity throughout service. The linear cumulative Palmgren-Miner's damage rule is widely used for damage assessment, despite its well-known limitations. The scatter of fatigue testing data suggests that a probabilistic characterization of the material behavior should be considered during fatigue life predictions. In this work, the inherent uncertainties of the fatigue phenomenon and the influence of a geometrical discontinuity are explored in the fatigue life estimation of a structural component subjected to random vibration profiles. The fatigue life estimated using the methodology proposed in this work presented a good agreement with testing results using Dirlik frequency domain counting methods.

Keywords: Fatigue, Frequency Domain, Damage, Notches, FEM

NOMENCLATURE AND ABREVIATIONS

$[A\ ]$: Multiaxial Power Spectral Density input loading

D_{p_x} : Damage associated to S-N curve with probability of survival px

$E[0]$: Expected number of upward zero crossings

$E[P]$: Expected number of positive peaks

f: Frequency (Hz)

$[G]$: Stress Response PSD matrix

PSD: Power Spectral Density

$p(S_i)$: Probability Density Function of Stress Amplitude

m_n: Spectral moments of the PSD of Stress

$R_{xx}(\tau)$: Autocorrelation Function

rms: Root Mean Square

$T_{exposure}$: Exposure time under determined PSD profile

$S_{xx}(f)$: Power Spectral Density – Excitation

TTF: Time to Failure

INTRODUCTION

Fatigue as a technical problem became evident around the middle of the 19th century. Wöhler performed systematic fatigue tests of smooth and notched railway axles in the 1850s [1]. In addition to the introduction of the S-N diagram, a plot of the number of cycles to failure at a given stress level, his work also led directly to the concept of a fatigue (or endurance) limit, which represents the theoretical maximum cyclic load a material can withstand indefinitely with low risk of fatigue failure.

Significant advances in the fatigue research were achieved with the mean stress effect studies by Gerber and Goodman, the development of fatigue safety diagrams by Haigh, investigation of reversed loading phenomena by Bauschinger, analysis of the notch effect on fatigue limit by Heyn, formulation of empirical laws to characterize fatigue limit by Basquin, the introduction of the crack growth energy balance by Griffith, life estimation under variable loading by Palmgren and the recognition of the statistical nature of fatigue by Weibull [1-2].

In evaluating and predicting the fatigue life of structures, the role of mathematical and statistical models is crucial due to the high complexity of the fatigue problem. An efficient estimation of the corresponding parameters represents one of the most difficult challenges for the problem assessment. The possible shortage of data is a common feature in the case of fatigue experimentation due to economic and time reasons.

The fatigue damage is traditionally determined from time histories of loading, usually in the form of stress or strain. This approach is satisfactory but requires extensive time records to describe a random loading process accurately. Alternatively, a frequency domain approach has a significant advantage in terms of computational time when finite element analysis is used. Random loading and responses are characterized using power spectral densities (PSDs).

The frequency-domain approach offers a significant advantage in terms of calculation time, which can be used to solve much larger and complex problems using finite element analysis. Therefore, this work aims to review and identify modeling parameters that increase the accuracy of the results compared to the time domain. The fatigue life of a 6061-T6 aluminum notched specimen is estimated for a random loading profile. The fatigue damage and life are estimated in frequency domains for different S-N curve percentiles. Also, the signal in the time domain was analyzed and the corresponding fatigue life was calculated. By correcting the S-N curve by the notch effect and modeling the damping behavior of the model, the frequency domain approach yields fatigue life prediction that is in close agreement with the experimental random vibration tests.

METHODOLOGY

The fatigue life in the time domain is estimated by applying the time-series load history to the component under evaluation; a transient structural analysis is performed to generate the stress histories at all discretized points of the structure, defined by the mesh size in the finite element modeling. Then, Rainflow cycle counting and linear cumulative damage, per Palmgreen-Miner's rule, are applied to estimate the component's total damage and fatigue life. On the other hand, in the frequency domain approach, illustrated in Figure 1, the dynamic characterization of the discretized structure is made by its frequency response functions (FRFs), which are calculated based on a unit load excitation (e.g. for a PSD given in unit of acceleration g^2/Hz, the structure is harmonically excited with one (1) g acceleration at the correspondent excitation direction).

The PSD input load multiplies the square of FRFs stress tensor resulting in the stress response spectrum (PSD). Figure 1 illustrates an input PSD and the corresponding PSD of one stress tensor component. The fatigue life is then estimated by calculating the spectral moments of the PSD of stress, probability density function (PDF), and duration of the vibration profile. The obtained number of cycles subjected to the structure is compared to the material S-N curves via linear cumulative damage. The following section describes the basic formulation to analyze a random time signal and identify the corresponding PSD. Then, the Dirlik method to estimate the probability density function of the stress amplitudes is described with the respective formulation dependent on the PSD of the stress.

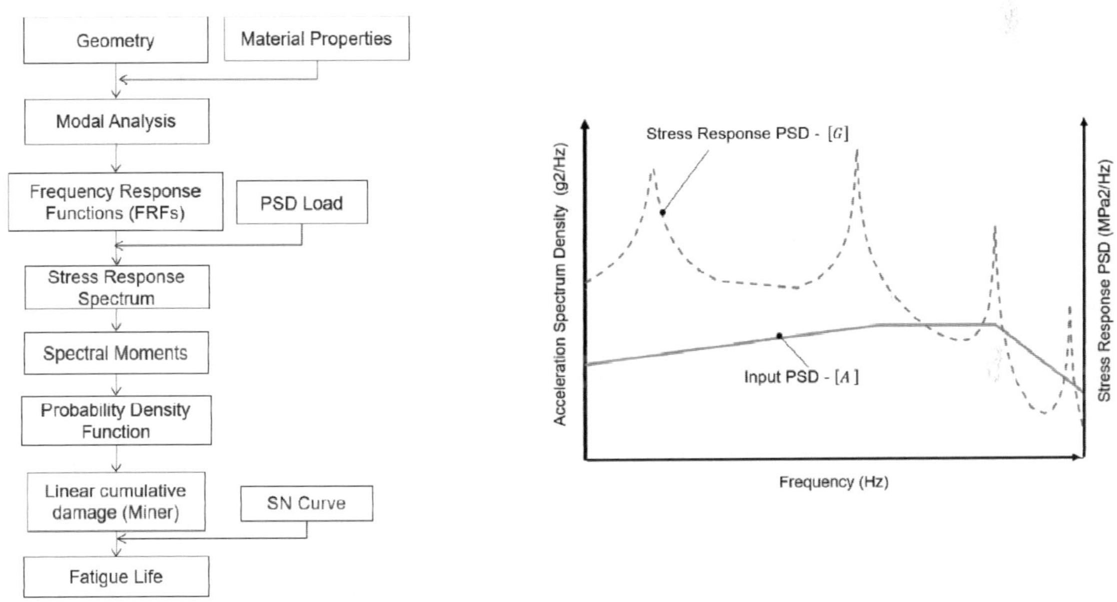

Figure 1: Flowchart for fatigue life estimation based on the frequency domain and illustration of input and output PSDs.

A periodic time history may be represented by the summation of a series of sinusoidal waves of various amplitude, frequencies, and phases. This is the basis of the Fourier series expansion. The Fourier transform pair allows transformation between the time and frequency domains. The Fourier transform is defined as [3]:

$$X(f) = \int_{-\infty}^{\infty} x(t)e^{-i2\pi ft}\,dt \tag{1}$$

and the inverse Fourier transform is:

$$x(t) = \int_{-\infty}^{\infty} X(f) e^{+i2\pi ft}\, dt \tag{2}$$

The autocorrelation function that defines how a signal is correlated with itself, with a time separation τ can be written as

$$R_{xx}(\tau) = \int_{-\infty}^{\infty} x(t)x(t+\tau)dt = E[x(t)x(t+\tau)] \tag{3}$$

The autocorrelation and the power spectral density (PSD) are related by the Fourier transform pair.

$$R(\tau) = \int_{-\infty}^{\infty} S_{xx}(f) e^{-j\omega t}\, d\omega \tag{4}$$

$$S_{xx}(f) = \int_{-\infty}^{\infty} R(\tau) e^{+j\omega t}\, d\tau \tag{5}$$

The one-sided PSD $G_{xx}(f)$ defined for $0 \leq f \leq \infty$, is:

$$G_{xx}(f) = 2S_{xx}(f) \tag{6}$$

By definition, a random time history cannot be periodic; however, if the time history is taken from an ergodic stationary Gaussian random process, then it may be expressed in the frequency domain. A random process is said to be stationary if the probability distributions of the ensemble remain the same for each time instant [3]. If the ensemble probability distribution function is Gaussian, then the process is known as a Gaussian random process. A stationary process is called ergodic if the statistics taken from one sample are the same as those obtained for the ensemble. For nonstationary processes, the statistics obtained from a sampled time history would not be representative of those of the whole random process as they would be continuously changing.

The PSD gives a statistical representation of a stationary random process in the frequency domain [4-5]. Spectral moments are calculated from the one-sided PSD by:

$$m_n = \int_0^{\infty} f^n G(f) df = \sum_{k=1}^{m} f_k^n G_k(f) \delta f \tag{7}$$

The mean square amplitude of the time history is the area under the PSD curve, and therefore the root mean square (RMS) is obtained by Eq. (8):

$$\sigma_{rms} = \sqrt{m_0} \tag{8}$$

The expected number of upward zero crossings is given by

$$E[0] = \sqrt{\frac{m_2}{m_0}} \tag{9}$$

The expected number of positive peaks can be written as:

$$E[P] = \sqrt{\frac{m_4}{m_2}} \quad (10)$$

Figure 2 depicts a random stress variation and shows the representation of $E[0]$ and $E[P]$. Dirlik's approach [3] uses Equation 9-11 to obtain a closed-form equation for the pdf of the stress amplitude as described in the next section.

Dirlik's model and probabilistic linear cumulative damage

The particular case of narrowband random fatigue loading was extended to arbitrary stress cycles in the 1960s, using cycle versus amplitude counting methods. The de-facto standard that emerged was called the Rainflow cycle method, attributed to Matsuishi and Endo [6]. Research showed that the Rainflow cycle method combined with the linear damage model gave the most accurate fatigue life estimates [7]. This technique is a time-domain method.

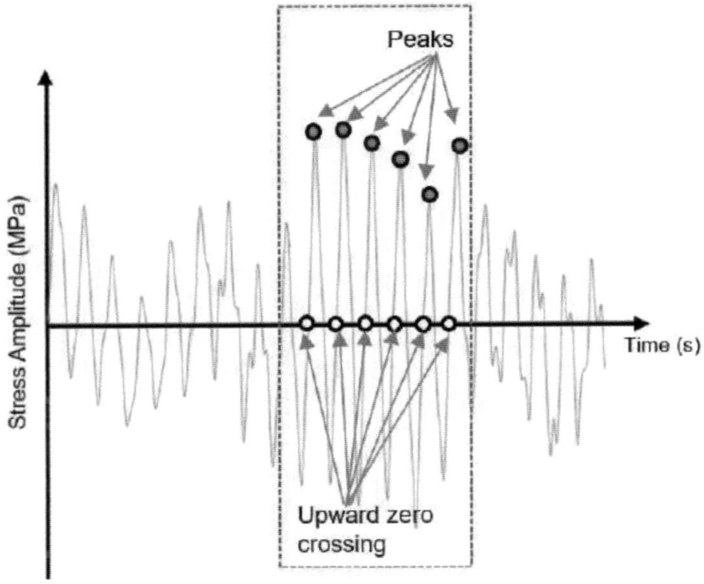

Figure 2: Representation of expected number of peaks and upward zero crossings.

The difficulties with the computation of the rainflow cycles distributions from time-domain data have led to the development of frequency-domain techniques. The seminal work by Rice [8] and research since has shown that the PDF of peak response can be determined from moments of the frequency domain response power spectral density (PSD) and that the peak PDF is a combination of a Gaussian and a Rayleigh PDF.

In 1985, Dirlik developed a practical closed-form solution for the wideband random problem., He proposed a pdf that considered the sum of two Rayleigh and one exponential distribution generated from an extensive Monte Carlo simulation [9]. Although apparently more complicated than some alternative methods, it is only a function of the four moments of area of the PSD of the stress. This method is widely applicable and constantly outperforms all the other available methods for wideband vibration problems [3-4].

The Dirlik's pdf of stress amplitude developed in [9] is given by:

$$p(s) = \frac{\frac{D_1}{Q} e^{\left(\frac{-s}{2Q\sqrt{m_0}}\right)} + \frac{D_2}{R^2} \frac{s}{2\sqrt{m_0}} e^{\left(\frac{-s^2}{8R^2 m_0}\right)} + D_3 \frac{s}{2\sqrt{m_0}} e^{\left(\frac{-s^2}{8m_0}\right)}}{2\sqrt{m_0}} \tag{11}$$

where the parameters D_1, D_2, D_3, Q, a R are defined as a function of the PSD moments as shown from Eq. (12) through Eq.(17)

$$x_m = \frac{m_1}{m_0}\sqrt{\frac{m_2}{m_4}} \tag{12}$$

$$D_1 = \frac{2(x_m - \gamma^2)}{1 + \gamma^2} \tag{13}$$

$$R = \frac{\gamma - x_m - {D_1}^2}{1 - \gamma - D_1 + {D_1}^2} \tag{14}$$

$$D_2 = \frac{1 - \gamma^2 - D_{1+}{D_1}^2}{1 - R} \tag{15}$$

$$D_3 = 1 - D_1 - D_2 \tag{16}$$

$$Q = \frac{1.25(\gamma - D_{3+}D_2 R)}{D_1} \tag{17}$$

The number of cycles at a specific value of stress *i* are obtained for a given time of vibration exposure *T* by

$$n_i = p[S_i]dS \times E[P] \times T \tag{18}$$

The damage is then calculated generally using the 50% or P50 percentile S-N curve. However, due to the scatter in the material data, a probabilistic linear cumulative is proposed to characterize the variability of the damage estimation using the different percentile fatigue life curves.

The probabilistic linear cumulative damage is then defined as:

$$D_{p_x} = \sum_{i=1}^{k} \frac{n_i}{N_{i_{p_x}}} \tag{19}$$

where x is the percentile curve. The damage is now computed not only for the commonly used median percentile but for all confidence level intervals.

The time to fatigue failure could also be calculated considering the time which the component or structure is exposed to the vibration profile relative to the damage at determined life percentile, as shown in Eq. (20):

$$T_{failure_{px}} = \frac{T_{exposure}}{D_{px}} \tag{20}$$

EXPERIMENTAL STUDY AND FINITE ELEMENT MODEL

A notched 6061-T6 aluminum specimen illustrated in Figure 3 is subjected to a random vibration profile in the vertical direction(Z) in two steps. The first PSD profile was defined with a specific time duration; subsequently, the component is subjected to a second PSD profile with an increased root mean square acceleration where the component was tested to failure. Based on the cumulative damage process, the objective of the first test is to pre-damage the structure. Both PSD profiles are shown in figure 4. Three samples were tested simultaneously to measure the fatigue life variations among the specimens. Also, the specimens mounted on the shaker table can be seen in Figure 4.

Both applied PSD profiles are shown Figure 4. The first PSD was applied during 7800 s. A DSA-4K - GW Gearing & Watson Electronics Limited electrodynamic shaker and three Bruel & Kjær DeltaTron 4519-003 accelerometers were used to excite and measure the acceleration responses at the notches on the test specimens. Both PSD profiles were defined to create a broadband excitation with stress responses below the yield point of the AL 6061-T6 material.

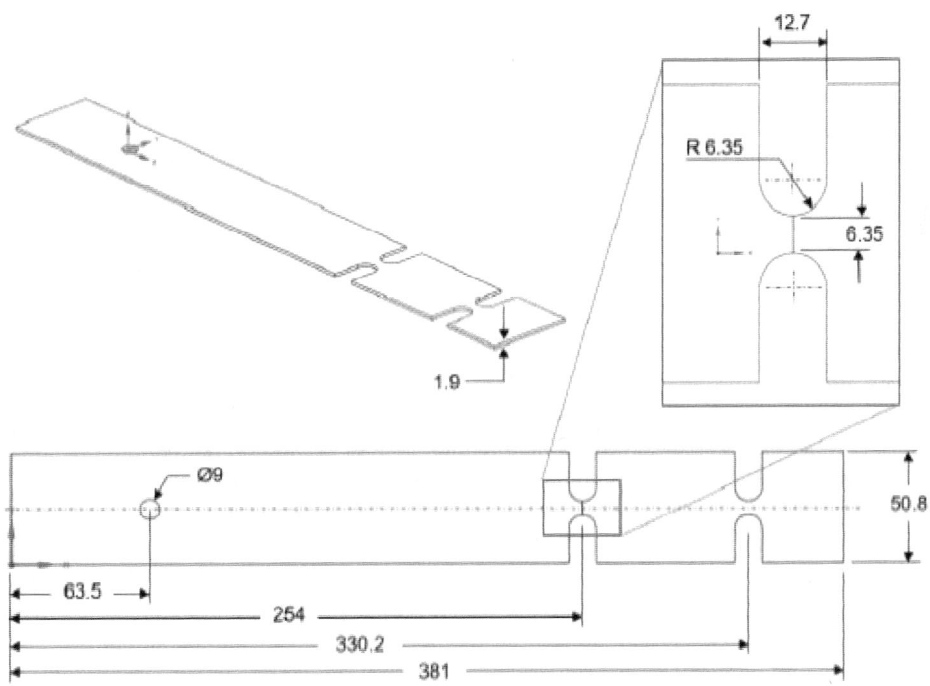

Figure 3: Main dimension of tested sample.

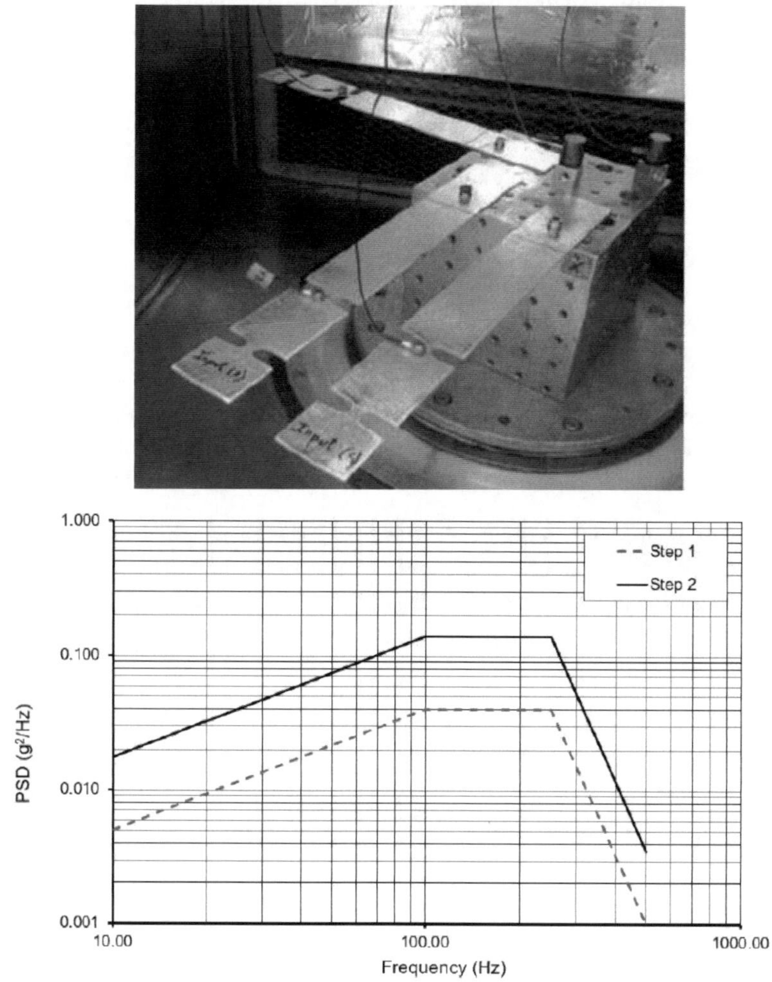

Figure 4: Acceleration PSD base excitation input and samples mounted on the shaker table.

Figure 5 shows the mesh of the finite element model used in the study. A mesh independence study was carried out to verify the sensibility of the results to the mesh refinement. The final adopted mesh size has square elements of 0.125 mm close to the notch area. The dynamic characterization of the structure was carried out by performing the modal analysis solving the natural frequencies and vibration modes of the system. The boundary conditions were defined to emulate the fixation of the specimen in the shaker fixture. A shell model with the element formulation SHELL181 was used in Ansys software. Seven modes were identified in the frequency range up to 500Hz, which is the upper limit of the vibration profiles. The lower and higher natural frequencies were f_1 =17.29 Hz and f_7 =453.8 Hz, respectively. These modes are used to obtain the frequency response functions (FRFs) at the modal frequency response step for a unit harmonic acceleration input.

To properly define the transient response and obtain the FRFs, the damping must be defined. In this work the damping was modeled using the half power bandwidth method obtained from accelerometer data. The Rayleigh proportional damping was used in the dynamic analysis. Details of the calibration procedure for the constants representing the mass (α=4.188) and stiffness (β=8.49e-7) proportional damping coefficients can be found in [11].

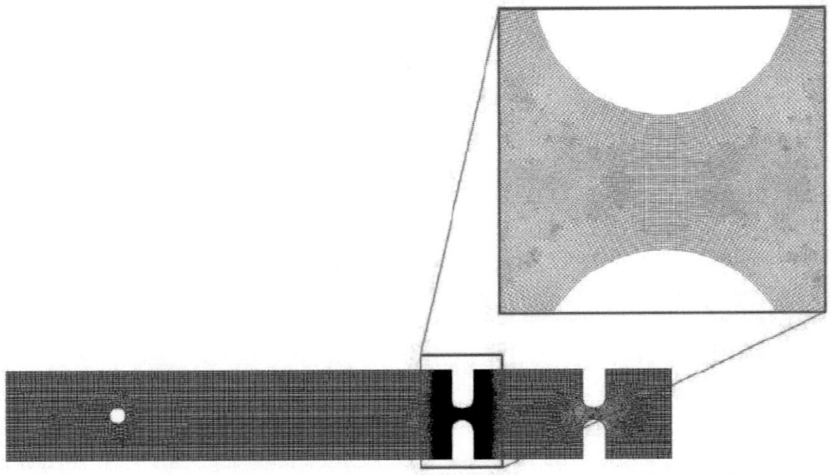

Figure 5: Finite element model and zoom at the notch area to see the mesh refinement.

The material properties, engineering stress-strain curve and unnotched fatigue data, for 6061-T6 aluminum are reported in Pascualinotto [11,12]. The tensile stress-strain curve and fatigue S-N curve were taken from the military handbook [13] and the results correspon to unnotched specimens. The fatigue strength reduction factor k_f is calculated using the volumetric approach to correct the S-N curve for the notched condition, as shown in [11]. The fatigue strength reduction factor k_f is then applied to the unnotched S-N percentile curves.

Based on the assumption that the fatigue failure requires a physical volume to occur, the volumetric approach is a macro-mechanical method that uses elastic-plastic distribution and stress gradient evolution to predict the fatigue reduction factor kf, based on the idea of fatigue damage accumulation in a specific region near notch tip. The fact that the fatigue tests are generally affected by a large scatter indicates that the stress distribution near the notch tip is different between specimens, and, consequently, the first derivative of the stress distribution function is considered to obtain the fatigue concentration factor. The obtained kf parameter was 1.14.

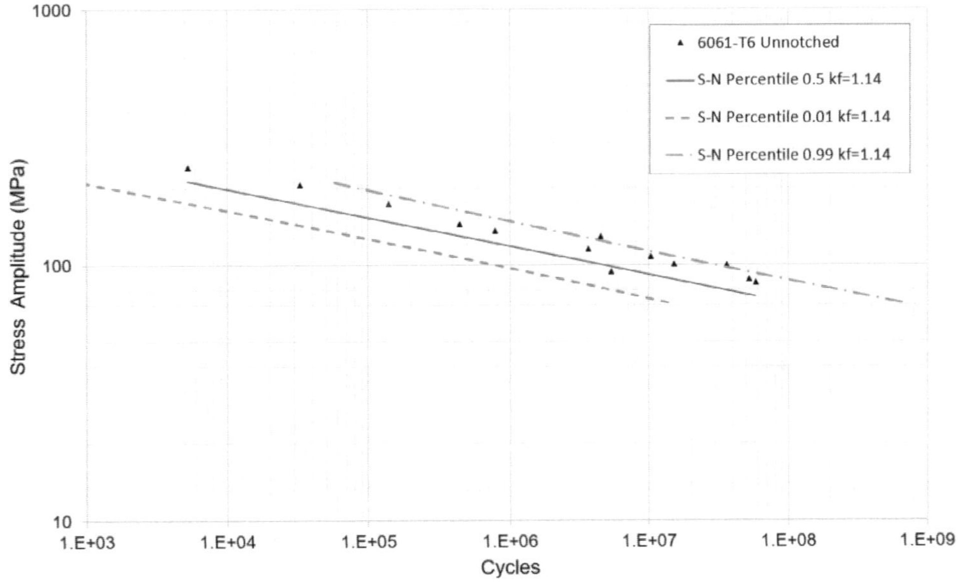

Figure 6: Finite element model and zoom at the notch to see the mesh refinement

The corrected P-S-N curves in a log-log scale are illustrated in Figure 6. It is noticeable the impact of the notch on the fatigue strength of the specimens..

Table 1 displays the results of the experimental random fatigue. The failure was defined when the complete fracture (separation) has occurred based on the assumption that takes most of the fatigue life to initiate a dominant crack. None of the samples failed during the application of the first loading PSD profile. As can be seen in the table, the failure occurred after more than 1500 seconds of duration of the second PSD function for all three samples. The mean time to failure was calculated as 9667 seconds.

Table 1. Experimental fatigue results.

Specimen	STEP 1	STEP 2	Total Time to failure (s)	Mean time to failure (s)
	Exposure time (s)	Exposure time (s)		$\mu \pm \sigma$
Sample 1	7800	1772	9572	
228Sample 2	7800	2324	10124	9667±228.5
Sample 3	7800	1534	9334	

RESULTS AND DISCUSSION

The equivalent von Mises stresses plot is illustrated in Figure 5. The high-stress concentration at the notch is notable, which has a higher likelihood of failure. The stress response spectrum is obtained at the critical area of the part. Figure 6 shows the stress response spectrum obtained at the critical point. The numerical results show predominantly axial stresses, with peak responses at the first and second modes. Shear stresses are significantly lower for the uniaxial excitation case. The critical plane method was used to combine the stress response PSD components, $\{\sigma_x(f), \sigma_y(f), \tau_{xy}(f), \}$, as embedded feature in Ansys software. The equivalent von Mises σ stress plot is used here not as the failure criteria. A comparison with the fatigue limit could result in a completely wrong interpretation of the fatigue phenomenon. Instead, it is used to identify the areas with higher stress concentration that are likely to fail.

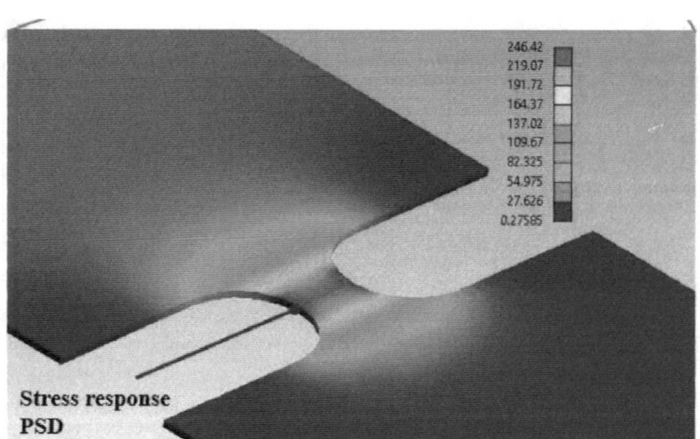

Figure 7: von Mises stresses at the notch area

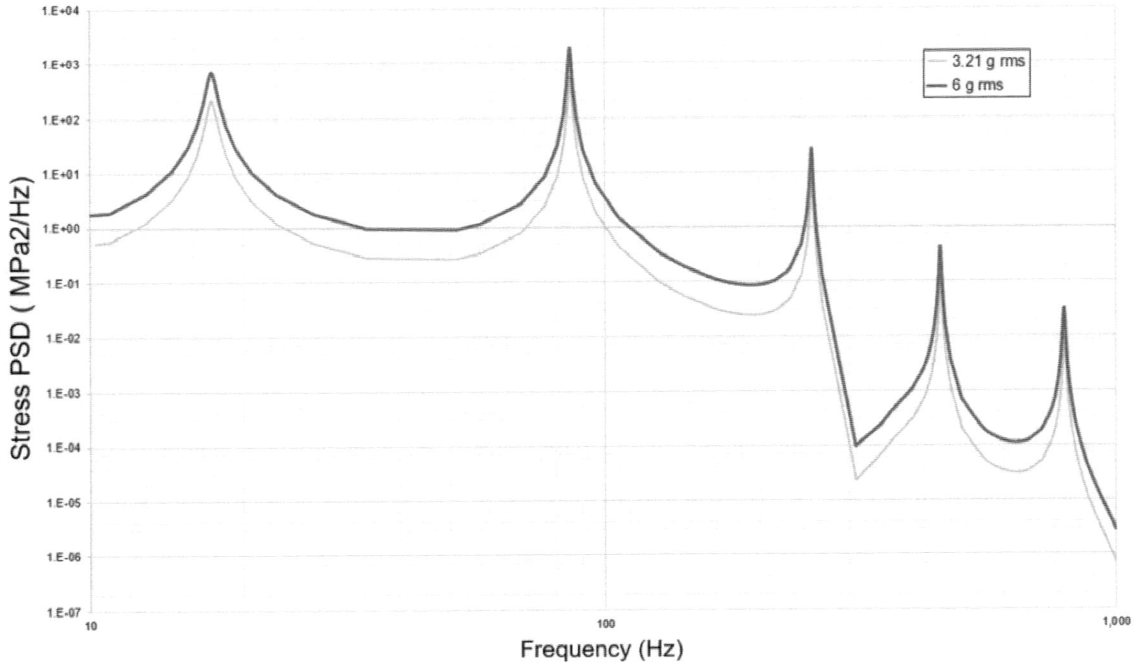

Figure 8: Stress Response PSD

The number of cycles is estimated from the combined stress response PSD, based on the probability density function (Dirlik), input PSD exposure time, and expected positive peaks. The PSD has no time duration as it is simply a representation of the frequency content and energy of the time signal. Thus, the total number of cycles must be scaled by the period that the structure is exposed to the vibration profile. Similar to the definition of the acceleration spectrum magnitudes, the time durations were defined to create a cumulative damage process in the first step not leading to failure, followed by a sufficient exposure time to the second profile leading to failure. The exposure time was defined as 7800s for the 3.21 $grms$ PSD input and 2500s for the six (6) $grms$ PSD. The fatigue life was estimated based on the linear cumulative damage calculated for the defined S-N curve percentiles. The damage is then calculated from the stress PSD for the different percentile curves.

The linear cumulative damage rule was calculated based on the material fatigue testing data's S-N percentiles 0.01, 0.5, and 0.95. The time domain approach was also used to calculate the failure time. Acceleration time histories were synthesized following the acceleration spectrum densities specifications by using the inverse Fourier transform, Eq. 2, and setting random values for the phase of each periodic signal that is used to define the time signal. The synthesized acceleration time signals were applied as a transient base excitation to the notched specimen at the fixed support defined in the modal analysis. The time transient analysis is computationally expensive analysis as the stress response tensor is calculated for each time step increment which for the total signal length of 25s gives a total of 131,095 acceleration records. Details of the time domain analysis can be found in [12].

The number of cycles at each stress level is then calculated from the stress response PSD, given the duration of the exposure to the vibration profile. The usage of the 0.95 percentile would rate the damage as only 14% compared to the experiment. The upper limit of the S-N curve, which corresponds to p=0.99, is non-conservative for fatigue predictions. Therefore, it is not recommended to be used for design purposes. The results in Table 1 show good agreement between testing and analytical results when the meantime to failure from the test results is compared to the results using the percentile 0.5 of the S-N curve. As one would expect, the predictions using p=0.01 S-N are highly conservative in all cases. Finally, a summary of damage and time to failure is summarized in Table 2 for both methods. A good agreement of both methods, time and frequency domain, is observed in table 2 when compared to the mean failure time from the fatigue experiments. There are various sources of variation when modeling fatigue in a random environment, however, using the linear cumulative damage with p=0.5 produced results in good agreement with testing results.

Table 2. Fatigue Life and Damage relative to the S-N curve percentile.

S-N curve percentiles	Damage (PSD_1+PSD_2)	Life (repeats) (PSD_1+PSD_2)	Frequency domain Dirlik (s)	Time domain Time to failure (s)	Experiment Mean time (s)
0.01	6.5	0.15	8146	8129	
0.5	1.35	0.74	9835	9647	9667
0.95	0.14	6.84	No-failure	No-failure	

Despite the well-known limitations of the linear cumulative damage, the proper dynamic characterization of the structure via a representative damping modeling and the consideration of geometrical factors (notch) which affect the fatigue behavior were found in this work to influence the fatigue life prediction under random vibration loading.

CONCLUSION

In this work, the fatigue life estimation of a notched structural component under random vibration was calculated using a probabilistic linear cumulative damage model and frequency domain technique. There are various sources of variation when modeling fatigue in a random environment. However, using the linear cumulative damage with p=0.5 produced results in good agreement with testing after the appropriate characterization of the effect of the notch and modeling of the damping. The fatigue strength reduction in the presence of a notch was estimated based on the volumetric approach. The proper correction of the base S-N curve (p=0.5) influenced the fatigue life estimation of approximately 69%, after that, including the effects of geometric discontinuities is critical to the accuracy of the prediction. Another fact contributing to the good agreement of the predictions is the primarily uniaxial behavior of the stress tensor, which approximates the material response to the one obtained in the standard S-N test.

REFERENCES

[1] Suresh, S.; Fatigue of Materials, Cambridge University Press (Cambridge), 1991.

[2] Dowling, N. E.; Mechanical Behavior of Materials, Prentice Hall (Upper Saddle River, NJ), 1999.

[3] Yung-Li Lee, Jwo Pan, R. Hathaway, M. Barkey. Fatigue Testing and Analysis: Theory and practie. Elsevier, 2005.

[4] A. Halfpenny, *A frequency domain approach for fatigue life estimation from Finite Element Analysis*. In M. D. Gilchrist, J. M. DulieuBarton, and K. Worden, editors, DAMAS 99: Damage Assessment of Structures, volume 167-1 of Key Engineering Materials, pages 401–410, 1999.

[5] Halfpenny, A, *Dynamic Analysis of Both On and Offshore Wind Turbines in the Frequency Domain*. PhD thesis, University College London, 1998.

[6] Matsuishi, M. and Endo, T.; *Fatigue of metals subjected to varying stress*. Japan Society of Mechanical Engineers, 1968.

[7] Dowling, N. E.; *Fatigue failure predictions for complicated stress-strain histories*. Journal of materials, vol 7, pp71-87, 1972.

[8] Rice, S.O.; *Mathematical Analysis of random noise*. Bell Syst. Tech. J., vol 23, pp.282-332, 1944.

[9] T. Dirlik. *Application of Computers in Fatigue Analysis*. PhD thesis, The University of Warwick, 1985.

[10] R. D Pollak, Analysis of methods for determining high cycle fatigue strength of a material with investigation of *Ti-6AL-4V gigacycle fatigue behavior*. PhD thesis, Air Force Institute of Technology, 2005.

[11] Vagner Pascualinotto Jr. Fatigue life estimation of a notched component using frequency domain technique and probabilistic linear cumulative damage model. Master thesis, University of São Paulo, Brazil, 2021.

[12] Vagner Pascualinotto Junior, Diego Felipe Sarzosa Burgos. Fatigue Life Estimation Using Frequency Domain Technique and Probabilistic Linear Cumulative Damage Model. Pressure Vessels and Piping Conference. PVP2020-21536,2020.

[13] MIL-HDBK-5G, Military Handbook – Metallic Materials and Elements for Aerospace Vehicles Structures, Vol. 1 and 2, 1994.

SNAME
THE INTERNATIONAL COMMUNITY
FOR MARITIME AND OCEAN PROFESSIONALS

CAPSIZE OF LIFTBOAT IN TRANSIT

William P. Stewart
Stewart Technology Associates

Proceedings of the 27th Offshore Symposium, February 22nd, 2022, Houston, Texas
Texas Section of the Society of Naval Architects and Marine Engineers

ABSTRACT

The 2021 capsize of the liftboat Seacor Power, with loss of 13 lives, has focused attention on the afloat stability of liftboats in transit. This paper presents a numerical method of simulating the dynamic capsize of liftboats in waves using time domain analysis and a commonly used computer program. Capsize of a similar vessel to the Seacor Power is predicted in 6-ft quartering and beam seas even without wind. Model tests are proposed to help calibrate calculation techniques and define limiting metocean conditions more reliably for safe transit of liftboats.

Keywords: Capsize, Liftboat, Seacor Power; AVCO V; Sub-harmonic Response; GZ Curve, Froude-Krylov

Introduction

In 1989 a Class 105 (meaning leg length 105 feet) liftboat, the *AVCO V*, capsized with the loss of 10 lives. The *Seacor Power* was a Class 250 liftboat with an extra 15 feet added to its legs, making it a Class 265 (meaning leg length 265 feet) vessel. Figure 1 compares the outboard profiles of the 105 and 250 classes.

The two capsize incidents occurred about 50 miles apart, off the Louisiana Coast, south of Houma. In the *AVCO V* capsize [1] the waves were described as "6 and 7 feet" high 30 minutes before the capsize, and building. At the point of capsize, opinions in [1] put the waves as "about 6 to 8 feet high". The master estimated the seas to be 15 feet high at the point of capsize at 0505 on July 31, 1989. The wind speed was estimated at 30 knots, "gusting to 50 knots". The water depth at the *AVCO V* capsize location was 27 feet.

In the *Seacor Power* capsize [2] the wind speed increased suddenly in a rain squall. Vessels in the area reported winds in excess of 80 knots from the north with rapidly building seas. From testimony given in the August USCG Hearing [3] less than 30 minutes before the capsize, the wind had been around 18-kt from between south and south-southeast with 3-4 feet seas from the same direction. The vessel was making about 3 knots (from AIS transmission) and heading southeast. In a period of less than 2 minutes the wind changed direction to north (coming from) and its speed was speed registered on the vessel anemometer at 79 mph (67 kt). After about 2 minutes the wind speed dropped (to possibly 40-kt) and a white out occurred (extremely heavy rain and almost zero visibility) for 5-10 minutes. Immediately following the white out the wind speed was noted to be 30-45 mph (26-38 kt) and the waves were estimated to be 3-6 feet, coming from the north and rapidly increasing.

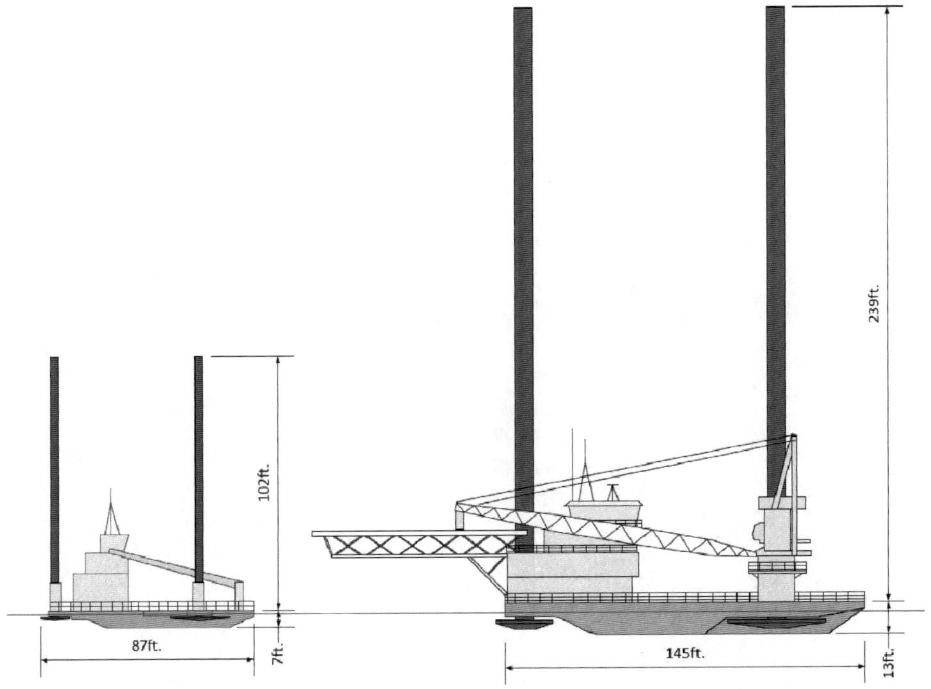

Figure 1 – Comparison of Class 105 and Class 250 Liftboat Afloat Profiles

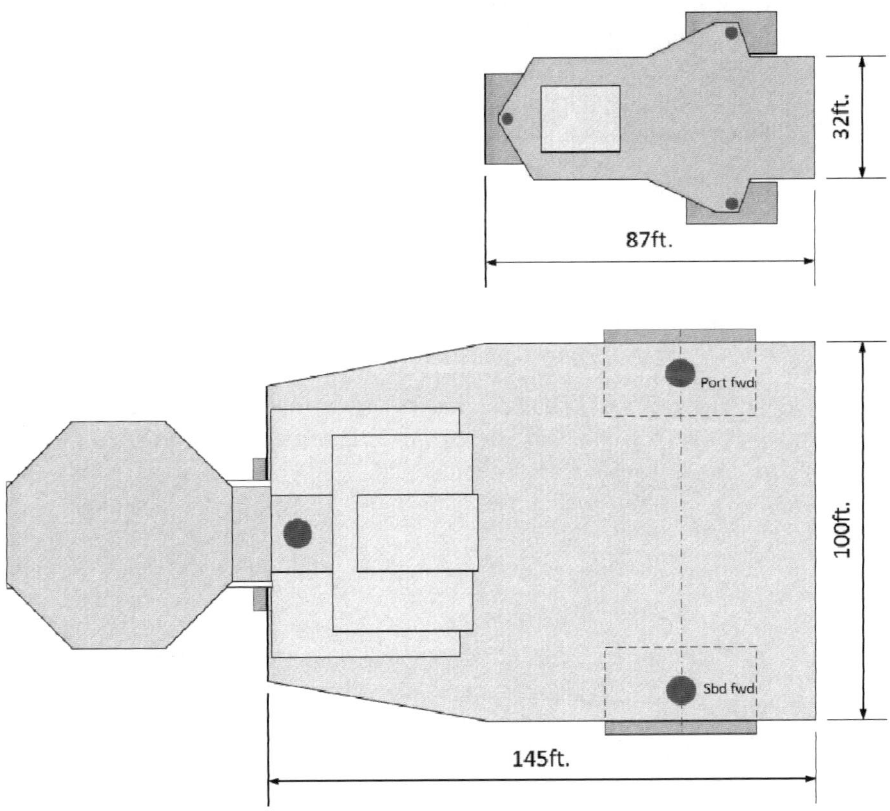

Figure 2 – Plan View Comparisons of the *AVCO V* and *Seacor Power*

During the white out the captain decided to stop and elevate the hull. The crew simultaneously began jacking the legs down turning the vessel into the wind. The turn was to port and the capsize occurred to starboard during the turn. The vessel had the wind and waves from the north on its port side at the time of the capsize. The water depth at the *Seacor Power* capsize location was 54 feet.

From testimony and from AIS transmission during the white out, the vessel speed had been increased by the wind from the north to about 5.0 knots. It is likely to have reduced to less than 4.0 knots during the turn.

Figure 3 Locations of Capsizes, *AVCO V* and *Seacor Power*

Tube Raft Hull Model - Hydrostatics

This paper introduces a tube raft. The liftboat hull is built with a tube raft made from an assembly of OrcaFlex [4] line elements. Each tube has a discrete solution for its buoyancy and Froude-Krylov force. Rafts of tubes are assembled to occupy the same volume as each block of the liftboat hull. Hull geometry blocks are shown in Figure 4.

The target segment length selected for each tube is 7.5 ft. Thus, the longest upper tubes have 19 segments each. The tubes are 6.5 ft OD and very stiff in flexure. All line elements are attached to an OrcaFlex 6D buoy, which represents the vessel hull. Figure 5 shows hull geometry blocks filled with tubes.

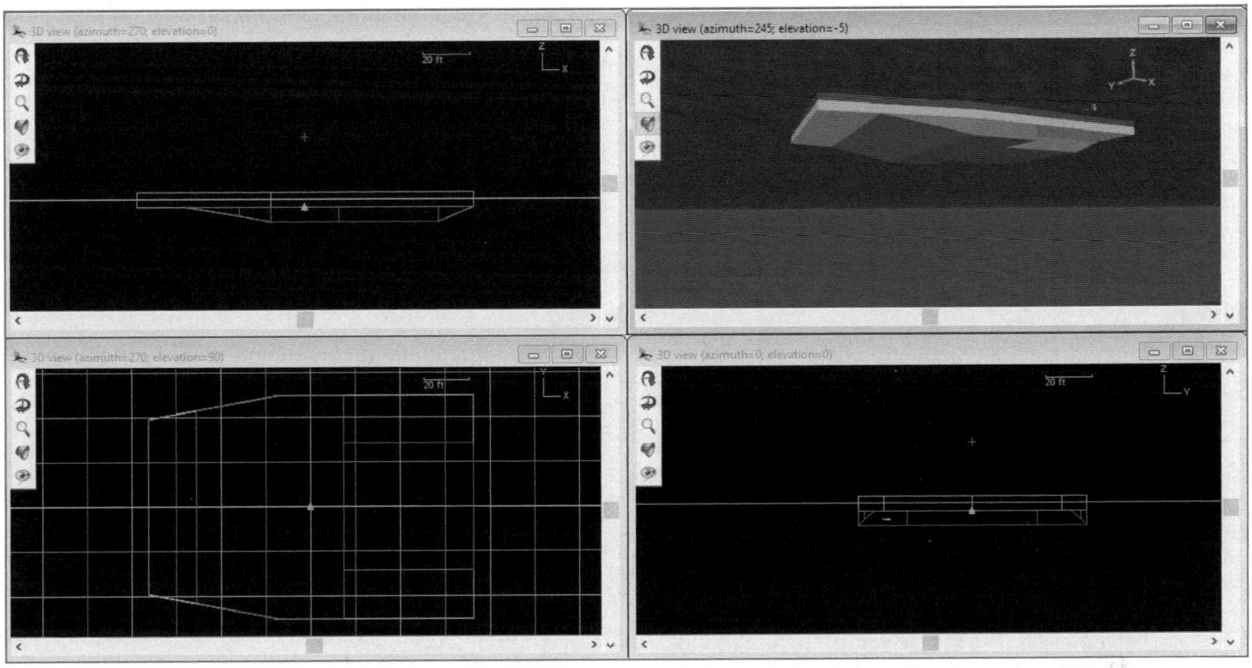

Figure 4 – Hull Geometry blocks Shown with OrcaFlex Shapes (green upper, and red lower hull block)

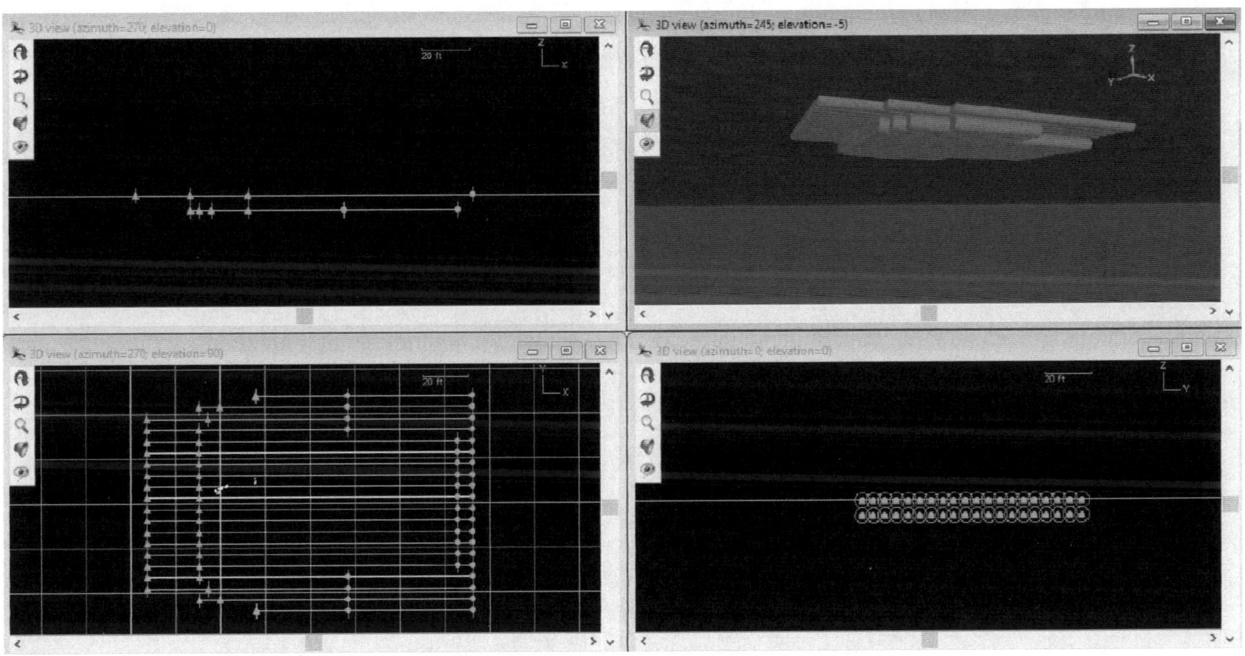

Figure 5 – Hull Geometry made with OrcaFlex Line Elements (Tubes)

The *Seacor Power* hydrostatics table from the vessel's Marine Operations Manual (MOM) was made public in Reference 4. The OrcaFlex tube raft model was adjusted in order to get matching displacement vs draft data to within 0.2% in the operating depth range of interest (from 9 to 10 ft draft). The comparison is shown in Figure 6.

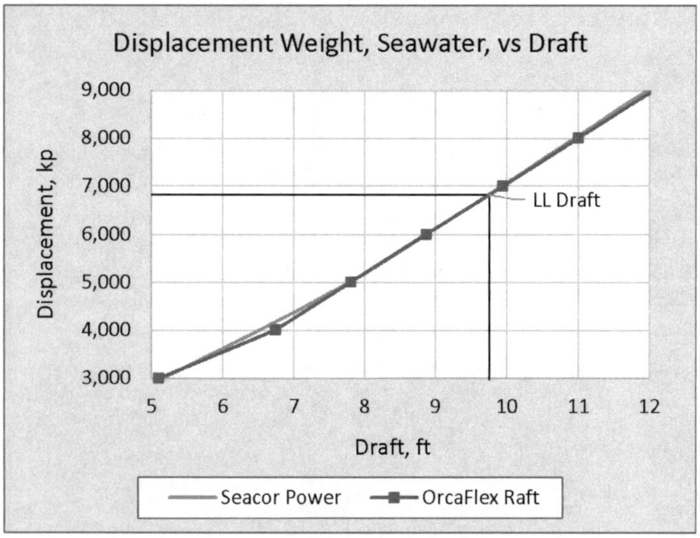

Figure 6 – Displacement Weight vs Draft Comparison, Zero Trim

Class 250 Liftboat Weight Distribution & Afloat Condition

The weight distribution used for the Class 250 liftboat (250-ft leg length) described in this paper is similar to that of the *Seacor Power* and is shown in Table 1. The *Seacor Power* had a similar weight distribution but had 265-ft leg lengths and slight differences in its hull, when it capsized.

Table 2 shows the Class 250 tube raft vessel afloat condition, with draft, heel, trim, and allowable KG, based on the *Seacor Power* allowable KG, as shown in Figure 7.

Components of Weight	Weight each	Weight total	LCG from center	LCG from Fr.0	TCG	VCG from Keel	LCG from LCB
3 Legs, 7.5' OD	479.4 kp	1438 kp	2.97 ft	69.53 ft	0.00 ft	127.00 ft	11.95 ft
2 Fwd Pads	157.1 kp	314 kp	36.70 ft	35.80 ft	0.00 ft	4.00 ft	45.68 ft
1 Aft Pad	78.6 kp	79 kp	-64.50 ft	137.00 ft	0.00 ft	4.00 ft	-55.52 ft
Lightship Hull, w/o helideck or cranes		3124 kp	-9.21 ft	81.71 ft	2.32 ft	9.00 ft	-0.23 ft
2 120' Crane Booms	22.4 kp	45 kp	-27.50 ft	100.00 ft	0.00 ft	38.00 ft	-18.52 ft
2 Crane assemblies w/cabs, bearimgs, motors, etc.	130.0 kp	260 kp	36.70 ft	35.80 ft	0.00 ft	40.00 ft	45.68 ft
Helideck		300 kp	-107.50 ft	180.00 ft	0.00 ft	26.50 ft	-98.52 ft
Total Variable Load (max = 529.4 LT at max 13' above deck)		1186 kp	35.00 ft	37.50 ft	-6.00 ft	26.00 ft	43.98 ft
Summary Load Condition		**6745 kp**	**-0.07 ft**	**72.57 ft**	**0.02 ft**	**39.02 ft**	**8.91 ft**
		3060 MT	-0.02 m	22.12 m	0.01 m	11.89 m	2.72 m

Table 1 – Vessel Weight Distribution Used in this Paper

Stability Results for Class 250 Loaded Condition	**T, Draft**	**W, Weight**	**LCB* from Fr.0, ft**	**LCB Bow, ft**	**KML, ft**	**KMT, ft**	**KG corr =VCG -FS**	**GMT = KMT - Corr KG**
Draft at LCF	9.66 ft	6745 kp	68.30 ft	81.48 ft	298.74 ft	112.22 ft	39.12 ft	73.10 ft
Heel	0.02º	Stbd Down		Allowable KG (70-kt wind)			**39.71 ft**	OK
Trim	0.94º	Bow Up		Allowable KG (80-kt wind)			**38.06 ft**	NOT OK
Drafts and Freeboards at Stern and Bow	**Stern**	**Bow**		*Allowable KG data for Seacor Power*				
Drafts	10.85 ft	8.47 ft					32.62 ft	z for Orca
Freeboards	2.15 ft	4.53 ft						
CFR 174.185: Check Stern Freeboard > 20 in	OK		*LCB* matched to tube raft model in this paper*					

Table 2 – Afloat Condition, Draft, Heel & Trim, Showing Bow and Stern Freeboards

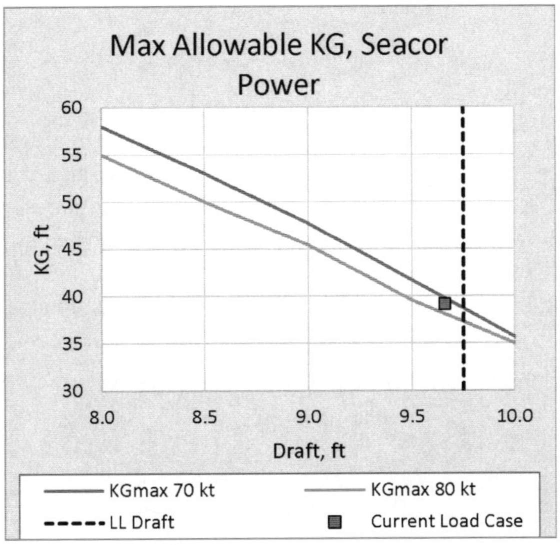

Figure 7 – Allowable KG

The maximum allowable KG data for the *Seacor Power* was presented in Reference [3] and is shown plotted in Figure 7 together with the actual draft and KG data for the vessel in this paper, from Table 2.

It was not clear from testimony given in [3] whether the *Seacor Power* Allowable KG data presented was for the vessel before or after the legs were extended from 250 ft to 265 ft.

Wind Forces & Heeling Moments

Table 3 presents a summary of wind areas, forces and heeling moments developed for the vessel studied in this paper using the standard industry "building block" method. Selected shape and height coefficients are shown. Results are given for the upright condition for a 9.66-ft draft and 70-kt wind speed.

Separately a large angle heeling moment curve is developed (by rotating the 3-dimensional vessel with zero trim and using projected areas) and applied to find the wind area ratio as discussed later.

Component	Length (or diameter)	height	area, A	VCA above CB	Cs	Ch	F	Vmom from CB
3 Legs from 200 to 250	23.40 ft	42.34 ft	991 ft^2	213.0 ft	0.85	1.37	19.11 kpf	4070 kpf-ft
3 Legs from 150 to 200	23.40 ft	50.00 ft	1170 ft^2	166.8 ft	0.85	1.30	21.41 kpf	3572 kpf-ft
3 Legs from 100 to 150	23.40 ft	50.00 ft	1170 ft^2	116.8 ft	0.85	1.20	19.77 kpf	2309 kpf-ft
3 Legs from 50 to 100	23.40 ft	50.00 ft	1170 ft^2	66.8 ft	0.85	1.10	18.12 kpf	1211 kpf-ft
Fwd 2 Legs from Crane Top to 50	15.60 ft	4.13 ft	64 ft^2	52.0 ft	0.85	1.00	0.91 kpf	47 kpf-ft
Aft Leg from Above Deck House to 50	7.80 ft	26.23 ft	205 ft^2	130.3 ft	0.85	1.00	2.88 kpf	375 kpf-ft
Deck House Level 1	50.00 ft	10.10 ft	505 ft^2	13.2 ft	1.00	1.00	8.36 kpf	111 kpf-ft
Deck House Level 2	53.00 ft	9.70 ft	514 ft^2	23.1 ft	1.00	1.00	8.51 kpf	197 kpf-ft
Deck House Level 3	37.10 ft	10.10 ft	375 ft^2	33.0 ft	1.00	1.00	6.21 kpf	205 kpf-ft
Deck House Level 4	28.70 ft	10.90 ft	313 ft^2	43.5 ft	1.00	1.00	5.18 kpf	225 kpf-ft
Fwd 2 Leg Towers	29.30 ft	18.40 ft	539 ft^2	17.4 ft	1.00	1.00	8.93 kpf	155 kpf-ft
Fwd 2 Crane Main Assemblies	37.80 ft	23.50 ft	888 ft^2	38.32 ft	1.00	1.00	14.71 kpf	564 kpf-ft
2 Main Crane Booms	244.46 ft	5.50 ft	1345 ft^2	35.87 ft	1.00	1.00	22.27 kpf	799 kpf-ft
2 Upper Crane Assemblies	8.52 ft	48.10 ft	410 ft^2	50.62 ft	1.50	1.10	11.20 kpf	567 kpf-ft
2 Crane Hook Assemblies	6.40 ft	7.50 ft	48 ft^2	38.17 ft	1.50	1.00	1.19 kpf	46 kpf-ft
Rigging cables, masts, misc	250.00 ft	1.00 ft	250 ft^2	64.00 ft	1.50	1.10	6.83 kpf	437 kpf-ft
Helideck	87.30 ft	10.80 ft	943 ft^2	31.00 ft	0.90	1.00	14.05 kpf	436 kpf-ft
Deck Cargo	70.00 ft	30.00 ft	2100 ft^2	30.00 ft	1.10	1.00	38.26 kpf	1148 kpf-ft
Hull	145.00 ft	3.34 ft	484 ft^2	6.50 ft	1.00	1.00	8.02 kpf	52 kpf-ft
TOTALS	**145.00 ft**	**92.99 ft**	**13483 ft^2**	**70.05 ft**	**1.00**	**1.06**	**236 kpf**	**16525 kpf-ft**
			1253 m^2	21.35 m	1.00	1.06	**1049 kN**	**22405 kN-m**

Table 3 – Class 250 Wind Areas, Forces, Heeling Moments at 9.66 feet Draft, 70-kt Wind Speed

Heel and Trim with 70kt wind on Beam or Bow	
Heel	1.93º
Trim	1.48º
Freeboard at Stern with 70kt Wind on Bow	17.6 in

Table 4 – Heel, Trim & Stern Freeboard with 70-kt Wind

Large Angle Stability

The method used to produce conventional large angle stability data, for tube raft liftboat in this paper, uses OrcaFlex as follows. For free trimming without twist, two pairs of winches are used at fixed elevations, one above and one below the free-floating vessel. The winches are positioned far to each side of the vessel and have long (weightless) winch wires that apply a horizontal pull to points on the vessel above and below the center of flotation. The heeling moment is applied slowly, in the time domain, and very slowly rotates the vessel, which is free to trim and to yaw, or twist, very slightly. No buoyancy of above deck structures is included. The method is illustrated with two annotated screen shots in Figure 8.

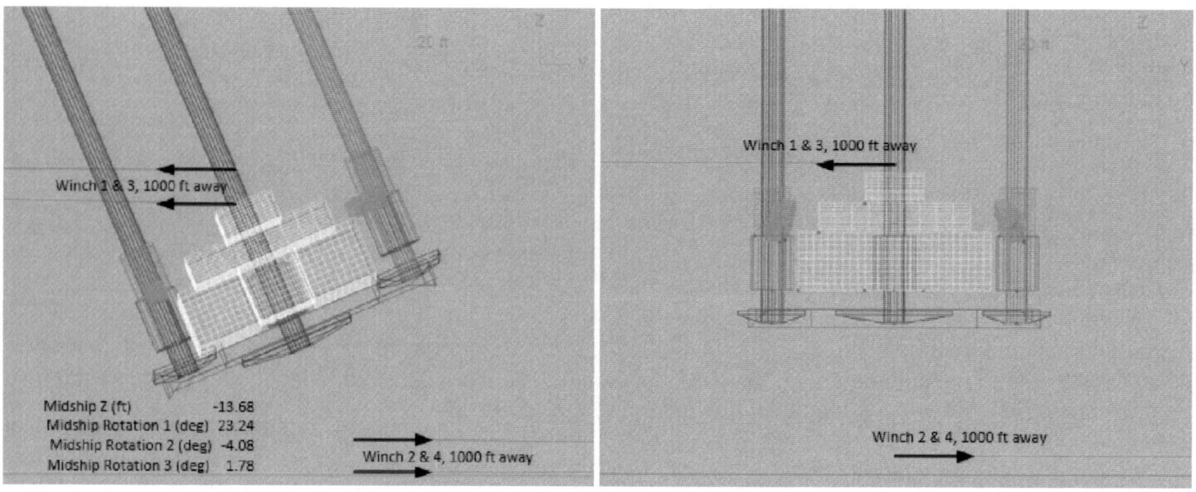

Figure 8 – Inclining with Winch Pull, Free to Trim & Yaw, Left. Before Inclining, Right

For this paper, the winch loads are increased from zero by steadily reducing the winch wire lengths over a period of 400 seconds. This length of time is sufficient to make inertial effects negligible during the rotation, effectively resulting in a continuous static solution. The adjustment length is set such that the vessel just reaches the limit of positive stability at time equals 400 seconds. Initially simulations are run until the vessel capsizes. The capsize can be observed at real time in the time domain. Just before capsizing, the applied moment from the winches has reduced to zero, the heel angle is 23.4°, the trim is 4.3° (stern down) and the yaw, is 1.9°.

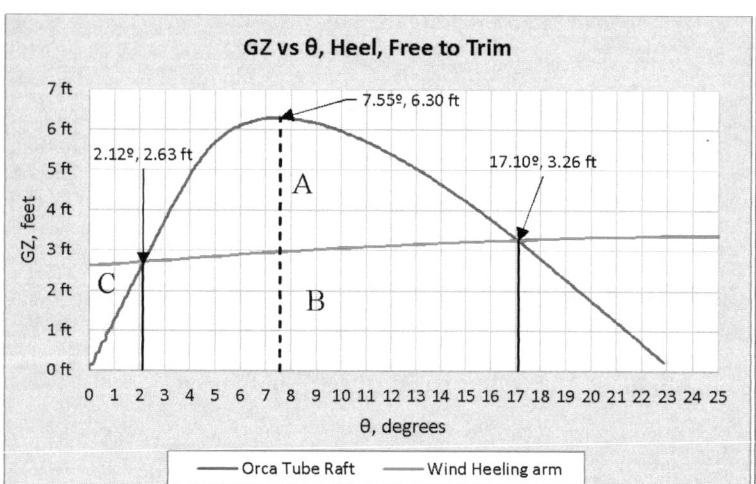

Tube Raft GZ Curve Summary & Checks	
Displacement al LL Draft	6,745 kp
Wind Speed	70 kt
Upright Wind O/T Moment, kpf-ft	16,525 kp-ft
Area Ratio, (A + B)/(B + C)	1.58
CFR Check Area Ratio > 1.4	OK
θ at maz GZ	7.55º
max GZ	6.30 ft
Area under righting arm to max GZ	31 deg.ft
Area to max GZ > 15 ft.deg?	OK
1st Intercept	2.1º
2nd Intercept	17.1º
CFR stability range check > 10º	OK
Residual righting energy, Area A	29 deg.ft
CFR Check Area A > 5 ft.deg	OK
Metacentric height, GM	71.57 ft

Figure 9 - GZ Curve for Tube Raft Heel, Free to Trim (more data in table on right)

The heeling moment created by the winch loads is divided by the vessel weight to give the righting arm, GZ, which is plotted against the angle of vessel rotation at every time step, to create the GZ curve shown in Figure 9.

The 70-kt wind heeling arm curve is calculated separately and plotted with the GZ curve in the above figure. The resulting area ratio is 1.58, which is 13% above the 1.40 minimum ratio required to satisfy the CFR [5] liftboat stability regulations. The value for θ at maximum GZ is 7.55° and the initial GM is 71.6 feet.

Using only two winches (one above and one below), the same approach allows both free trimming and free yawing. The result for a GZ righting arm curve is similar and is shown in Figure 10. The area ratio is slightly reduced to 1.55, θ at maximum GZ is increased to 7.87° and the initial GM is decreased to 65.7 feet.

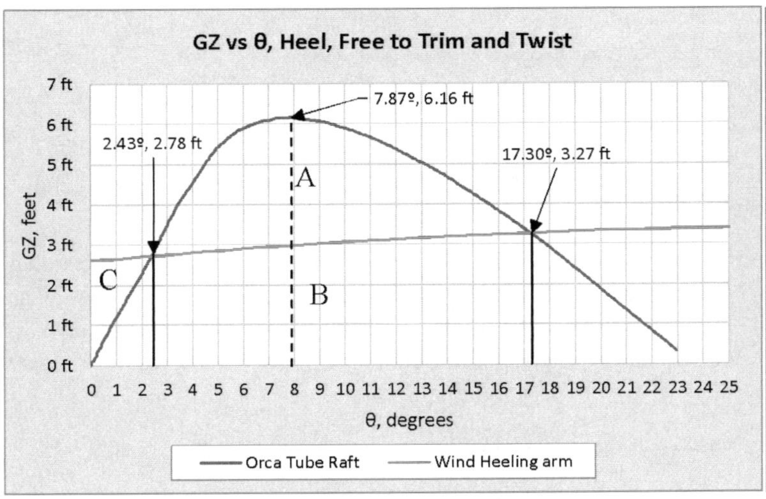

Tube Raft GZ Curve Summary & Checks	
Displacement al LL Draft	6,745 kp
Wind Speed	70 kt
Upright Wind O/T Moment, kpf-ft	16,525 kp-ft
Area Ratio, (A + B)/(B + C)	1.55
CFR Check Area Ratio > 1.4	OK
θ at maz GZ	7.87º
max GZ	6.16 ft
Area under righting arm to max GZ	31 deg.ft
Area to max GZ > 15 ft.deg?	OK
1st Intercept	2.4º
2nd Intercept	17.3º
CFR stability range check > 10º	OK
Residual righting energy, Area A	28 deg.ft
CFR Check Area A > 5 ft.deg	OK
Metacentric height, GM	65.30 ft

Figure 10 – GZ Curve for Tube Raft Heel, Free to Trim and Twist (more data in table on right)

The two-winch moment allows the vessel to twist about the initially vertical axis between the winch wire attachment points (50 ft above and below the vessel midships center). This is similar to the free twist method described by van Santen [6].

It should be noted that the stern deck edge begins to submerge at a heel angle of 2.7°. The bow deck edge begins to submerge at a heel angle of 5.5°, at which point the stern deck edge is submerged by 2.25 feet. At 10° of heel, the stern deck edge is submerged by 7.15 feet and the bow deck edge is submerged by 3.4 feet. The GZ curve is close to linear only up to 4°.

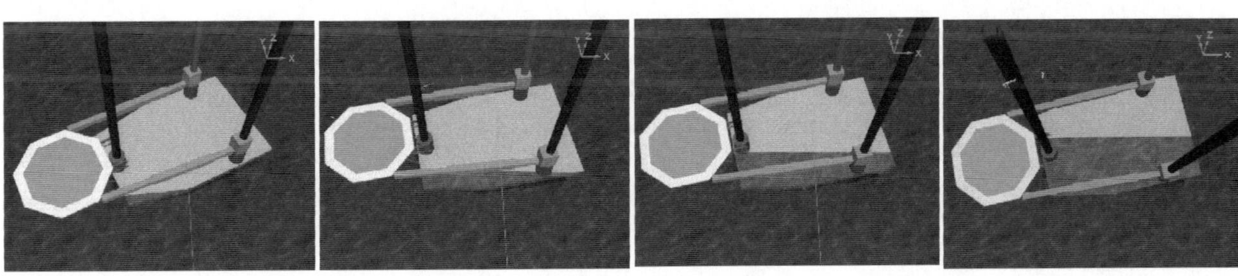

Figure 11 – Free Twist Heeling, 3.5°, 5.3°, 8.7°, 23.5°, Free Trim Range -0.2° to 0.4°, Water Surface Shown

In comparisons of other tube raft hydrostatic models and other large angle stability software, the resulting GZ curves have been found to be accurate to within less than a 2% error.

Figure 12 presents GZ curves for the tube raft trim, in both directions (bow down and stern down) about midship's axis, starting from the 0.9° stern down condition summarized in Table 2.

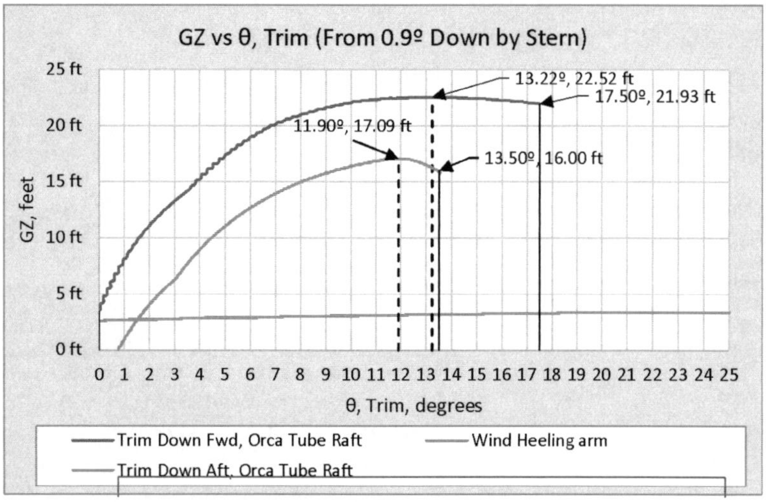

Figure 12 - GZ Curves for Tube Raft Trim, Both Directions

It is important to note that the large angle data this section is presented to give a clear picture of the conventional intact static stability of the liftboat that is subsequently presented with time domain dynamic analysis in waves. The dynamic analysis method, with a tube raft and OrcaFlex, does not use this conventional hydrostatic stability data, in contrast to most widely used seakeeping software. It is also important to note that the GZ heeling curve is linear only up to around 4° of heel angle.

Class 250 Liftboat Structural Mass Moments of Inertia

Structural Inertia Calculations	Ixx total	Iyy total	Izz total
3 Legs, 7.5' OD	68%	47%	30%
Lightship Hull, inc. pads, helideck, crane booms	25%	28%	63%
2 Crane assemblies w/cabs, bearimgs, motors, etc.	3%	2%	0%
Helideck	1%	16%	2%
Total Variable Load (max = 529.4 LT at max 13' above deck)	2%	8%	5%
Totals (kp.ft^2)	**1.35E+07**	**2.28E+07**	**1.47E+07**

Table 5 – Structural Inertias with Totals and Percentage Contributions

The mass moments of inertia for the vessel in this paper are summarized in Table 5. These data are attached to the Orcaflex 6D buoy located at the cg of the tube raft model.

Hydrodynamic Added Mass and Added Moments of Inertia

Fixed frequency terms for added mass and inertia for the tube raft heave, roll and pitch freedoms are shown in Table 6.

Hydrodynamic Mass in heave	21,200 kp
Hydrodynamic Inertia in roll	1.19E+07 kip.ft^2
Hydrodynamic Inertia in pitch	5.70E+07 kip.ft^2

Table 6 – Added Hydrodynamic Mass & Inertia

The hydrodynamic mass and inertia terms are selected from the in-house version of the author's program STA BARMOT. This is a barge motions program used for calculating displacement RAOs, partly based on model test data and partly based on diffraction and Froude-Krilov calculations.

Natural Periods and Motion Decay in Still Water

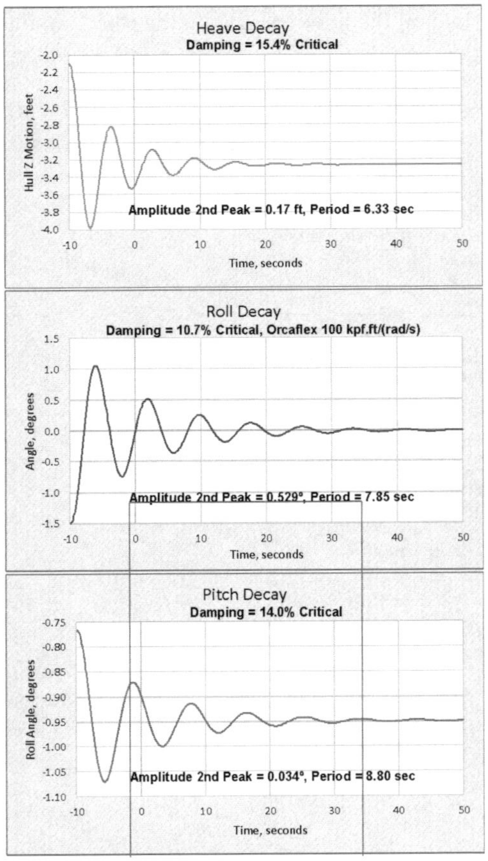

Figure 13 shows decay curves of small amplitude pitch, roll and heave motions of the OrcaFlex tube raft vessel in still water, 60 feet deep. The natural periods of pitch, roll and heave are estimated from the decay curves and shown in Table 7.

The restoring forces are not discretely input but are a result of the combined forces at each time step, on each segment of each tube, or OrcaFlex line element, in the theoretically undisturbed wave.

The constant hydrodynamic damping terms for the OrcaFlex tube raft (applied at a point close to the center of buoyancy in still water at an even keel) are adjusted from experience to obtain the expected form of decay for the liftboat vessel. The damping forces and moments selected are directly proportional to the velocity or angular velocity of the vessel relative to earth, to simulate radiation damping, see Reference [4] for more details. The selected hydrodynamic damping terms are shown in Table 7.

Figure 13 – Heave, Roll & Pitch (green, blue, and red) Decay Curves in Still Water

Characteristic	Heave	Roll	Pitch
Natural Period	6.38 sec	7.82 sec	8.73 sec
Hydrodynamic Damping	250 kpf/(ft/s)	1.00E+05 kpf.ft/(rad/s)	4.00E+05 kpf.ft/(rad/s)
Equivalent %age Critical Damping	15.40%	10.70%	14.00%

Table 7 – Natural Periods (estimated from decay curves) and Hydrodynamic Damping

The natural periods of the moored vessel, for response in waves, are also found from OrcaFlex modal analysis and are shown in Table 8, below.

Characteristic	Surge	Sway	Yaw
Natural Period	81.68 sec	81.64 sec	24.66 sec
Characteristic	**Heave**	**Roll**	**Pitch**
Natural Period	6.22 sec	7.73 sec	8.66 sec

Table 8 b- Modal Periods Computed by OrcaFlex

Response to 1-ft Regular Waves in 60-ft Water Depth

The RAO definition used for the following charts is the response range divided by the wave height, noting that the response and the wave may both be non-sinusoidal, but periodic, and may both be about non-zero mean values. As will be shown later, the periodic responses of roll and pitch may be at twice the wave period, or subharmonic.

Heave, roll and pitch response results from time domain analyses in regular stream function (10th order) 1-ft high waves at attack angles 0° (from the stern), 45° and 90° are summarized in Figure 14. The model is held in 60-ft water depth, on a 4-line mooring with one attachment point at the centerline on the bow and one on the stern. Long light horizontal lines at ±45° from each end of the vessel are used.

The load condition shown in Table 2 is used, with an initial 0.94° of stern down trim. The vessel is analyzed with zero forward speed. The following RAO chart axes are kept the same for each response type (heave, roll, pitch) for each wave attack angle, thereby enabling easier comparison of values with different wave heights.

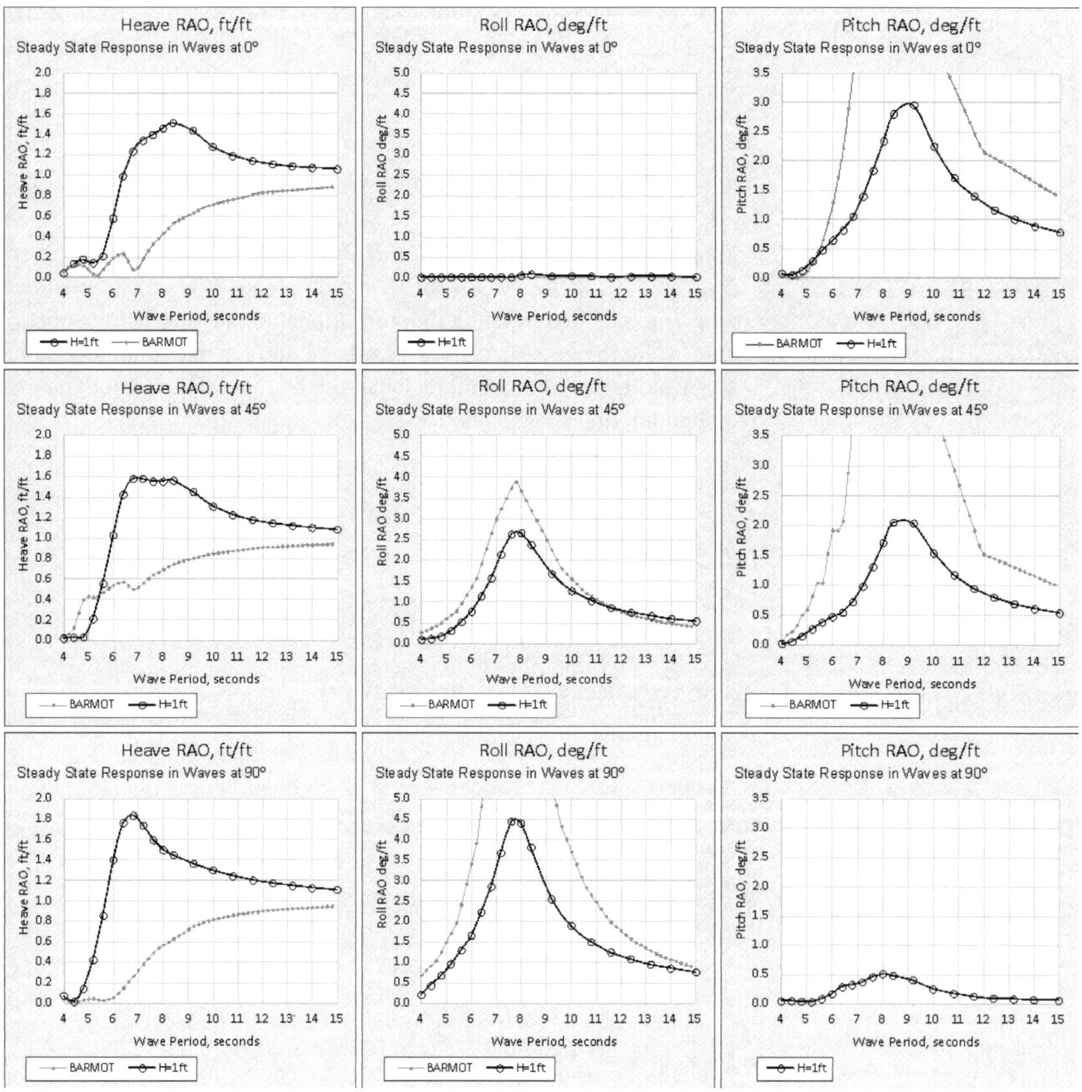

Figure 14 - Steady State Response Ranges at 1-ft Wave Height, Tube Raft Comparison with STA BARMOT

The steady state response ranges (plotted as RAOs) in 1-ft waves, for heave, roll and pitch for the OrcaFlex tube raft, are shown for wave attack angles of 0°, 45° and 90° in Figure 14. Results from STA BARMOT are shown for comparison, but it must be noted that the Class 250 vessel dimensions, with L/B = 1.44 and B/T = beam/draft = 10.4, are well outside the range for which the program has been calibrated.

Each segment of each tube in the tube raft experiences hydrostatic and hydrodynamic forces in waves. The standard OrcaFlex Morison equation is used with drag force coefficient, Cd, set to zero or 0.1 and the added mass coefficient, Ca, set to zero. The inertia coefficient, Cm is set to the default value of 1.0 in the normal directions, thus giving the

Froude-Krylov pressure force on each tube. The added mass terms for the tube raft are separately specified as described in a previous section. Diffraction forces are not calculated.

For the 1-ft waves and the 0° wave direction, the heave RAO reaches a peak value of 1.5 (ft/ft) at 8.4 seconds then decreases towards 1.0 at 15 seconds. The BARMOT values underestimate the heave peak (close to the pitch natural period). For 1-ft waves and the 90° direction the heave RAO reaches a peak value of 1.8 at 6.8 seconds (the heave natural period) and then decreases towards 1.0 at 15 seconds.

For 1-ft waves at 0° the pitch RAO peaks at 2.9 (deg/ft) at 8.7 seconds, the pitch natural period. For 1-ft waves at 90° the roll RAO peaks at 4.5 (deg/ft) at 7.7 seconds, the roll natural period. BARMOT predicts similar RAO behavior with lower amplitudes.

RAOs from Responses to Regular Wave up to 9-ft in 60-ft Water Depth

In larger waves the form of the displacement RAOs change significantly, as the liftboat deck edges become submerged and the restoring forces become far from linear. Additionally, the response in stern seas and head seas are notedly different, partly due to the above and below water hull shape and partly due to the trim angle.

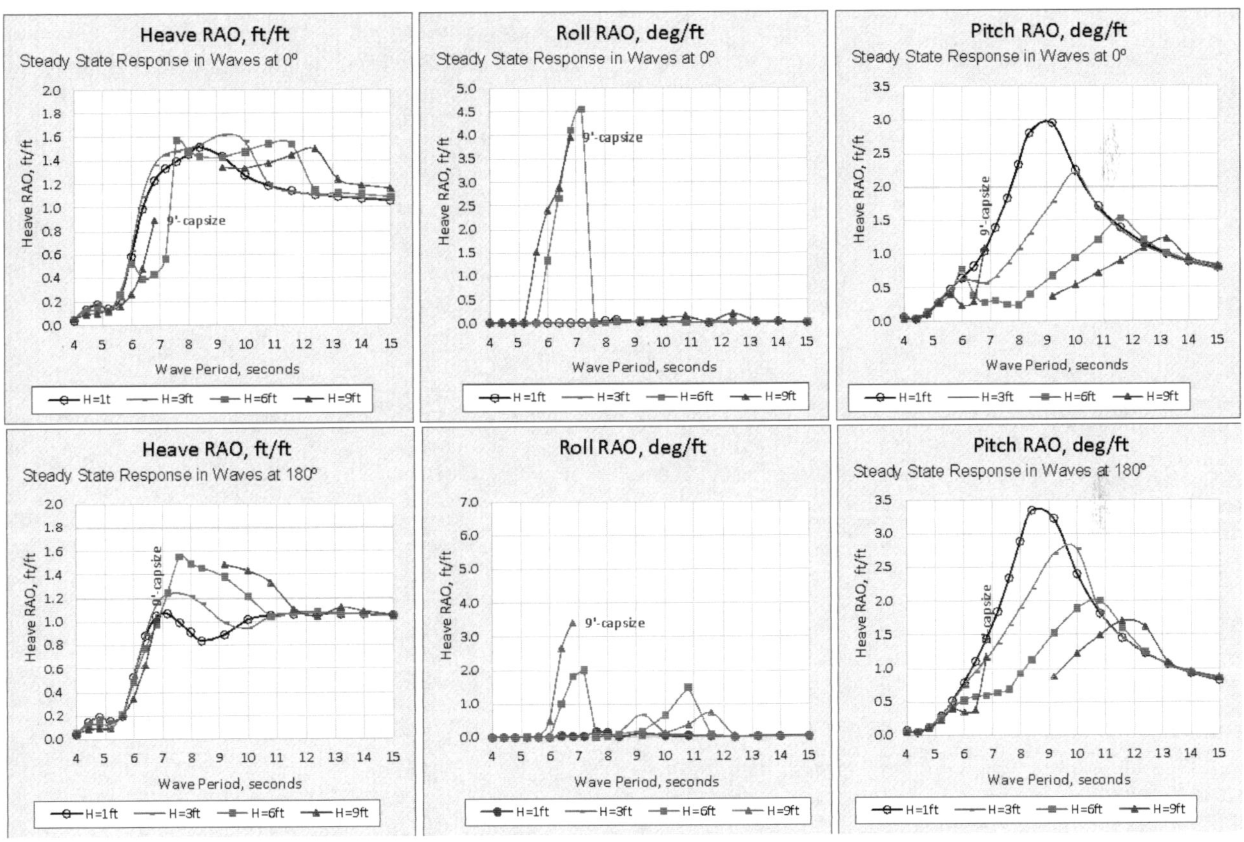

Figure 15 – Comparison of Tube Raft RAOs for 0° and 180° Angles, Wave Heights 1-ft to 9-ft

Figure 15 presents a comparison of tube raft RAOs in stern (0°) and head (180°) seas with 1-ft, 3-ft, 6-ft, and 9-ft wave heights. In head seas (180°) the heave RAO for 1-ft waves peaks at 1.1 (ft/ft) at 7.0 seconds, then decreases to 0.8 at 8.5 seconds, then rises to just over 1.0 and stays at this value from 11 to 15 seconds. In stern seas the RAO for 1-ft waves, as described previously, peaks at 1.7 at 8.4 seconds then decreases towards 1.0 at 15 seconds.

The heave RAOs generated from responses to waves larger than 1-ft have similar values up to around 7 seconds and then have peak values occurring at longer periods. Peak values are all around 1.6 for stern seas.

The pitch RAOs for 1-ft waves are similar for beam and head seas, both peaking around 8.7 seconds, with head seas having a slightly larger value of 3.3 (deg/ft). The pitch RAOs generated from higher waves show decreasing peak values at progressively longer periods, all decreasing to a value of 0.8 at 15 seconds.

The vessel capsizes in 9-ft regular waves in both stern and head seas. Capsize occurs in both seas with periods from 7.2 to 8.4 seconds. The data point at the last period before capsizing is marked on the charts with the word **capsize**. The response curve is continued at the next period for which a steady state response is found, until 15 seconds wave period. Significant submergence of the deck is seen in the capsize cases.

Surge sway and yaw motions are computed but detailed responses are not reported in this paper. The yaw motions, restrained by the mooring system, keep the selected vessel heading from deviating by more than ±4° in up to 6-ft waves, and to within ±8° in 9-ft waves.

Some of the differences in the RAO curves generated with different wave heights are attributed to the increased influence of the water depth on the hydrodynamic characteristics of the wave as height increases (as predicted by the 10th order stream function theory being used in this paper). Larger differences are attributed to the non-linear hydrostatic restoring forces as increasingly larger parts of the deck become submerged during the steady state responses observed.

The unexpectedly large roll response in stern and head seas, with 6-ft and 9-ft wave heights, at periods from 5 to 8 seconds is discussed later.

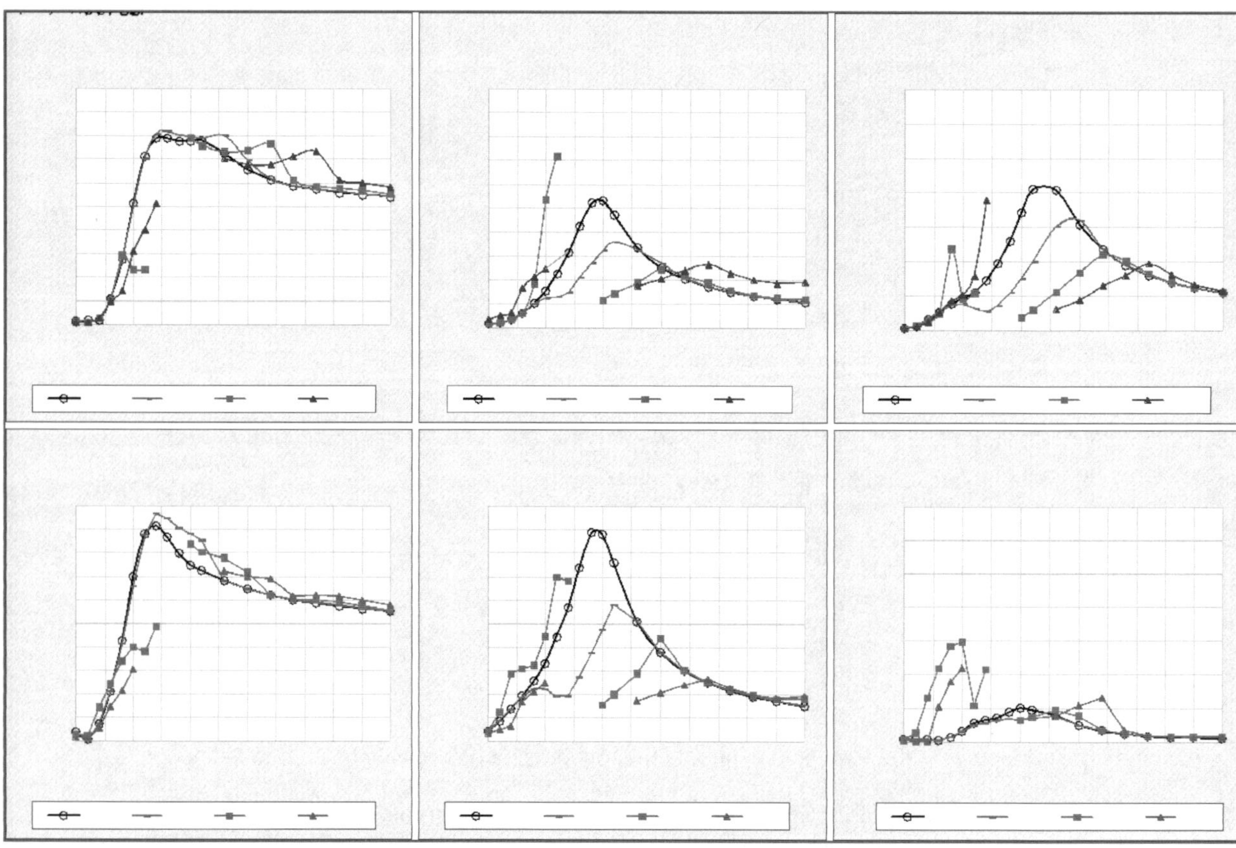

Figure 16 – Comparison of Tube Raft RAOs for 45° and 90° Angles, Wave Heights 1-ft to 9-ft

Figure 16 shows that the vessel capsizes in 6-ft and 9-ft regular waves from 45° and 90°. The maximum period range for capsize is from 6.4 to 8.4 seconds.

The heave RAOs for beam waves (90°) generated from responses to waves larger than 1-ft, have similar values to the 1-ft wave heave RAO. The roll RAOs for 45° and 90° directions, for 3-ft, 6-ft, and 9-ft waves, show similar &

180characteristics to the pitch RAOs for larger waves in the 0° direction, with peak values occurring at longer periods at the wave height increases.

MAX RESPONSES, NO WIND, 1-FT TO 9-FT WAVES, 0° & 180°, 60-FT WATER DEPTH

In this section the vessel maximum heave, roll and pitch steady state responses are presented (not RAOs, as in the previous section) to regular waves with heights 1-ft, 3-ft, 6-ft, and 9-ft, from directions 0°, and 180°. The results are summarized in Figure 17, below.

The data point at the last period before capsizing is marked on the charts with the word **capsize**. The response curve is continued at the next period for which a steady state response is found, until 15 seconds wave period. Significant submergence of the deck is seen in the capsize cases.

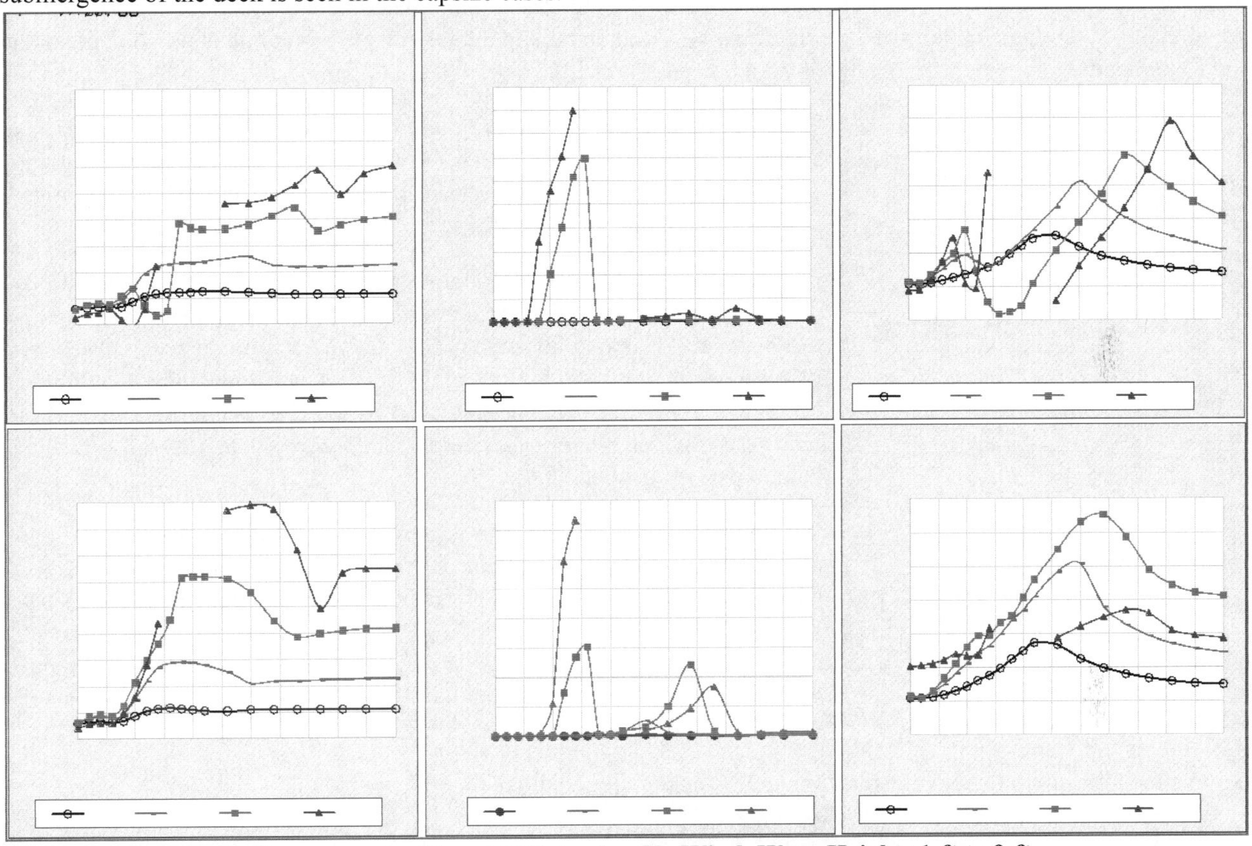

Figure 17 - Maximum Responses for 0° and 180° Angles, No Wind, Wave Heights 1-ft to 9-ft

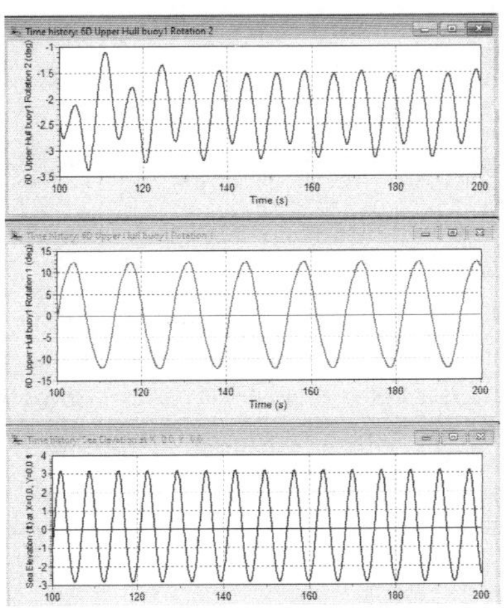

Figure 18 – Steady State Pitch, Roll, Wave Elevation, 6.8 sec, 6-ft

In 6-ft and 9-ft waves, for both wave directions, a large amplitude steady state roll response becomes periodic at exactly twice the wave period, in the wave period range leading up to the roll natural period of 7.8 seconds. In some cases, the pitch response also "locks" into this twice-wave-period response.

Figure 18 illustrates a condition where the roll response (red curve) is at twice the wave (dark red) period with the pitch (blue curve) remaining the same as the wave period. The wave direction is 0° and the wave height is 6-ft.

In 1-ft and 3-ft waves, the roll response is periodic at the wave period in all cases and no twice-wave-period responses are found.

In these stern and head seas, the unexpectedly large roll, has similarities with parametric rolling of large ships which is well described by the ABS in [7].

Parametric Roll Resonance In Longitudinal Waves

In parametric roll resonance, when a ship moves in longitudinal waves, the roll response is periodic at, or close to, the ship's natural roll period. It is caused by the periodic change of stability as the roll restoring moment is increased and decreased during the passage of a wave. An increase in the waterplane area occurs when wave crests are simultaneously passing the flared parts of the bow and stern, and the wave trough is at midships. A decrease in the waterplane area occurs when the wave crest is at midships [7].

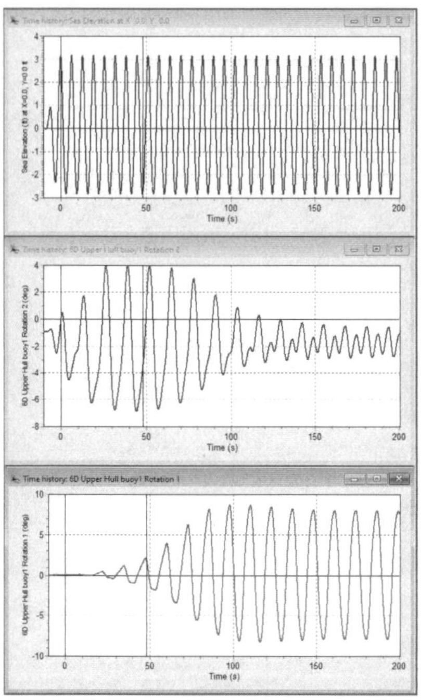

Figure 19 – Build Up to Steady State, Wave Elevation, 6.8 sec, 6-ft

The large amplitude, twice-wave-period roll, and pitch response of the Class 250 liftboat described in this paper is not considered to be associated with these same changes of the waterplane area (and hence changes in roll stiffness) during the passage of waves in the longitudinal direction.

In most large amplitude roll responses in stern and head waves, the liftboat pitch motion begins with a response that locks into the twice-wave-period mode and roll is negligibly small. The pitching amplitude then reduces, and its response period becomes the same as the wave, while the roll amplitude becomes large and sub-harmonic, at twice the wave period. These twice-the-wave sub-harmonic periods are longer than the roll and pitch natural periods in still water.

In Figure 19 the wave elevation is shown in dark brown and has a period of 6.8 seconds. The blue pitch response initially builds to +4° to -7° (representing an amplitude of 5° about a mean of -1°, which is the vessel initial trim) and has a period of 13.6 seconds. The pitch then decays to around 1° amplitude with a period of 6.8 seconds. The red roll response begins with negligible amplitude and grows to an 8° amplitude at twice the wave period as the pitch reduces.

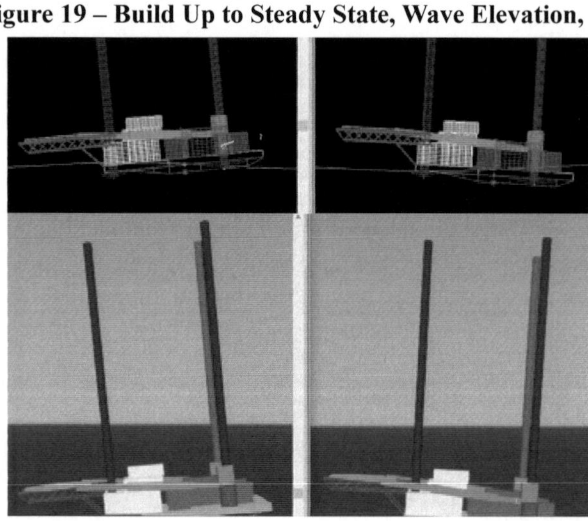

Figure 20 – Deck Submergence in 6-ft, 9-sec Head Waves

In 8 seconds and longer period waves, the heave, roll and pitch responses all quickly reach a steady state at the wave period. In longitudinal seas, roll is close to zero, as expected.

Exceptions where small amplitude roll motion is seen at longer periods in head seas are associated with small mean angles of yaw (between 1° and 3°) permitted by the mooring system. The roll response is at wave period in these cases.

Capsize occurs in 9-ft regular waves from both the 0° and 180° directions. Capsize does not occur in these directions with 6-ft waves.

The phenomenon of sub-harmonic (period-doubling) motion of other ocean structures (not free floating) has been reported elsewhere and confirmed with model tests [8].

Max Responses, No Wind, 1-ft to 9-ft Waves, 45° & 90°, 60-ft Water Depth

In this section the vessel maximum heave, roll and pitch steady state responses are presented to regular waves with heights 1-ft, 3-ft, 6-ft, and 9-ft, from directions 45°, and 90°. The results are summarized in Figure 21, below.

Capsize occurs in both 6-ft and 9-ft waves from 45° and 90°.

In wave periods less than the roll and pitch natural periods, both roll and pitch show large amplitude periodic responses at twice the wave period. The roll and pitch response period ratios are presented for all steady state responses from time domain simulations in Table 9. This table also summarizes all conditions where capsize occurs.

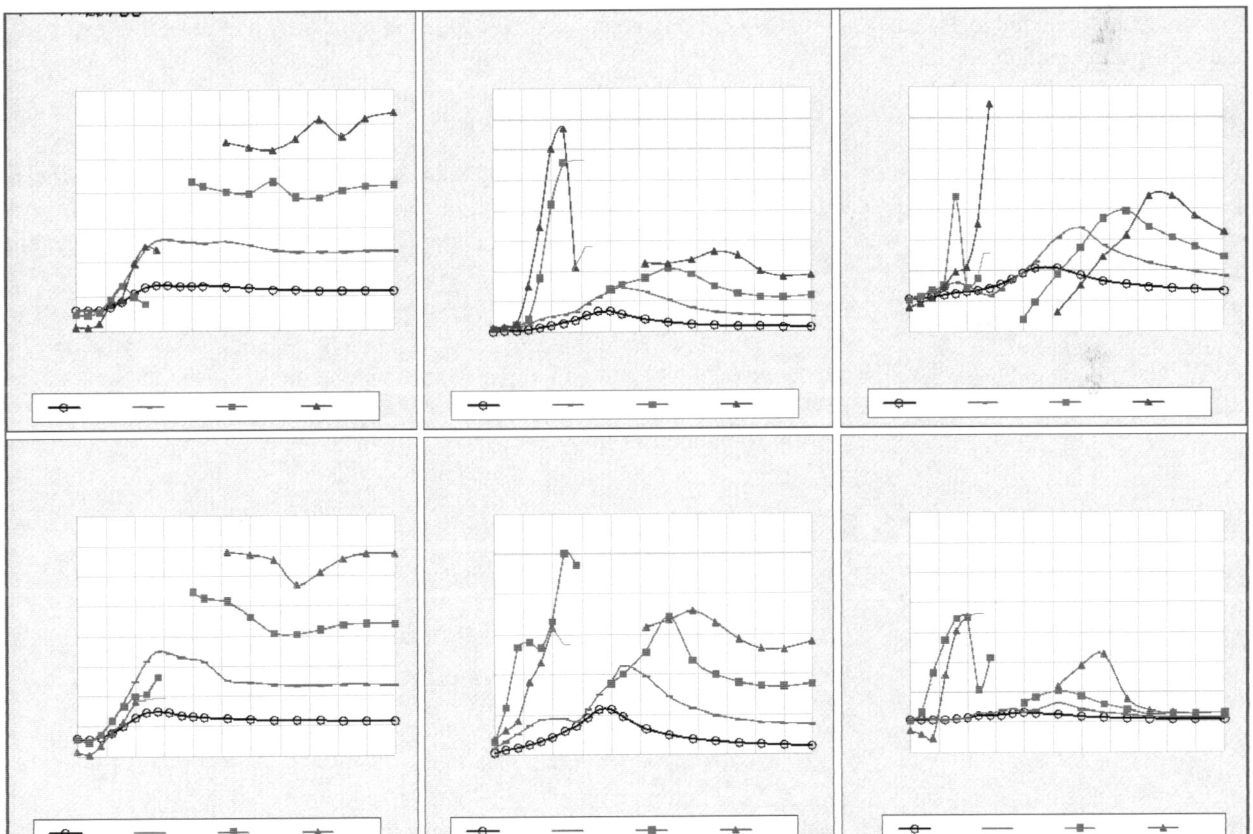

Figure 21 - Maximum Responses for 45° and 90° Angles, No Wind, Wave Heights 1-ft to 9-ft

Angle, Height	Response Period Ratio	4.0 s	4.4 s	4.8 s	5.2 s	5.6 s	6.0 s	6.4 s	6.8 s	7.2 s	7.6 s	8.0 s	8.4 s	9.2 s	10.0 s	10.8 s	11.6 s	12.4 s	13.2 s	14.0 s	15.0 s
0°, 6ft	Roll Ratio	1.0	1.0	1.0	1.0	2.0	2.0	2.0	2.0	2.0	1.0	1.0	1.0	1.0	1.0	1.0	1.0	1.0	1.0	1.0	0.5
	Pitch Ratio	1.0	1.0	1.0	1.0	1.0	1.1	1.0	1.0	1.0	0.7	1.0	1.0	1.0	1.0	1.0	1.0	1.0	1.0	1.0	1.0
0°, 9ft	Roll Ratio	1.0	1.0	1.0	2.0	2.0	2.0	2.0	2.0	0.0	0.0	0.0	0.0	1.0	1.0	1.0	1.0	1.0	0.9	1.0	0.5
	Pitch Ratio	1.0	1.0	1.0	1.0	1.2	1.0	1.1	1.5	0.0	0.0	0.0	0.0	1.0	1.0	1.0	1.0	1.0	1.0	1.0	1.0
45°, 6ft	Roll Ratio	1.0	1.0	1.0	1.0	2.0	2.0	2.0	0.0	0.0	0.0	1.0	1.0	1.0	1.0	1.0	1.0	1.0	1.0	1.0	0.5
	Pitch Ratio	1.0	1.0	1.0	1.0	2.0	1.0	1.0	0.0	0.0	0.0	1.0	1.0	1.0	1.0	1.0	1.0	1.0	1.0	1.0	1.0
45°, 9ft	Roll Ratio	1.0	1.0	1.0	2.0	2.0	2.0	1.7	2.0	0.0	0.0	0.0	0.0	1.0	1.0	1.0	1.0	1.0	1.0	1.0	0.5
	Pitch Ratio	1.0	1.0	1.0	1.0	1.0	1.1	1.3	2.0	0.0	0.0	0.0	0.0	1.0	1.0	1.0	1.0	1.0	1.0	1.0	1.0
90°, 6ft	Roll Ratio	1.0	1.0	2.0	2.0	2.0	2.0	1.7	2.0	0.0	0.0	1.0	1.0	1.0	1.0	1.0	1.0	1.0	1.0	1.0	0.5
	Pitch Ratio	1.0	2.0	2.0	2.0	2.0	2.0	1.2	1.6	0.0	0.0	1.0	1.0	1.0	1.0	1.0	1.0	1.0	1.0	0.5	0.5
90°, 9ft	Roll Ratio	1.0	1.0	1.0	1.4	2.0	2.0	0.0	0.0	0.0	0.0	0.0	0.0	1.0	1.0	1.0	1.0	1.0	1.0	0.5	0.5
	Pitch Ratio	1.0	1.0	0.7	2.0	2.0	2.0	0.0	0.0	0.0	0.0	0.0	0.0	1.0	1.0	1.0	1.0	1.0	0.5	0.5	0.5
180°, 6ft	Roll Ratio	1.0	1.0	1.0	1.0	1.0	1.2	2.0	2.0	2.0	1.0	1.0	1.0	1.0	1.0	1.0	1.0	0.5	1.0	1.0	0.5
	Pitch Ratio	1.0	1.0	1.0	1.0	1.0	1.0	1.0	1.0	1.0	1.0	1.0	1.0	1.0	1.0	1.0	1.0	1.0	1.0	1.0	1.0
180°, 9ft	Roll Ratio	1.0	1.0	1.0	2.0	2.0	2.0	2.0	2.5	0.0	0.0	0.0	0.0	0.9	1.0	1.0	1.0	1.0	1.0	0.5	0.5
	Pitch Ratio	1.0	1.0	1.0	1.0	1.0	1.0	1.0	1.4	0.0	0.0	0.0	0.0	1.0	1.0	1.0	1.0	1.0	1.0	1.0	1.0

Table 9 – Capsize Periods, with Roll and Pitch Steady State Response Period Ratios, No Wind

Table 9 is arranged with wave periods in the top row and wave direction and height in the left column.

- Cells with a light red color indicate roll or pitch steady state response period ratios of greater than 1.3. Values of 2.0 indicate response at twice the wave period.
- Cells with dark red color indicate capsize.
- Cells with green color indicate response period ratios of less than 0.9.

No results are included for 1-ft and 3-ft waves where all responses were found to be at the wave period, as expected, and no capsize was found.

Max Responses, Regular Waves with 50-kt & 70-kt Wind Heeling

Time domain steady state responses for all runs in waves without wind were repeated with wind heeling moments from 50-kt wind and 70-kt wind. The heeling moments were applied as a constant local moment and did not have any variation with roll angle or wave direction. The moment was applied as if the wind was always coming from the 90° direction. Results are summarized in Figure 22 and Figure 23.

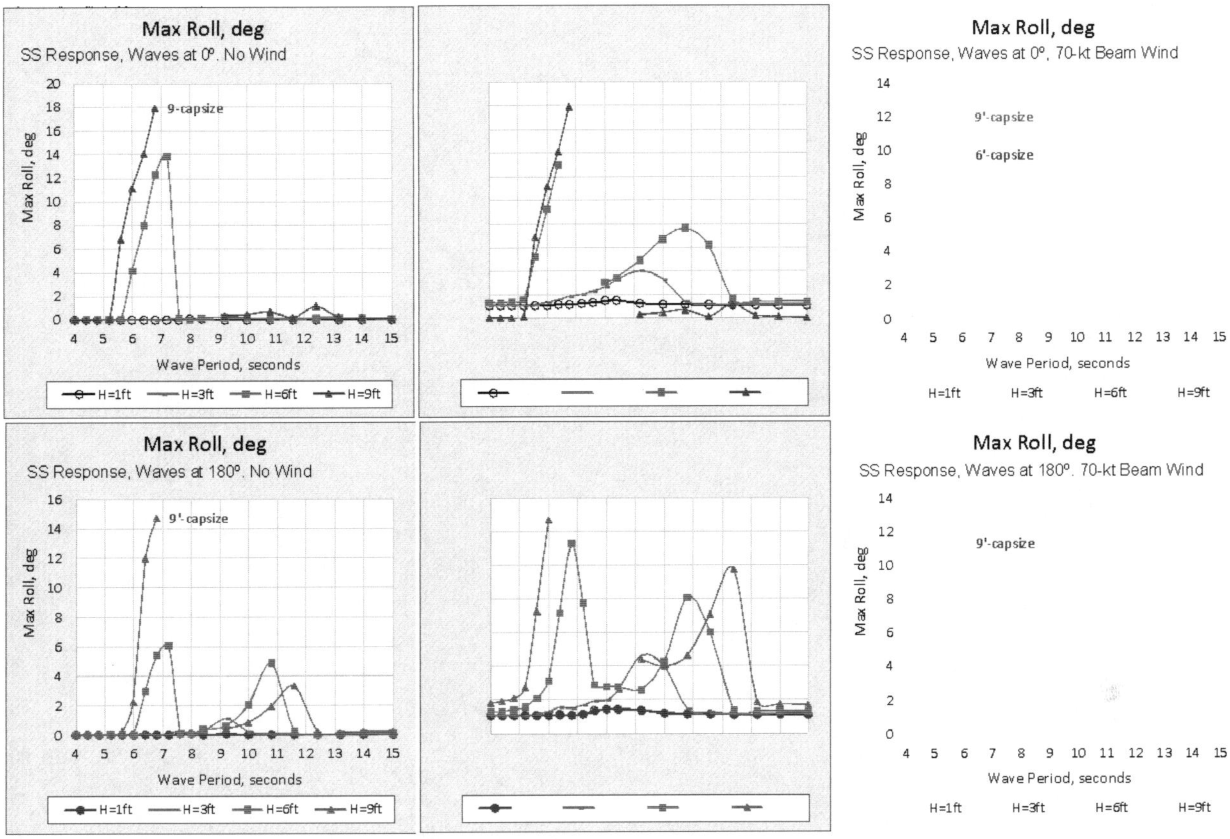

Figure 22 – Comparison of Roll Responses in Beam Seas with 0-kt, 50-kt and 70-kt Wind on Beam

Figure 22 shows capsize occurs in stern seas in both 6-ft and 9-ft waves with 50-kt and 70-kt beam winds. Capsize occurs in head seas, only in 9-ft waves with 50-kt and 70-kt beam winds. The range of wave periods with capsize is increased up to 13.2 seconds in stern seas with 70-kt beam winds.

Figure 23 shows capsize in both 6-ft and 9-ft waves at 45° and 90°, with and without beam winds. The range of wave periods with capsize with 9-ft waves and 70-kt beam winds is increased up to 13.2 seconds with waves from 45° and up to 10.8 seconds with beam waves.

Note that all results are without forward speed and without any rudder or other turning forces. Class 250 liftboats rarely travel at more than 5 knots in still water and reduce speed in waves.

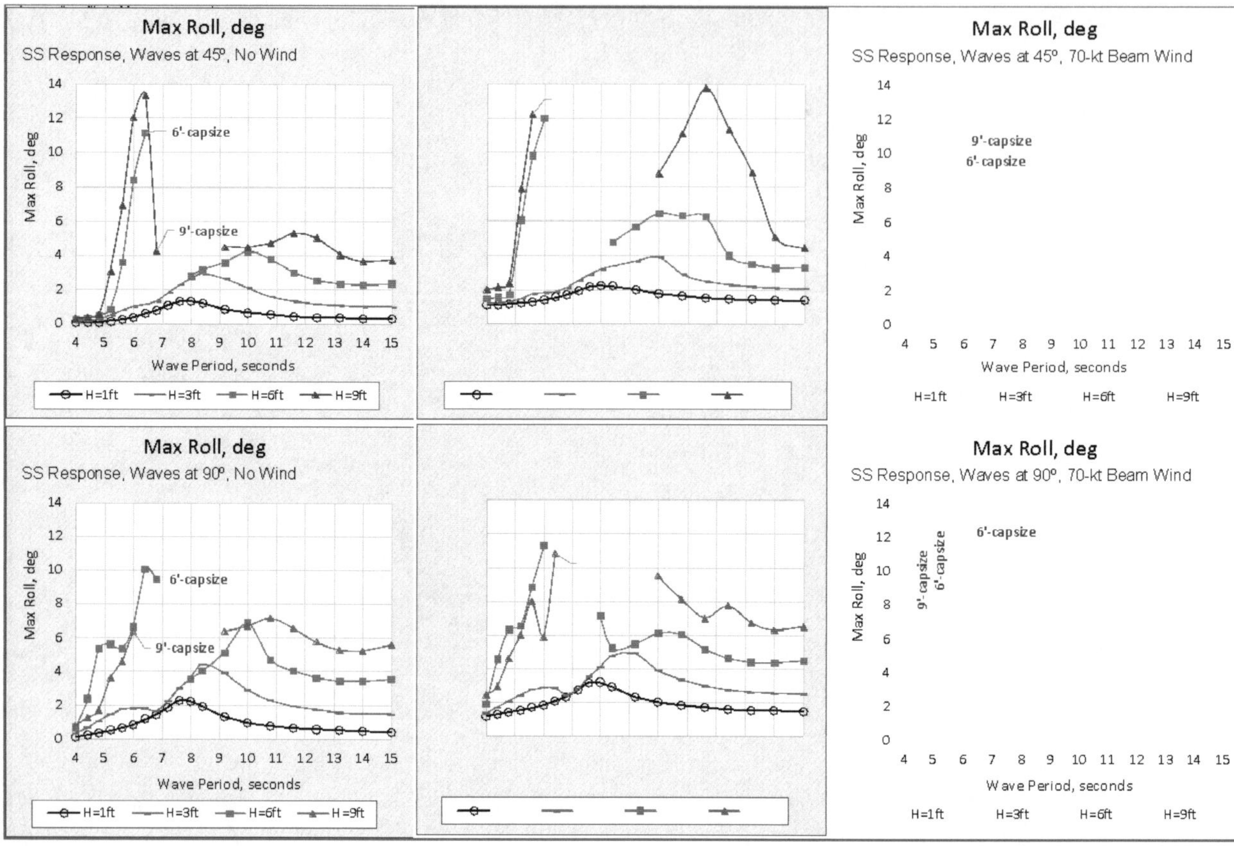

Figure 23 – Comparison of Roll Responses in 45° Waves with 0-kt, 50-kt and 70-kt Wind on Beam

Angle, Height	Response Period Ratio	4.0 s	4.4 s	4.8 s	5.2 s	5.6 s	6.0 s	6.4 s	6.8 s	7.2 s	7.6 s	8.0 s	8.4 s	9.2 s	10.0 s	10.8 s	11.6 s	12.4 s	13.2 s	14.0 s	15.0 s
0°, 6ft	Roll Ratio	1.0	1.0	1.0	1.0	1.9	2.0	2.0	0.0	0.0	0.0	1.0	1.0	1.0	1.0	1.0	1.0	1.0	0.5	0.5	0.5
	Pitch Ratio	1.0	1.0	1.0	1.0	1.0	1.0	1.0	0.0	0.0	0.0	1.0	1.0	1.0	1.0	1.0	1.0	1.0	1.0	1.0	1.0
0°, 9ft	Roll Ratio	1.0	1.0	1.0	2.0	2.0	2.0	2.0	2.0	0.0	0.0	0.0	0.0	1.0	1.0	1.0	1.0	1.0	0.9	1.0	0.5
	Pitch Ratio	1.0	1.0	1.0	1.0	1.2	1.0	1.1	1.5	0.0	0.0	0.0	0.0	1.0	1.0	1.0	1.0	1.0	1.0	1.0	1.0
45°, 6ft	Roll Ratio	1.0	1.0	1.0	2.0	2.0	2.0	0.0	0.0	0.0	0.0	0.0	1.0	1.0	1.0	1.0	1.0	1.0	1.0	1.0	1.0
	Pitch Ratio	1.0	1.0	1.0	1.0	1.0	1.8	0.0	0.0	0.0	0.0	0.0	1.0	1.0	1.0	1.0	1.0	1.0	1.0	1.0	1.0
45°, 9ft	Roll Ratio	1.0	1.0	1.0	2.0	2.0	0.0	0.0	0.0	0.0	0.0	0.0	0.0	0.0	1.0	1.0	1.0	1.0	1.0	1.0	1.0
	Pitch Ratio	1.0	1.0	1.0	1.0	2.0	0.0	0.0	0.0	0.0	0.0	0.0	0.0	0.0	1.0	1.0	1.3	1.0	1.0	1.0	1.0
90°, 6ft	Roll Ratio	1.0	1.5	2.0	2.0	2.0	2.0	0.0	0.0	0.0	0.0	1.0	1.0	1.0	1.0	1.0	1.0	1.0	1.0	1.0	0.5
	Pitch Ratio	1.0	2.0	2.0	2.0	2.0	1.0	0.0	0.0	0.0	0.0	1.0	1.0	1.0	1.0	1.0	1.0	1.0	0.5	0.5	0.5
90°, 9ft	Roll Ratio	1.0	1.0	1.0	2.0	2.0	1.6	1.5	0.0	0.0	0.0	0.0	0.0	0.0	1.0	1.0	1.0	1.0	1.0	1.0	0.5
	Pitch Ratio	1.0	1.0	2.0	2.0	2.0	2.0	1.9	0.0	0.0	0.0	0.0	0.0	0.0	1.0	1.0	1.0	1.0	1.0	0.5	0.5
180°, 6ft	Roll Ratio	1.0	0.5	1.0	1.0	1.0	1.2	2.0	2.0	1.5	1.0	1.0	1.0	0.5	1.0	1.0	1.0	1.0	1.0	1.0	0.5
	Pitch Ratio	1.0	1.0	1.0	1.0	1.0	1.0	1.0	1.2	1.0	1.0	1.0	1.0	1.0	1.0	1.0	1.0	1.0	1.0	1.0	1.0
180°, 9ft	Roll Ratio	1.0	1.0	1.0	1.7	2.0	2.0	0.0	0.0	0.0	0.0	0.0	0.0	1.0	0.8	1.0	1.0	1.0	1.0	0.5	0.5
	Pitch Ratio	1.0	1.0	1.0	1.0	2.0	1.4	0.0	0.0	0.0	0.0	0.0	0.0	1.0	1.0	1.0	1.0	1.0	1.0	1.0	1.0

Table 10 – Capsize Periods, with Roll and Pitch Steady State Response Period Ratios, 50-kt Wind

Angle, Height	Response Period Ratio	4.0 s	4.4 s	4.8 s	5.2 s	5.6 s	6.0 s	6.4 s	6.8 s	7.2 s	7.6 s	8.0 s	8.4 s	9.2 s	10.0 s	10.8 s	11.6 s	12.4 s	13.2 s	14.0 s	15.0 s
0°, 6ft	Roll Ratio	1.0	1.0	1.0	1.0	2.0	2.0	0.0	0.0	0.0	0.0	0.0	1.0	1.0	1.0	1.0	1.0	1.0	0.5	0.5	0.5
	Pitch Ratio	1.0	1.0	1.0	1.0	1.0	1.0	0.0	0.0	0.0	0.0	0.0	1.0	1.0	1.0	1.0	1.0	1.0	1.0	1.0	1.0
0°, 9ft	Roll Ratio	1.0	1.0	1.0	2.0	2.0	1.1	0.0	0.0	0.0	0.0	0.0	0.0	0.0	0.0	0.0	0.0	0.0	0.0	1.0	0.5
	Pitch Ratio	1.0	1.0	1.0	1.0	1.0	1.0	0.0	0.0	0.0	0.0	0.0	0.0	0.0	0.0	0.0	0.0	0.0	0.0	1.0	1.0
45°, 6ft	Roll Ratio	1.0	1.0	1.0	2.0	2.0	0.0	0.0	0.0	0.0	0.0	0.0	0.0	1.0	1.0	1.0	1.0	1.0	1.0	1.0	1.0
	Pitch Ratio	1.0	1.0	1.0	1.0	2.0	0.0	0.0	0.0	0.0	0.0	0.0	0.0	1.0	1.0	1.0	1.0	1.0	1.0	1.0	1.0
45°, 9ft	Roll Ratio	1.0	1.0	1.0	2.0	2.0	0.0	0.0	0.0	0.0	0.0	0.0	0.0	0.0	0.0	0.0	0.0	0.0	0.0	1.0	1.0
	Pitch Ratio	1.0	1.0	1.0	2.0	2.0	0.0	0.0	0.0	0.0	0.0	0.0	0.0	0.0	0.0	0.0	0.0	0.0	0.0	1.0	1.0
90°, 6ft	Roll Ratio	1.0	1.0	2.0	2.0	0.0	1.8	0.0	0.0	0.0	0.0	0.0	0.0	1.0	1.0	1.0	1.0	1.0	1.0	1.0	0.5
	Pitch Ratio	1.0	1.0	2.0	2.0	0.0	2.0	0.0	0.0	0.0	0.0	0.0	0.0	1.0	1.0	1.0	1.0	1.0	1.0	0.5	0.5
90°, 9ft	Roll Ratio	1.0	1.0	1.0	0.0	0.0	0.0	0.0	0.0	0.0	0.0	0.0	0.0	0.0	0.0	0.0	0.7	0.7	1.0	1.0	1.0
	Pitch Ratio	1.0	1.0	2.0	0.0	0.0	0.0	0.0	0.0	0.0	0.0	0.0	0.0	0.0	0.0	0.0	1.0	1.0	1.0	1.0	0.5
180°, 6ft	Roll Ratio	1.0	1.0	1.0	1.0	2.0	2.0	1.0	1.0	1.0	1.0	1.0	1.0	0.5	0.5	1.0	1.0	1.0	1.0	1.0	0.5
	Pitch Ratio	1.0	1.0	1.0	1.0	1.7	2.0	1.1	1.0	1.0	1.0	1.0	1.0	1.0	1.0	1.0	1.0	1.0	1.0	1.0	1.0
180°, 9ft	Roll Ratio	1.0	1.0	1.0	1.9	2.0	1.2	0.0	0.0	0.0	0.0	0.0	0.0	0.0	0.0	0.0	0.5	1.0	1.0	1.0	0.5
	Pitch Ratio	1.0	1.0	1.0	1.3	2.0	1.5	0.0	0.0	0.0	0.0	0.0	0.0	0.0	0.0	0.0	1.0	1.0	1.0	1.0	1.0

Table 11 – Capsize Periods, with Roll and Pitch Steady State Response Period Ratios, 70-kt Wind

Table 10 and Table 11 have the same format and arrangement as Table 9 with wave periods in the top row and wave direction and height in the left column. The dark red cells indicate conditions with capsize.

Influence of Roll Damping

Limited investigations into the effects of significantly increased or decreased roll damping were undertaken. With hydrodynamic damping (proportional to the vessel angular velocity relative to earth) significantly increased (to above an equivalent of 20% critical) the change from large amplitude sub-harmonic pitching to large amplitude sub-harmonic rolling in longitudinal waves is reduced or does not occur at all. With this very large damping, capsize is not found in 6-ft waves and 70-kt wind, but still occurs in 9-ft waves and 70-kt wind.

With roll damping reduced to an equivalent of around 5% critical, the roll amplitudes are increased and capsize is found in 6-ft waves where it did not occur with 10% critical damping. The sub-harmonic roll response remained common to 6-ft and 9-ft waves but did not occur in 1-ft or 3-ft waves.

Capsize in Irregular Waves

Capsize is found in irregular waves with 5-ft to 6-ft significant wave heights, with and without wind heeling forces. A full description is beyond the scope of this paper. Figure 24 provides a visual example of a capsize sequence in a JONSWAP following sea with Hs = 6-ft, Tz = 6 sec and γ =2.0, with a 50-kt wind on the starboard side, causing heeling to port. The port leg is colored blue. Twelve screen captures are shown each taken one second apart. Each capture shows a side elevation, an end elevation and a plan view with shaded graphics and perspective.

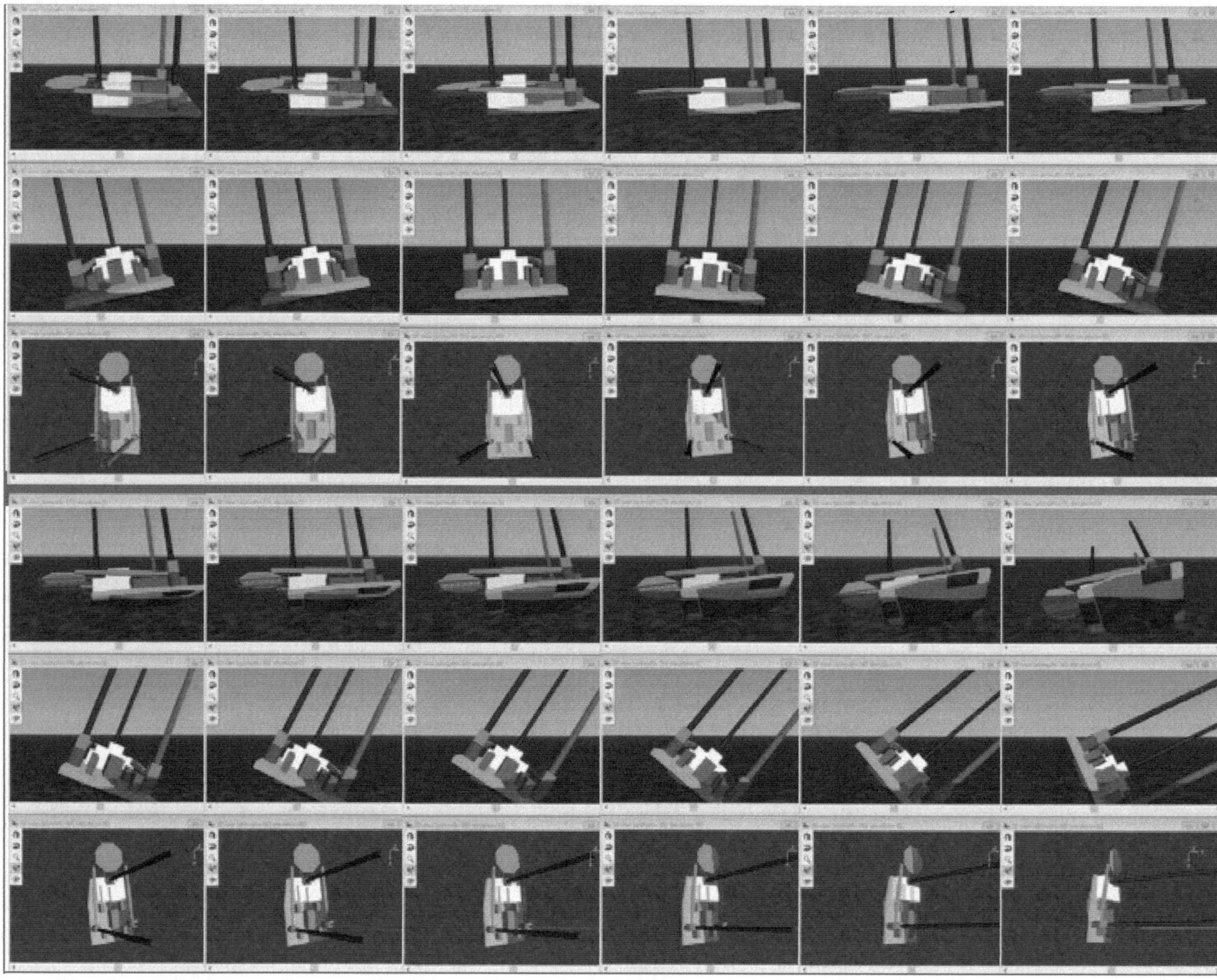

Figure 24 – Example Capsize Sequence, 1-Frame/sec, JONSWAP, Hs=6-ft, Tz=6 sec, γ=2.0, Direction 0°

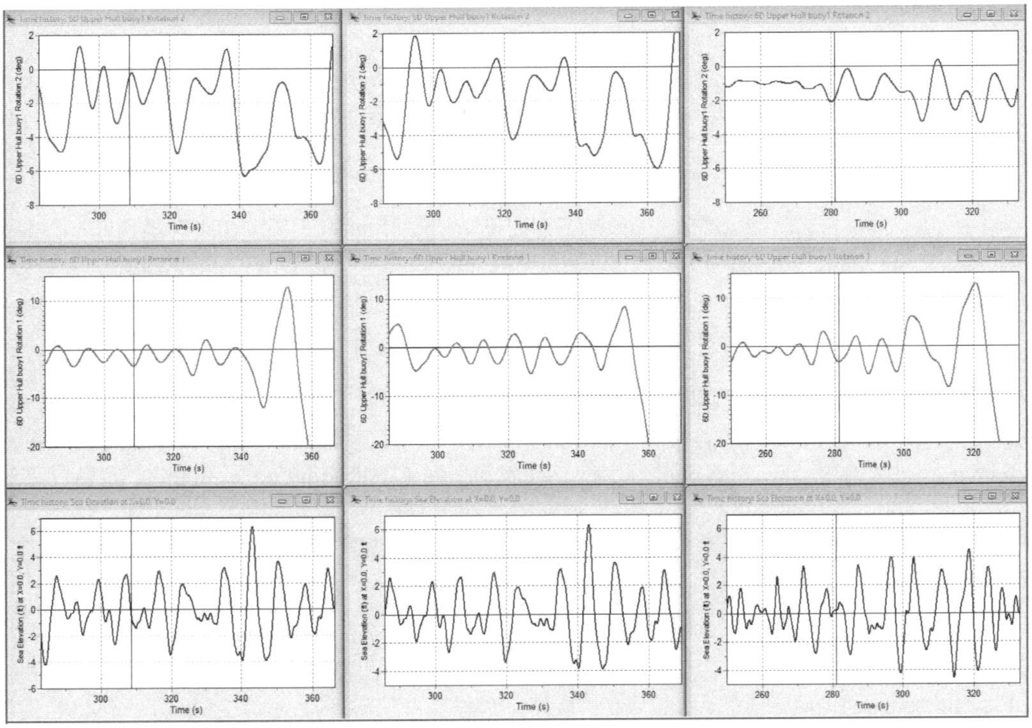

Figure 25 – Capsize Examples, Pitch, Roll & Wave Elevation Before Capsize

Figure 25 shows three sets of time histories for pitch, roll and wave elevation (blue, red, and brown) for three cases with a 50-kt wind all resulting in capsize. The cases are for wave attack directions of 0°, 45° and 90° (left, middle and right) with a JONSWAP wave spectrum of long-crested waves with Hs = 6-ft, Tz = 6 sec and γ = 2.0. The charts are drawn with the same vertical scales for each color. The red roll negative axis is limited to 20° to show more clearly the roll response in the 80 seconds before the capsize.

The capsize sequences in waves from 0° and 45° both begin during the passage of a relatively high wave with a 4-ft trough, 6.3-ft crest, and 6.9 second period from trough-to-trough. In both these cases the response before and during the capsize shows pitching with a longer period than the waves.

The capsize in 90° waves is not associated with a single high wave event but is preceded by a 20-second build-up of longer-than-wave-period pitching and rolling. The capsizes are not associated with a "Pure Loss of Stability" as defined by the IMO and which is a common reason for fishing vessel capsize on the crest of a wave [9].

No vulnerability criteria for capsize appropriate to vessels like liftboats appears in drafts of the IMO Second Generation Intact Stability Code [10] although the Dead Ship Condition, which applies to vessels in beam seas, suggests model tests.

Influence of Leg Lowering

A little over one minute before the capsize, the *Seacor Power* had begun lowering her legs at a rate of 6ft/minute. As the buoyant 7.5-ft OD legs and 40-ft x 23-ft pads lowered from beneath the upper hull, the heave, roll and pitch damping would have been increasing, and the vertical centers of gravity and of buoyancy would have been lowering. The pad of one leg can be seen lowered by about 10 feet in photos of the capsized vessel. The combined effect of lowering the legs by 10 feet would have been to slightly reduce the likelihood of capsize.

Effect of Buoyant Superstructure

Responses change significantly when the tube raft elements are extended above the barge deck to model the weathertight deck houses. Capsize nevertheless occurs. Full results are beyond the scope of this paper.

Model Testing

A well-planned series of model tests is advocated to investigate the dynamic afloat stability of liftboat-type hulls, with relatively low freeboard/beam and draft/beam ratios and relatively high centers of gravity, partly as a result of their long legs. The author originally proposed such a series of tests when he chaired the SNAME Liftboat Committee over 20 years ago.

Summary and Conclusions

1. Liftboats have wide beams, shallow drafts, and low freeboards. They experience water on deck and deck edge submergence in relative low wave heights. This makes them difficult to model with most seakeeping software.
2. Two cases of intact liftboat capsize in waves with loss of life highlight the need to provide better guidance for liftboat capsize avoidance in waves.
3. The Class 250 liftboat, in the nearly full load condition as presented in this paper, is prone to capsize in 6-ft waves from the beam and quartering directions, without any wind.
4. It is shown how a liftboat vessel hull can be modelled as a tube raft built from an assembly of general purpose OrcaFlex line elements.
5. The tube raft model is shown to have accurate large angle stability properties in still water.
6. The model permits unlimited submergence at any angle in a time domain analysis while reliably predicting hydrostatic restoring forces.
7. The model is demonstrated to have natural heave, roll, and pitch natural decay curves in still water with the expected natural frequency and damping properties of the liftboat vessel.
8. Dynamic response to non-linear steam function (and other) waves can be simulated in shallow water, although the presence of the vessel does not alter the wave particle kinematics.
9. The model includes Froude-Krylov forces within the whole of the wetted surface (entire volume) and constant hydrodynamic added mass terms for 6DOF. Diffraction forces are not included.
10. Time domain analysis of the model demonstrates that vessel RAOs change their form with increasing wave height and are not linear with wave height.
11. The model illustrates deck submergence (or deck in water) in waves, but the solutions are of unknown accuracy.
12. The model is able to realistically simulate capsize and identify wind and wave conditions that will result in capsize.
13. The time domain simulations run relatively quickly, with a batch of twenty 10-minute real time analyses taking 5 minutes computer time.
14. Large amplitude sub-harmonic rolling, and pitching are observed at twice the wave period for periods leading up to roll and pitch natural periods in still water. This motion has similarities with but is not the same as parametric rolling.
15. Model tests are needed to define limiting metocean conditions more reliably for safe transit of liftboats.

References

[1] NTSB, Marine Accident Report, Capsizing and Sinking of the U.S. Liftboat M/V AVCO V, Gulf of Mexico, July 31, 1989

[2] NTSB, Preliminary Report, Marine, DCA21MM024, Capsizing and Sinking of the US-flagged Liftboat Seacor Power, April 13, 2021.

[3] USCG Marine Board Hearing, Capsizing and Loss of Life on the Seacor Power, April 13, 2021, USCG and NTSB Joint Formal Public Hearing and Examination, Houma, LA, August 2 – 13, 2021.

[4] OrcaFlex, Finite Element Computer Software for Static and Dynamic Analysis of Ocean Structures, Developed by Orcina Ltd., UK.

[5] US Code of Federal Regulations, Title 46 / Chapter I / Subchapter S / Part 174 / Subpart H - Special Rules Pertaining to Liftboats, Source: CGD 82-004 and CGD 86-074, 62 FR 49355, Sept. 19, 1997.

[6] Santen, JA van, Problems met in stability calculations of offshore rigs and how to deal with them, 13th International Ship Stability Workshop, 2013.

[7] ABS GUIDE FOR THE ASSESSMENT OF PARAMETRIC ROLL RESONANCE IN THE DESIGN OF CONTAINER CARRIERS, 2019.

[8] Orszaghova J, Wolgamot H, Draper S, Eatock Taylor R, Taylor PH, Rafiee A. Transverse motion instability of a submerged moored buoy. Proc. R. Soc. A 475:20180459, 2019.

[9] Matsuda, A., Terada, D., Hashimoto, H., The characteristics of capsizing phenomena of Japanese fishing vessels, Proceedings of ISSW 2017.

[10] Naval Surface Warfare Center, Carderock Division, Development of Second Generation Intact Stability Criteria, NSWCCD-50-TR-20011/065, December 2011.

SNAME
THE INTERNATIONAL COMMUNITY
FOR MARITIME AND OCEAN PROFESSIONALS

ROLE OF CLASSIFICATION IN THE FLOATING WIND ASSURANCE

Piotr Szalewski
DNV, Approval Center Houston

Robert Galinski
DNV, Director Offshore Wind North America

Proceedings of the 27th Offshore Symposium, February 22nd, 2022, Houston, Texas
Texas Section of the Society of Naval Architects and Marine Engineers

ABSTRACT

By 2050, global installed floating wind capacity is expected to reach 264 GW. Project sanction will require levelized cost of energy (LCOE) being significantly below current levels, but also to demonstrate that the life, property, and the environment are safeguarded.

Most class societies were established in late 1800's as independent organizations addressing technical risks in shipping industry. In 1980's classification model has been successfully extended to the offshore oil and gas, resulting in significant safety improvements and cost savings. Most recently, the same model has been successfully applied to offshore fish farms.

During the last decades classification has been transforming towards fully risk and data driven solutions. By using fleet-wide knowledge, significant cost savings are generated and passed to the asset owner and operator. This makes classification ideally suited to floating wind.

The paper discusses areas of class society involvement in certification of floating wind foundations, including mooring system and wind turbine interface. Risk based integrity management in harsh environments the floating wind platforms are expected to operate are addressed. Potential impact is illustrated using examples from the past where such benefits were realized, including using fleet data to improve efficiencies and reduce costs. Required adaptation of classification concepts to specifics of the floating wind are also discussed.

The paper also shares examples of class society role and activities beyond certification and assurance. We hope this will help appreciate the overall value of classification provides and encourage the industry to tap into benefits from its broad knowledge and experience.

Keywords: SNAME, Offshore Symposium, Classification, Floating Wind, Assurance, Certification, Design, Fabrication, Installation, Integrity Management, Mooring, Hull

INTRODUCTION

By 2050, global installed floating wind capacity is expected to reach 264 GW. Project sanction will require levelized cost of energy (LCOE) being significantly below current levels, but also to demonstrate that the life, property, and the environment are safeguarded.

Most class societies were established in late 1800's as independent technical organizations insurance risks in maritime industry. In 1980's classification model has been successfully extended to the offshore oil and gas, resulting in significant safety improvements and cost savings. Most recently, the same model has been successfully applied to offshore fish farms.

During the last decades classification has been transforming towards fully risk and data driven solutions., By using fleet-wide knowledge, significant cost savings are generated and passed to the asset owner and operator. This makes classification ideally suited to floating wind.

Classification societies contributions go far beyond certification. Leading industry collaborations, assessment of new concepts and technologies, shelf state verification, software development, independent analysis, data platforms and integrity management solutions are most relevant to floating wind. On multiple projects they have been used in synergy with classification, maximizing the benefits of both.

So, let's first look into the cost of floating wind first, and next move into discussion on how this cost can be minimized. The main highlights include:

- Efficient maritime execution model
- Assessments of novel concepts and new technologies
- Strong set of technical standards and software
- Classification as a building block to shelf-state compliance
- Efficient, risk-based data driven integrity management
- Interface with other industries

COST OF FLOATING WIND

Floating offshore wind is currently about four times more expensive than other wind concepts, hence has the greatest cost saving potentials compared to onshore wind and bottom fixed offshore turbines, where majority of cost optimization has already happened [1]. As floating wind technology maturity lags other wind concepts, the drivers of change are also quite different, focused mainly on non-turbine investment (estimated 94 USD/MWh or 42% reduction) and OPEX (49 USD/MWh or 22%), see Figure 1. Classification activities concentrate around the activities with anticipated cost reductions.

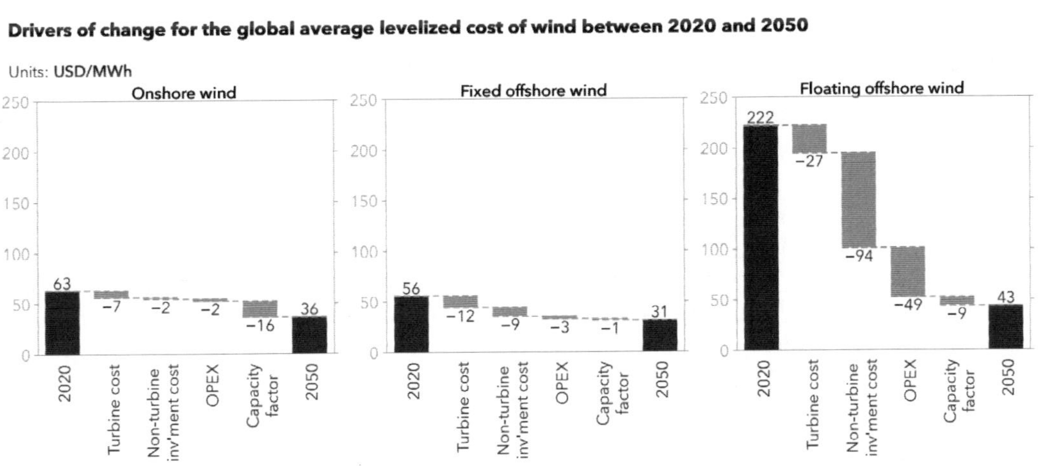

Figure 1: Drivers of change for the global average levelized cost of wind between 2020 and 2050. [2]

EFFICIENT MARITIME EXECUTION MODEL

Fabrication of floating wind foundation creates substantial opportunities for EPCI companies, shipyards, raw material and equipment manufacturers. However, new type of projects come with significant risks and challenges, including even higher pressure on project costs compared to shipbuilding and offshore industries [3]. Proven maritime execution model means smooth project execution with predictable and familiar acceptance criteria implying reduced cost with required quality which reduces initial CAPEX and future OPEX.

Maritime execution model solves many of the concerns through standardized material certification, standardized equipment and package specifications, standardized project documentation and construction requirements, and lean and efficient project supervision. Steel mills do not need to adjust production lines, welders do not need to be re-qualified, scope required by vendors is well understood, products already have type approval certificates, and expected fabrication quality remains the same hence productivity has already been optimized. Both owner's and third-party assurance process is well defined and understood by all stakeholders. Classification provides all of these benefits for maritime and offshore industries, and project tracker record shows substantial efficiency and cost savings can be achieved with standardized approach.

For new construction, EPCIC and EPC contracting models prevail in the offshore industry. The EPCIC is typically a turnkey solution where a major engineering contractor is responsible for all project aspects and delivers a final product based on owner's specification and build to classification requirements, though additional compliance requirements may still be specified. In EPC model, the owner would typically retain overall project responsibility and separately contract and actively oversee FEED, EPC, installation, and commissioning activities. Classification may still be used as basis, though some owners choose to rely solely on set of project specific standards with significant portion being the owner's own standards.

Several industry studies clearly showed the project cost of EPCIC model is about 50-60% of that of EPC. Figure 2 shows a comparison performed by an oil major for a dozen of recent FPSO projects. Higher cost of the EPC model may be attributed to added interface complexity, removing of key efficiency incentives, less competition and larger cost markups by sub-contractors.

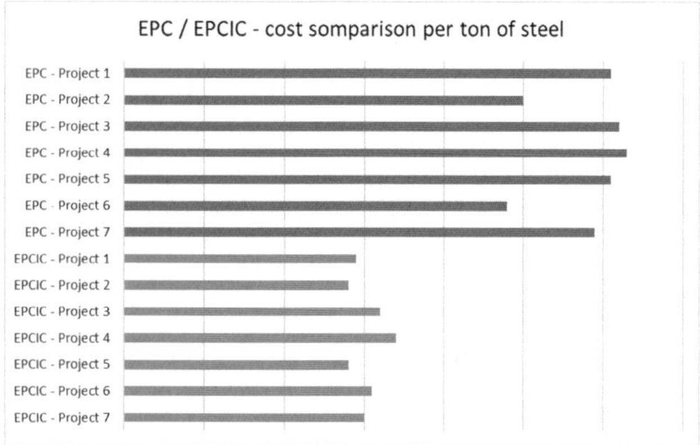

Figure 2 – EPC / EPCIC – project cost comparison, owner's own study of recent FPSO projects.

Similar conclusions are drawn from numerous naval construction projects built to class rules versus navy specifications. One of the benefits is using off the shelf materials which generally leads to savings both in project cost and duration. Other examples include standardization of subsea forgings [4], which reduced lead times, enhanced stock keeping and allowed for interchangeability of materials.

Commercial floating wind projects are expected to be mass produced. The yard/fabricator will go through an extensive optimization of the fabrication process continuously incorporating learnings into subsequent units of the same design, including more substantial changes like adjusting block erection sequence or even design optimization of subsequent units. Some production mistakes will also happen, and in-yard repairs will need to be performed. It is important that production is done to a recognized constructions standard such as IACS Rec.47 [5] which is very familiar to the fabricators. Design codes are calibrated to the building tolerances prescribed therein, what has been proven by in-service performance of structures built with this approach. Classification surveyors follow a risk based survey scheme to verify that the fabrication quality follows design documentation and is according to the agreed construction standard. This is done efficiently with independent inspections focusing on challenging construction activities and critical structural details, where potential deficiencies are more likely and may have higher consequences. Standard building activities are normally well covered by fabricator's quality system, hence third party activities can be limited to audits and random unscheduled inspections. Risk based approach will also mean that the survey activities will be greatly reduced for subsequent identical units. This is no different than long ship series where class societies have accumulated substantial experience to achieve the right balance between scope and value.

Owners may also choose to follow an enhanced fabrication testing scheme such as described in DNV FAB notation. That may be important for units intended for harsher environments, or if the owner intends to reduce in-service inspection. Class involvement may also include review of coating procedures (COAT notation), or application of enhanced fatigue methodologies (FMS notation) [6]. Enhanced design and fabrication standards may increase initial project costs, but will be offset by savings in future inspection, maintenance and repair costs A balance giving lowest overall investment cost can found using cost models based on probabilistic RBI methodology described in DNV-RP-C210 [7]. The cost approach has proven beneficial in multiple life extension projects of offshore installations and can be used to find the most cost effective floating wind integrity management program. The cost models are also able to incorporate factors like developer risk tolerance or maximum time allowed for in-situ inspections which may be driven by other field activities such as. turbine maintenance.

In addition to meeting fabrication and installation standards, accurate information about the achieved quality level will also be needed, including relative quality of individual units and structural joints. This is important to be able to correctly choose representative inspection locations – high confidence in this selection will mean minimized scope of inspection and therefore reduced cost. This approach requires consistency in quality assurance and accurate record keeping across different vendors and fabrication sites. Construction monitoring class notation such as DNV CMON [8] may help achieve that. A construction monitoring document will be developed based on design analysis, construction and fatigue improvement methods, complexity, building tolerances, experience from similar structures and more. Fabrication details such as joint preparation, alignment, material defects, welding procedure, NDT testing and any remedial actions are recorded in the document and follows into in-service phase.

Overall, a well-proven shipbuilding and offshore construction approach is a win-win for all stakeholders: owner, fabricators, designers, equipment vendors and others. Use of class scheme, with minimal owner specific requirements, establishes a standard solution: each stakeholder knows exactly what is expected of them and where are their responsibilities and scope boundaries. For contractors that means they can accurately price their products and do not need to incorporate risk premiums to cover unnecessary uncertainties. For owners it means trust in quality product they ordered.

Bringing confidence in new concepts and technologies

Floating wind technology is currently in early phases. Most concepts are on drawing boards, some have advanced into a scaled prototype, full-scale floaters have only been installed with few units in a field. The industry has come quite far to demonstrate that the technology does work, but significant innovations are still required to make it economically feasible. That includes development of new floating concepts, and optimization and scaling up of existing concepts. Existing technologies would need to be extended to new applications, crossed over from different industries or developed from scratch.

Project stakeholders need to trust that life, environment and property are protected during the entire project lifecycle. Lack of confidence introduces additional risks, which translate to monetary risk premiums, delays or may even turn into a project showstopper. These risks need to be managed starting in early project phases where cost effective solutions can still be found on the drawing board, at lowest possible cost.

Class methodologies such as Approval in Principle (AiP) and Technology Qualification (TQ) are widely used in shipbuilding and offshore to assess feasibility of proposed solutions of limited maturity. For floating wind, AiP would typically apply to a new floating concept including design of hull, mooring and relevant interfaces to turbine, seabed and power export. TQ's focus would be more on an individual technology such as use of polyester lines instead of chain in marine growth zone, export cable design, novel design of mooring connectors, motion based mooring tension monitoring and more.

AiP is a high-level concept review to evaluate new concepts and identify potential showstoppers. All relevant phases of the project are covered including design, fabrication, installation, integrity management. Interrelations between the phases are considered, as e.g. planned in-service inspection and maintenance philosophy plays significant role in required fatigue safety factors. Challenging aspects are determined, and follow-up actions agreed. Provided no showstoppers are identified, AiP certificate is issued along with a detailed report describing the basis for review, findings, and recommendations, allowing for additional confidence that the design will perform well once deployed.

Despite being an early assessment, TQ is used to assess and qualify new technologies. It is also applicable to evaluate extensions of existing technologies in size, new application, new environment and so on. Technology application needs to be considered and evaluation performed in a higher level context. As a first step, technology maturity is evaluated (Technology Assessment). A risk study is then performed to understand failure modes and associated risks, what leads to mitigation actions and selection of qualifications methods to demonstrate require performance (Technology Qualification Plan). The methods may include numerical studies, small- or full-scale testing, prototyping and more. Once performance is demonstrated, Technology Qualification Report is issued.

For both the AiP and TQ, the outcome allows the concept proponent and other parties to gain understanding and confidence of concept maturity and what further steps are required to advance it and ultimately achieve compliance with agreed set of technical standards. In simpler terms, achieving AiP and TQ means that it is realistic to expect that the concept or technology may be successfully deployed. Risk approach makes the assessment realistic by focusing on items of high and medium risk categories and defers low risk items to later project phases. With cross-industry expertise, the independent society experts are ideal to perform peer-to-peer assessment, adequately challenge the proponent and correctly assign risk category.

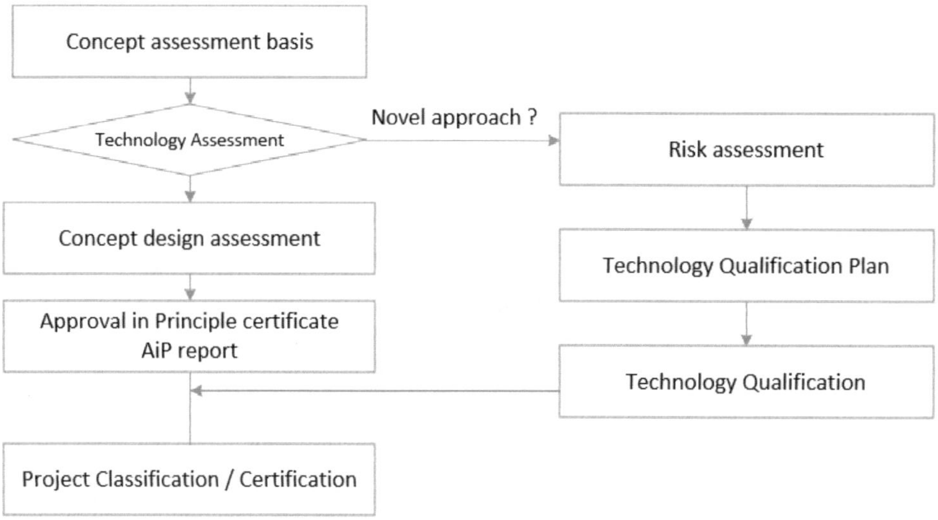

Figure 3 – Concept assessment process

As the project progresses into FEED and detailed design, full classification reviews are undertaken. The objective is to establish confidence in design analysis, what often cannot be sufficiently done by just reading design documentation. Independent analysis by class experts should be included, especially for novel concepts such as floating wind, where only limited experience is available. The independent calculations should as minimum reproduce designer results for governing load cases, and cover global performance, global strength, fatigue screening, stability and mooring. The independent analysis also establishes result consistency as the results can be compared with other assets. That helps achieving efficient risk based integrity solutions, discussed later in this paper.

Strong set of technical standards and software

Floating wind classification rules combine experience from other floating assets. They have been developed with industry involvement and tested on a multitude of new concepts. Technical provisions have been adapted from international standards from both oil and gas and offshore wind. Required level of safety was adjusted to risk level required in floating wind project, and safety factors calibrated accordingly. The rules are accompanied by recommended practices which describe accepted methodologies and design approaches.

Floating wind concepts are being developed using existing industry experience, so the floating wind standards build on relevant existing and proven norms. New features, challenges and differences have been comprehended and agreed by the industry based on numerous joint industry projects. The standards have been applied to numerous actual projects, with knowledge, experience and lessons learnt incorporated back into newer editions of the standards. Each update undergoes thorough industry hearing before publication. A journey of the DNV floating wind standards is summarized on Figure 4.

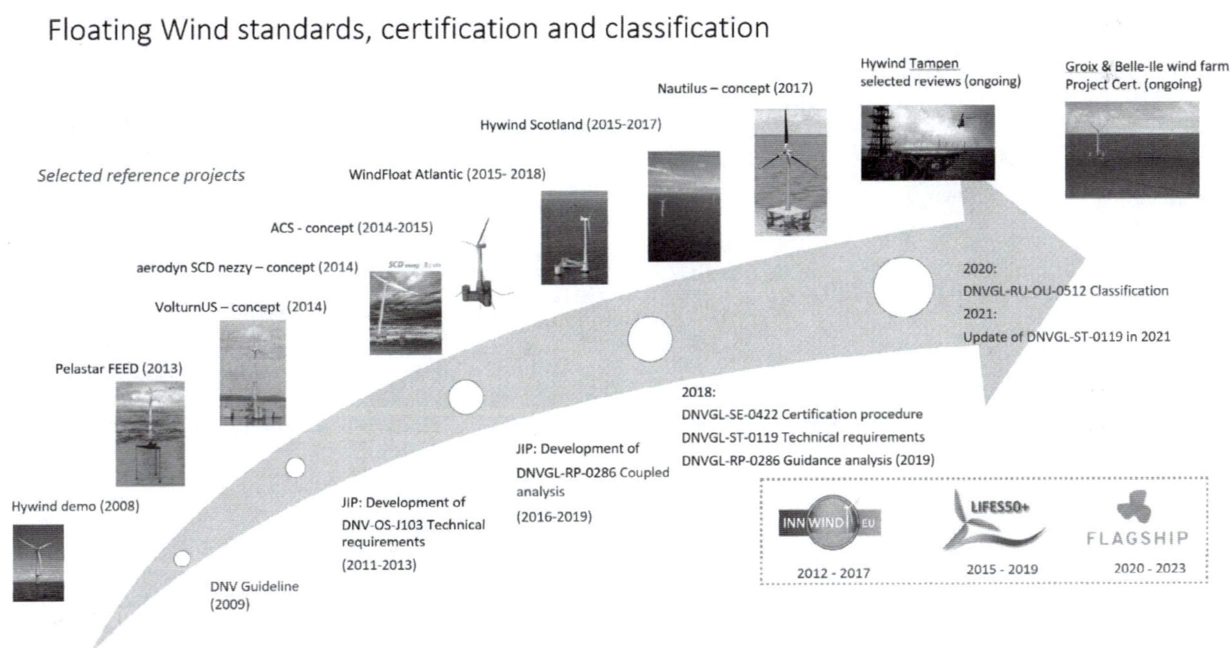

Figure 4 – Floating wind standard development journey.

First principle approaches have long been allowed by the classification standards, and will certainly be useful for floating wind design. This includes methods based on structural reliability analysis (SRA) and design assisted by testing. The challenge is to apply it in a consistent manner so the required level of safety can be demonstrated for the entire design life of the unit, and also establish basis for possible life extension. Class rules describe steps necessary to successfully implement such alternative approaches, and independent assurance can be provided by class experts ensuring the design meets safety intent defined by the standard.

Design methodologies for floating wind require significant computational efforts. While hull and mooring system may look like a familiar semi-submersible, TLP or barge, the interaction of the floater motions with wind turbine introduces additional analytical challenges. The motion of the floater influences the aerodynamic inflow conditions and introduces inertial loads to the turbine. In turn, the forces acting on the turbine influence motions of the floater. A fully coupled analysis need to be performed early in the design process to accurately represent these effects. A numerical model is developed including mooring system model and structural models of the floater, tower, blades, drivetrain, and turbine controller. Wind, wave and current loads need to be modelled with sufficient fidelity and cover an array of load cases representing both operational, survival and accidental scenarios [9].

Efficient design process requires a comprehensive software package specific to floating wind. The software also needs to facilitate postprocessing including code checking and fatigue damage summation. Software like DNV SESAM integrates floater analysis tools, with time domain simulation software Sima and turbine load generation program Bladed Turbine concept and controller data can be defined in the software or predefined concepts can be used. Code checking according to both DNV class standards, API, AISC, Eurocode, ISO and NORSOK are implemented. Fatigue summation in frequency and time domains are available, hence frequency domain may be used for preliminary studies, and the same model can be extended to time domain as design progresses. Figure 5 shows software workflow implemented in SESAM.

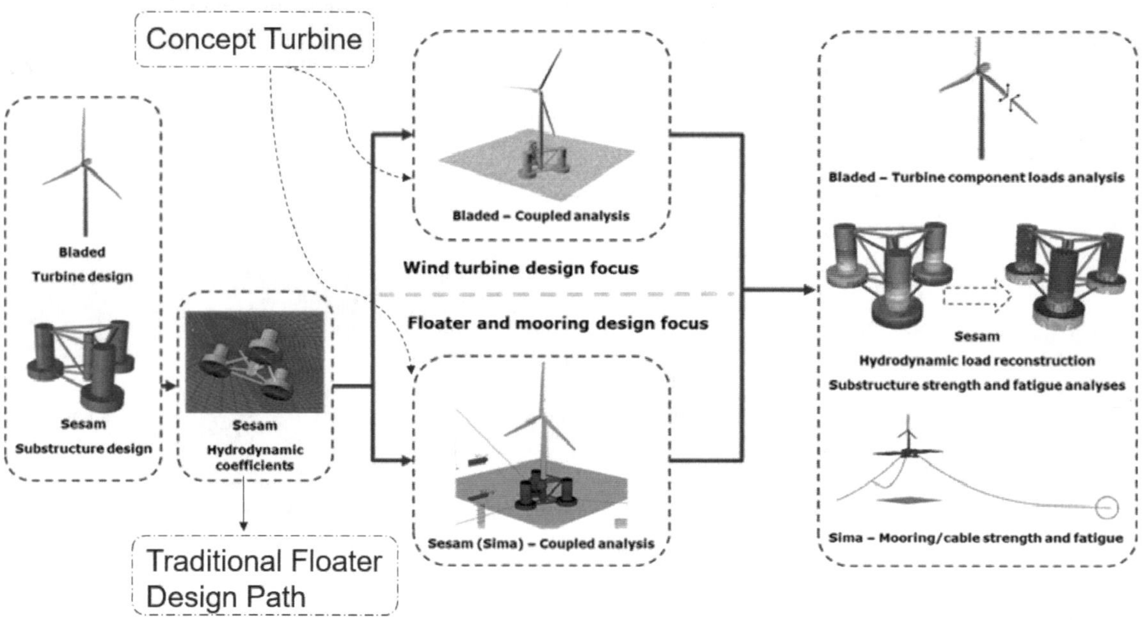

Figure 5 – SESAM implementation for floating wind design

Floating wind technology will advance in the future and the rules and software will need to be updated. As new concepts are designed and installed, it is important that lessons learnt are shared across the industry and incorporated into the standards. This is one of the areas where the independent third party organizations as class societies need to continue encouraging and leading industry collaborations. National regulators also need to be actively involved so they are up to date with state-of-the-art knowledge. Examples of such involvement is JIP aiming to update DNV-RP-C205 on environmental loads, where BSEE and BOEM participate along with other regulators and numerous industry partners. The project is planned for completion in 2023 and will result in updated recommendation focusing on close to shore infrastructures like floating wind. Other examples of ongoing industry collaborations include "DNV Floating offshore wind substations" JIP and Deepstar's projects on standardization of digital twins for mooring inspection [10].

Collaboration between class societies and several industry partners on software solution R&D is also key. One example is the "ImproveFLOW" project, which aims at facilitating more cost-effective designs of floating wind units and farms by developing methods and tools that will reduce uncertainty in the investment and planning phase, and improve accuracy, collaboration and efficiency in the design and engineering phases.

PATH TO SHELF-STATE COMPLIANCE

Floating wind projects will be installed in coastal waters and subject to permitting by a shelf state. A compliance with applicable laws needs to be documented and approved by local governments. We are yet to learn in detail what a practical implementation of these requirements will be in every country for floating wind projects. It appears that it will be somewhat similar to oil and gas floaters, where the industry has a lot of experience. The requirements and regulatory schemes may vary around the world, though classification rules are commonly accepted either as a sole basis or as a major element of the compliance programs.

In the UK, floating foundations should be designed to classification rules (hull) and ISO 19901-7:2013 standard (mooring, which is also fully incorporated in DNV class rules). Integrity management plan needs to be implemented with in-service monitoring. Third party verification by independent competent body is required to ensure quality of design, fabrication and operation. [11]. Similar requirement will likely be adopted by regulators in Asia, where DNV have recently been selected as certification bodies for offshore wind projects. [12]

In the US shelf floating wind installations are subject to US Code of Federal Regulations [13]. For floating foundations, applicable USCG requirements must also be met. Third-party verification is undertaken by a nominated CVA, who will perform independent assessment and certify to BOEM that the facility is designed to withstand the environmental and functional loads appropriate for the intended service life at the proposed location. CFR defines requirements to CVA such as relevant qualification, experience, technical capabilities, in-house technology, and the nomination is subject to BOEM's approval.

The CVA scope covers all components of the facility: wind turbine, support structure and electrical facilities including offshore power cables, and class societies are qualified to take that role. For floating wind that could combine the best experience from offshore wind and oil floaters in a self-contained integrated approach, with classification rules as a subset of a full verification scheme. An integrated class and verification approach have been executed for the first FPSO in the US GOM, and regulatory map have been subsequently developed by DNV [14]. DNV have also performed shelf state verification work of bottom fixed wind worldwide for over 30 years, including majority of US offshore wind projects.

Class activities may continue into operational phase, supporting integrity management and paving way to life extension. Experience from fleet risk management solutions of merchant ships, unit specific risk-based inspection programs for oil and gas floaters, and implementations of hull and mooring monitoring solutions are yet another value classification can bring to the picture.

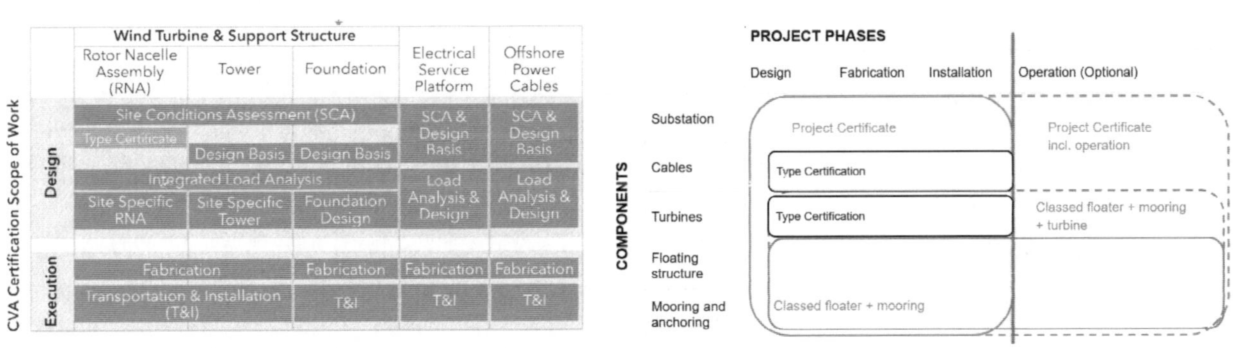

Figure 6 – CVA certification scope, based on bottom fixed installations (left) and possible areas of classification involvement for floating wind (right)

RISK BASED, DATA DRIVEN INTEGRITY MANAGEMENT

Data-driven, risk-based asset integrity management program is expected by the floating wind industry. The characteristics should include integrated wind farm approach, sensor-based monitoring, live risk profile of individual units and live inspection plan. There are numerous examples of individual solutions which have already been deployed. They can be easily adapted for floating wind and combined into a comprehensive integrity management solution. Figure 7 shows possible concept.

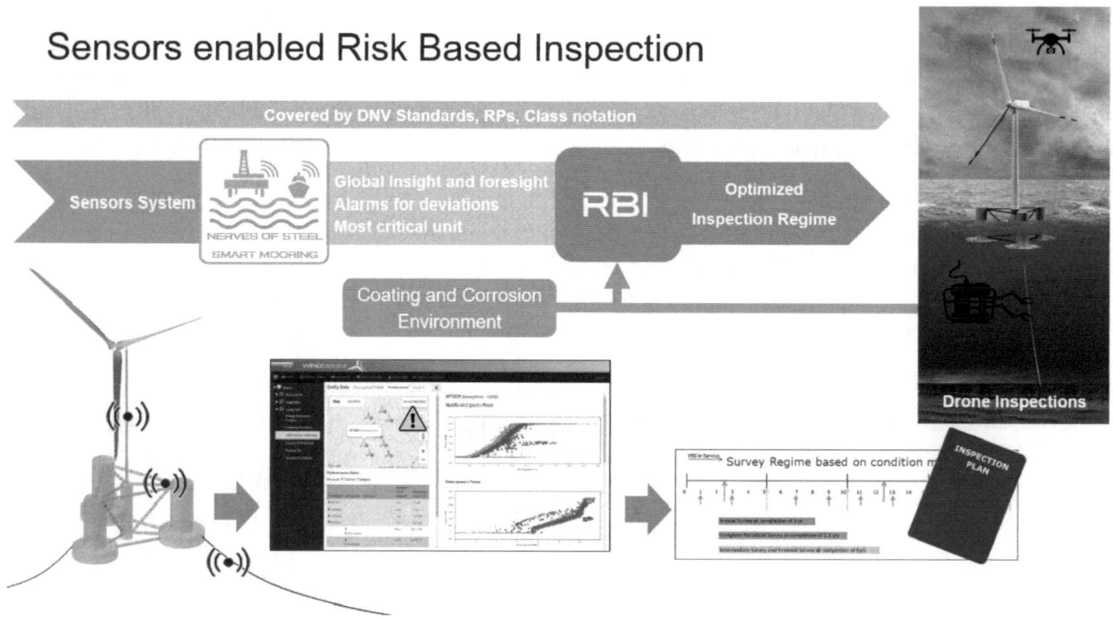

Figure 7 – Concept of sensor enabled risk based inspection for floating wind.

Unlike oil and gas floaters, the floating wind foundations will be installed in large groups. This creates numerous opportunities for synergies. An obvious approach is limiting inspection scope to representative floaters and details, and extrapolating conclusions to the entire wind farm. Inspection scope will need to be carefully determined based on relevant design results, fabrication and installation records. The scope will be regularly updated as monitoring data and inspection results become available. For the approach to work, data quality is essential. Uncertainties will need to be minimized, as they cause inspection scope to increase, possibly even to the point that implementation of a risk based solution is no longer feasible. Sources of uncertainties include inconsistent design methodologies, low confidence in design analysis, inconsistent fabrication quality, poor understanding of what quality was achieved for each unit / detail, installation variability, inaccuracies and misinterpretation of monitoring data, undocumented changes and repairs after installation, insufficient reporting of in-service inspection, inspection method accuracy and more. Through independent standardized approach, classification will help manage the uncertainties. Increased confidence level allows for reduction of inspection scope what translates to cost reductions. For full benefit, probabilistic methodologies such as the one described in DNV RP-C210 [7] will likely prevail in floating wind application. They have already been applied and accepted in lieu of prescriptive deterministic approach, for example in oil and gas life extension projects [15].

Sensor cost and reliability is a major challenge in implementing monitoring solutions. There is a trade-off between cost of sensors required to achieve desired confidence level and future savings inspection expenses. The optimum solution may differ by not only pure technical criteria, but also by operator's preference or risk tolerance. Numerous recent projects looked into optimum placement of minimum number of physical sensors for maximum monitoring benefit. "Nerves of steel" is an example of the advanced hull integrity monitoring system which was deployed for shuttle tanker fleet [16]. Sensor measurements were combined with other available data sets such as AIS and design fatigue analysis results to calculate live status of the structure. In some cases, useful insights may be available without any physical sensors– a digital twin system has already been in use for several years to establish risk index for the entire DNV vessel fleet based on analysis of available data sets and AIS data. Other example is "Smart mooring" providing mooring line tension measurements based on motion sensors installed on the floater. The solution is an approved alternative to physical sensor based mooring tension monitoring system, which was prone to malfunctions.

For floating wind it is expected that physical sensors will be installed, so there is a need to optimize their number and location to achieve monitoring objective. That may include combination of strain and motion sensors combined with other data sources like turbine monitoring data, weather data and so on. Sensitivities to future sensor failures and failure consequences need to be taken account. Further, accurate and consistent analytical prediction of design fatigue life is required to achieve meaningful insight with small number of sensors. In 2020 DNV introduced NUM and SENS qualifiers to FMS (fatigue methodology specification) class notation [6]. which integrates standardized fatigue evaluation methods with physical hull monitoring for integrity management. DNV-CG-0508 [17] was also published outlining planning and implementation of the hull stress monitoring system. The approach allows incorporation of owner risk tolerance and achieve desired balance between monitoring expense and inspection savings. As part of notation deliverables, DNV will setup independent analytical models, propose sensor locations, and establish required inspection scope to achieve target safety level. The model will be regularly updated once monitoring data becomes available.

Live risk profile will require comprehensive digital twins, which will aggregate data sources from multiple sources. Each additional data set generally enhances results fidelity, but also introduces potential errors. Quality assurance of digital twins is addressed by DNV-RP-A204 [18]. The scope covers assessment if digital twin meets stakeholder's expectations, establishing confidence in the data and computational models, and organizational readiness to implement and maintain digital twin. The methodology has been developed as industry collaboration and piloted with several companies in the oil and gas industry.

Remote inspection will save travel time and cost associated with inspection. It is important to realize that replacing physical presence of an engineer or surveyor should not necessarily mean that quality is compromised. Savings can be spent elsewhere, e.g. to add remote sensors, increase number of joints inspected and inspect inaccessible areas with drones and ROVs. Centralized inspection handling also means better consistency and better availability of subject matter experts. Inspection videos and measurement records are instantly recorded in a standardized way. Surveyors will still be sent to site if the remote findings are inconclusive.

Remote surveys have been in use for some time by class societies. It first started with remote monitoring of component testing and recently expanded into remote surveys on-board ships and offshore platforms. It gained more momentum in 2020 during the pandemic, where flag authorities have generally accepted remote techniques for statutory surveys. Some back-stepped meaning that the industry still needs to be demonstrate that the foremost objective is to improve safety, and direct and indirect cost savings come with it,

Risk based management programs have already been allowed as alternative to prescriptive based inspection requirements with fixed time intervals. An example is DNV POSMOOR [19], which fully accepts an approved mooring integrity management program based on best industry practices as alternative approach. The program shall comprehensively address risk of both the system and each individual component, describe strategy for risk mitigations (e.g. inspection, monitoring, replacement), and take into account risk introduced in each lifecycle phase (design, fabrication, installation and operation). Supporting documentation should combine theoretical studies calibrated to in-service findings. The program needs to be adjusted as the condition of the system or components change during its design life as outlined in both class and regulatory recommended practices [20], [21].

With technology enablers, live inspection plan will be possible to implement for floating wind foundations. However, that would require a step-change shift of the current status-quo which is heavily reliant on prescriptive inspection scope and intervals, and on in-person inspection. Novel approaches need to be accepted by the regulators and presenting compelling evidence may sometimes be a challenge. Industry needs to collaborate to establish a scheme and document that the proposal improves safety. It may come at the lower cost, but such should not be the main or the only driver. Class societies have historically been able to balance stakeholders' interests while improving safety in the maritime industry, hence are well suited to take on such role in floating wind journey.

Interfaces with Other Industries

Floating wind farms will be a part of larger ecosystem. Maritime infrastructure to support construction and maintenance is where class societies will obviously be involved. This part is relatively simple and a standard solution adapting well established concepts to offshore wind specification. What is less obvious and more challenging is the power export, storage and distribution, and some class societies have a footprint there. One current challenge for floating wind is power export cables, which have similarities to marine risers so existing assurance schemes can be implemented with updated technical requirements. Floating substations is another earlier mentioned example. Hydrogen generation, storage and transportation may develop into a viable option for floating wind in the future, and it is important it can be done safely and economically. Integration with oil and gas installations is already a reality in Equinor's Hywind Tampen, which is also seen a test bed for larger scale implementation. Offshore wind developments also have significant synergies and work well in tandem with other renewable energy sources such as solar, with offshore concept already in place. The industry can clearly benefit independent involvement from organizations with deep knowledge in maritime industry, but also having deep competence in cooperating industries a broad view of the overall ecosystem.

SUMMARY

A variety of examples have been provided to demonstrate how the floating wind industry can benefit from classification approach and other work done by class societies in all project phases: early concept, design, fabrication, installation, operation and integrity management. These contributions go far beyond a mandatory certificate and include efficient maritime execution model, assessment of novel concept and technologies, leading standardization efforts, software development and proven data-driven, risk-based integrity management. Independent third party position ensures impartial view and creates trust among all project stakeholders. It is now up to the industry how they tap into the experience.

ACKNOWLEDGEMENTS

We would like to thank you all colleagues in the industry and internally in DNV who shared their challenges with floating wind and inspired us to write this paper. Special thanks to Erik Henriksen, Chris Kubes and F.X. Sireta who read and commented on the manuscript. It would not be the same without your contributions.

Glossary

AiP – Approval in Principle

AIS – Automatic Identification System

BSEE – Bureau of Safety and Environmental Enforcement www.bsee.gov

BOEM – Bureau of Ocean Energy Management www.boem.gov

CVA – Certified Verification Agent

DNV – Det Norske Veritas www.dnv.com

EPC – Engineering, Procurement and Construction

EPCI – Engineering, Procurement, Construction and Installation

EPCIC – Engineering, Procurement, Construction, Installation and Commissioning

IACS – International Association of Classification Societies www.iacs.org.uk

LCOE – Levelized Cost of Electricity

POSMOOR – Position Mooring

RBI – Risk Based Inspection

SRA – Structural Reliability Analysis

TQ – Technology Qualification

References

[1] https://youtu.be/58EYcYbRKqk NREL webinar "Overview of Floating Offshore Wind" Feb 26, 2020.

[2] DNV Energy Transition Outlook 2021, A global and regional forecast to 2050. https://eto.dnv.com/2021

[3] Floating Wind: The power to commercialize. https://www.dnv.com/Publications/floating-wind-the-power-to-commercialize-192334

[4] https://www.drillingcontractor.org/dnv-unveils-recommended-practice-to-standardize-steel-forgings-for-subsea-applications-34908, DNV introduces steel forgings RP to increase subsea standardization, accessed Jan 14, 2022.

[5] IACS Recommendation No.47 "Shipbuilding and Repair Quality Standard

[6] DNV-RU-OU-0571 "Floating infrastructure installations", July 2021.

[7] DNV-RP-C210 "Probabilistic methods for planning of inspection for fatigue cracks in offshore structures, amended Oct 2021

[8] DNV-RU-SHIP Pt.6 Ch.9 Sec.6 "Construction monitoring", July 2021.

[9] DNV-RP-0286 "Coupled analysis of floating wind turbines" amended October 2021

[10] Matsumoto, Jaiswal, Sugimura, Honjo, Szalewski, „Mooring Integrity Management through Digital Twin and Standardized Inspection Data" OTC-31036-MS

[11] "Regulatory expectations on moorings for floating wind and marine devices" UK HSE August 2017.

[12] "DNV to Certify Wind Turbines in Korea" https://www.offshorewind.biz/2022/01/10/dnv-to-certify-wind-turbines-in-korea/, Jan10, 2022, accessed Jan 16, 2022

[13] https://www.law.cornell.edu/cfr/text/30/part-585 Code of Federal Regulations, 30 CFR Part 585 "Renewable Energy and Alternate Uses of Existing Facilities of the Outer Continental Shelf.

[14] DNV-SI-0003 "Verification for compliance with United States regulations on the outer continental shelf", amended July 2021.

[15] Szalewski, Malinowski, Lu "Demonstration target safety level for life extension of offshore structures, SNAME 24th Offshore Symposium, Houston 2019.

[16] DNV Industry Insights "Fleet-wide implementation of next generation hull integrity monitoring" May 11, 2021

[17] DNV-CG-0508 "Smart vessel", August 2021.

[18] DNV-RP-A204 "Qualification and assurance of digital twins", amended Sep 2021.

[19] DNV-RU-OU-0300, Rules for classification: Offshore units – Fleet in service, July 2021.

[20] DNV-RP-0308 Recommended practice for mooring integrity management.

[21] Oil & Gas UK, Mooring integrity guidelines, Issue 3, November 2014.

THE INTERNATIONAL COMMUNITY FOR MARITIME AND OCEAN PROFESSIONALS

Numerical Simulation of the effect of wave characteristics on PTO of Point Absorber Wave Energy Converter

Ahmed A. Hamada
Texas A&M University
Abigail Rolen
Texas A&M University
William McCullough
Texas A&M University
Mirjam Fürth
Texas A&M University

Proceedings of the 27th Offshore Symposium, February 22nd, 2022, Houston, Texas
Texas Section of the Society of Naval Architects and Marine Engineers

ABSTRACT

The ocean is currently an extremely large and under-developed source of renewable energy. The recent interest in the Blue Economy has led the scientific community to increase investigations in sustainable oceanic energy options, such as Point Wave Energy Converters (WEC). These devices harvest the wave energy using the excited oscillatory motion of the buoy, which is connected to a Power Take-Off system (PTO). During the last decades, the development of these devices has been boosted but they are still behind other renewable energy technologies. The Furthlab at Texas A&M University has showed that the spheroid buoy shape with a low length to diameter ratio is a good candidate shape to extract wave energy, by testing different buoy shapes and aspect ratios at a non-linear Stokes-II wave generation. This paper is the next step in our research and numerically investigates the effect of changing the wave characteristics, such as amplitude, frequency, and speed, on the power-generating ability of the spheroid buoy system. Three-dimensional Unsteady Reynolds-Averaged Navier-Stokes (URANS) simulations of the selected buoy were performed in OpenFOAM with the integration of a dynamic mesh module to handle the heave motion of the buoy. In addition, the PTO system was compensated with a forced oscillator mechanism of spring and damper. A comparison between the buoy's displacement and frequency responses, and power efficiency showed the optimal operating sea state to maximize energy output using the spheroid WEC. The results conclude that the best wave conditions to maximize the power extraction efficiency using a spheroid buoy with a diameter of 1 m and length of 0.5 m are wave length greater than 4 m, wave height less than 0.15 m, and wave speed between 0.07 and 0.12 m/s.

Keywords: Computational Fluid Dynamics, Ocean Energy Harvesting, Wave Characteristics, Point Wave Energy Converter, OpenFOAM

Introduction

Wave Energy Converters (WECs) are an emerging technology that aims to harvest renewable energy from the motion of ocean waves. With the increasing concerns about global warming, many countries are moving towards increasing their renewable energy usage, to decrease their carbon footprint. In particular, the U.S. and Canada are working to develop next-generation maritime or "blue" renewable energy devices, such as WEC, in response to the demand for "powering the Blue Economy" (NOAA 2021; Pacific Northwest National Laboratory 2022; Government of Canada 2021). The Blue Economy is defined according to The World Bank (2017) as "the sustainable use of ocean resources for economic growth, improved livelihoods, and jobs while preserving the health of ocean ecosystems." This has caused the scientific community focus on optimizing the performance of renewable energy sources, such as WEC. WEC works by harnessing the mechanical energy from the motion of ocean waves and transforming it into electricity. The focus of this paper is a buoy constrained to a single degree of freedom, called a Point Wave Energy Converter (PWEC). A PWEC harvests wave energy from the heave motion of the buoy (Wen et al. 2018). The buoy is connected to the Power Take-Off system (PTO), which acts as a damping system during the conversion of the mechanical motion of the buoy into electricity (López et al. 2017). By constraining the buoy to only move in heave direction, the impact of changes in wave properties can be directly observed.

Optimizing the buoy shape increases the PWEC's response motion, which increase the extracted power (Aderinto and Li 2018). However, the literature does not provide a single universally optimized buoy shape; it changed from study to study with the wave characteristics, as shown in Figure 1. Shadman et al. (2018) used the Design Of Experiment (DOE) method to optimize the buoy geometry of a PWEC near the coast of Rio de Janeiro, where the predominant wave period ranges between 7 and 13 s with an average significant height of 1.33 m. Their study concluded that a cylindrical buoy shape with a diameter of $D = 13.5$ m and a length of $L = 3$ m was optimal. A multi-objective optimization study by Kurniawan and Moan (2012) concluded that the elliptical shape was the best energy harvesting buoy shape when examining a uniform distribution of wave frequencies from 0.4 to 1.3 rad/s. Goggins and Finnegan (2014) found a truncated cylinder with a hemisphere attached at its base achieved the optimal heave velocity response in irregular waves over a frequency range of 0 to 0.2 Hz and an amplitude range between 0 and 2.75m. Further, a uni-symmetric shape with a bulbous body and wings, presented by McCabe et al. (2010); McCabe (2013), was the optimized buoy shape in irregular waves 40 km west of the Shetland Islands, where the frequency range is $[0-4]$ rad/s and the amplitude range is $[0.75-6.75]$ m. Pastor and Liu (2014a) analysed a buoys with a hemispherical bottom and a conical bottom over frequencies ranging between 0.2626 and 2.5849 rad/s and amplitudes between 0.25 and 3.75m. They found that the cone-cylinder buoy performed slightly better than the hemispherical buoy. Khojasteh and Kamali (2016) also found that the conical buoy outperforms the spherical in wave flume experimental trials on a 1/15.9 scale, with frequencies between 0 and 0.4Hz and amplitudes of 8 and 10.2 cm. In addition, Hamada and Fürth (2021) compared different buoy shapes (cylindrical buoy, cylindrical with a hemisphere buoy, cylindrical with a 45° conical buoy, hemi-ellipsoidal buoy, and spheroid buoy) to investigate the power extraction efficiency of these shapes on the Power Take-Off (PTO). They found that a spheroid shape was effective at harvesting ocean energy, due to the large waterplane area and low length-to-diameter ratio in waves with a 0.625Hz frequency and a 0.2m wave height.

The first Variable Shape PWEC (VSPWEC) was proposed by Zou and Abdelkhalik (2020). This buoy can change its shape depending on the incident wave. This is mainly aimed at eliminating the reactive power needed in the PTO unit and sequentially increases the energy harvesting efficiency. Shabara et al. (2021); Shabara and Abdelkhalik (2022) further investigated the performance of the VSPWEC using high-fidelity simulations, where an 8% increase in harvested power compared to the Fixed Shape PWEC (FSPWEC) was achieved. With the use of active shape optimization, optimal control algorithms and excitation wave estimation (Zou et al. 2021; Abdelkhalik et al. 2017), the VSPWEC can outperform FSPWECs due to their wider optimal operation ranges under different sea conditions and less complex PTO units.

It is clear that the wave characteristics greatly influence the optimal buoy shape, summarized in Figure 1. Therefore, it is essential to determine the range of optimal operation for each common shape of FSPWECs,

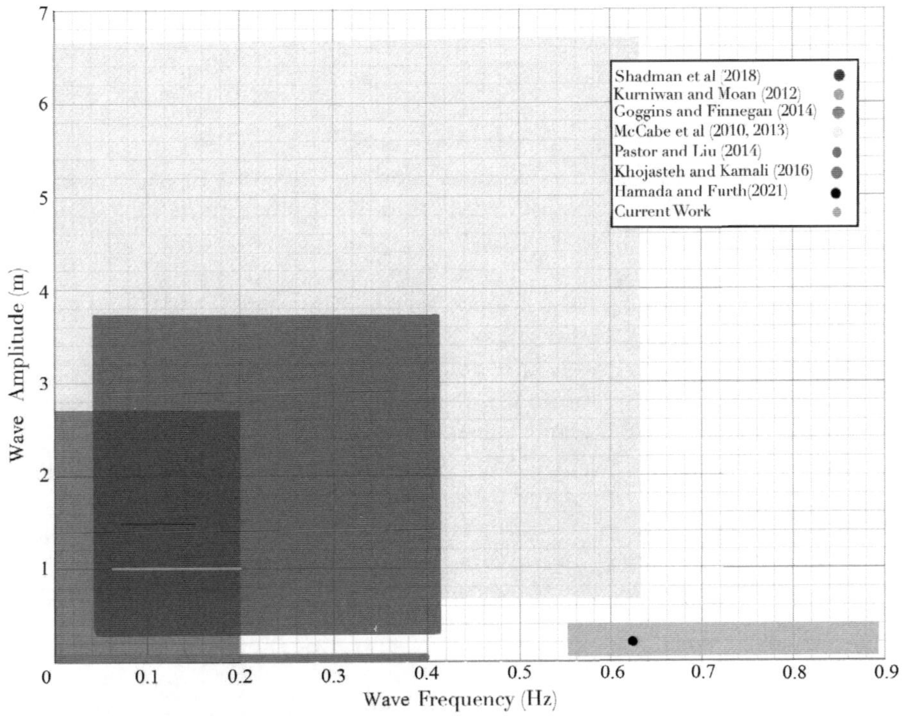

(a) Ranges of wave frequencies and wave heights tested in the aforementioned literature.

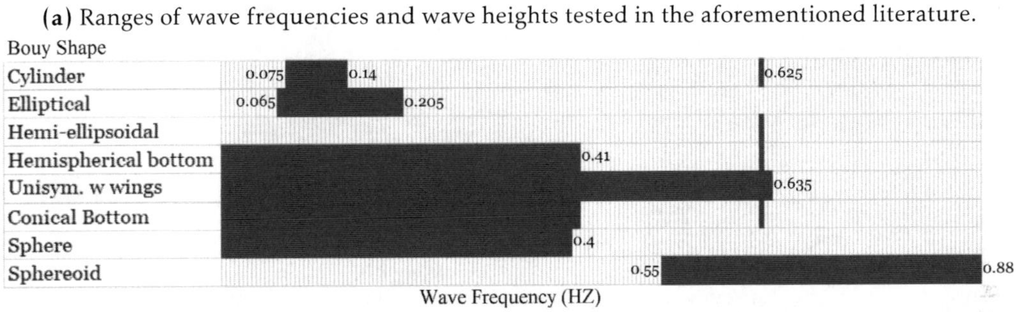

(b) Range of wave frequencies tested for different buoy shapes in the aforementioned literature.

Figure 1: Categorizing the literature according to the tested wave frequencies, wave heights, and buoy shapes.

starting with the spheroid buoy shape, as recommended by Hamada and Fürth (2021). The objective of this paper is to numerically determine the best operational conditions for energy harvesting using a spheroid buoy PWEC. This was achieved by studying the amplitude and frequency responses, and the power extraction efficiency of the buoy.

Computational Model

This section describes the computational model that was used to simulate the PWEC including the spheroid buoy shape and the PTO system in different sea states. In addition, the governing equations of the multi-phase flow, the equation of motion, and the equations of power extraction and efficiency were presented. The governing equations of the multi-phase flow were first used to solve the hydrostatic and hydrodynamic forces applied on the buoy due to the incoming waves. Then, the equation of motion was solved to determine the new heave position of the buoy, considering the effect of the PTO system. An iterative loop between these two sets of equations was marched with time. After this, the power model was implemented on the steady-state part of the results to obtain the efficiency of the spheroid buoy in the tested waves. In this section, the computational domain of the problem is presented briefly, including the numerical solver and the used grid. Further, more details about the convergent study were included in our previous work (Hamada and Fürth 2021).

Model Description

A spheroid buoy was partially submerged in a two-phase flow of water and air and was subjected to a Stokes wave (Stokes 1880). The buoy heaves along the z-direction due to the applied excitation force, F, on the buoy surface, from the wave motion. A low aspect ratio, L/D, of 0.5 was chosen for the buoy shape because the power extraction from the waves using a spheroid buoy shape increases as the aspect ratio decreases (Hamada and Fürth 2021). The buoy had a 1m diameter, a 0.5m length, and a mass of M_b, as shown in Figure 2. The centroid of the volume of the buoy was leveled with the undisturbed free surface of the water ($z = 0$), with the positive z-direction pointing upwards. The PTO system was modeled by a spring, K_{pto}, and a damper, C_{pto}, describing the buoy motion in the z-direction as a linear spring-mass-damper freely oscillating system. The PTO model's coefficients were selected to reflect typical values at the resonance frequency (Pastor and Liu 2014b). The properties of the fluids; water and air, were each kept constant. These values and the Stokes wave parameters are listed in Table 1. The wave flow was in the positive x-direction, and the wave velocity, height, and wavelength were varied for each case considering a deep water condition, as shown in Table 2. Furthermore, case 0 represents the case that was performed by Hamada and Fürth (2021) and was used as a reference case. The wave parameters (H_w, λ_w, and $\overline{U_1}$) were input parameters to

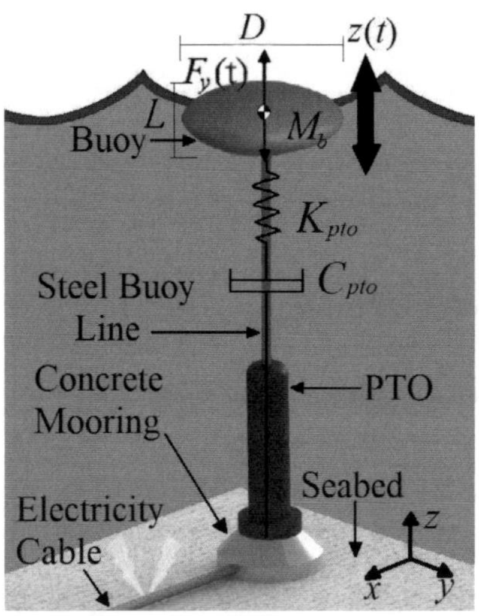

Figure 2: Schematic of the structure of the PWEC with a spheroid buoy shape. The buoy oscillates vertically, and the PTO is modeled with a spring and damper.

Table 1: System and fluid properties

Length to diameter ratio, L/D	0.5
Buoy diameter, D_b	$1\ m$
Buoy mass, M_b	$220\ kg$
Spring stiffness, K_{pto}	$1.6 \times 10^6\ N/m$
Damping coefficient, C_{pto}	$3000\ Ns/m$
Mean wave velocity, $\overline{U_1}$	$[0.03 - 0.15]\ m/s$
Wave height, H_w	$[0.05 - 0.8]m$
Wave length, λ_w	$[2 - 6]m$
Wave frequency, f_w	$[0.88 - 0.51]Hz$
Simulation final time, t_f	$20\ s$
Steady-state time range, t_{ss}	$12.5 - 20\ s$
Water Density, ρ_w	$1000\ kg/m^3$
Air Density, ρ_a	$1.2\ kg/m^3$
Water kinematic viscosity, ν_w	$1 \times 10^{-6}\ m^2/s$
Air kinematic viscosity, ν_a	$1.48 \times 10^{-5}\ m^2/s$

Table 2: Chosen wave characteristics for the simulated cases.

	Reference Case	Varied wave height			Varied wave length			Varied wave speed			
Case#	0	1	2	3	4	5	6	7	8	9	10
H_w (m)	0.2	0.05	0.1	0.4	0.2	0.2	0.2	0.2	0.2	0.2	0.2
λ_w (m)	4	4	4	4	3	5	6	4	4	4	4
$\overline{U_1}$ (m/s)	0.03	0.03	0.03	0.03	0.03	0.03	0.03	0.06	0.09	0.12	0.15

OpenFOAM. Further, they were chosen to reflect sea states between 1 and 3 on the Beaufort sea scale with no wind dependence (Beaufort 1805).

Governing Equations

Here, the *interDyMFoam* solver, implemented in the OpenFOAM® CFD package was used. *InterDyMFoam* is an extension of *interFoam*, which calculates the flow field for two incompressible fluids using a Volume of Fluid (VoF) approach. *InterDyMFoam* makes use of the optional mesh refinement- here, the mesh motion is refined close to the buoy. *InterDyMFoam* solves the Unsteady Reynolds-Averaged Navier-Stokes (URANS) equations (Ferziger et al. 2002), which are composed of the averaged continuity and momentum equations, and are expressed as:

$$\frac{\partial \overline{U_i}}{\partial x_i} = 0 \tag{1a}$$

$$\frac{\partial \overline{U_i}}{\partial t} + \overline{U_j}\frac{\partial \overline{U_i}}{\partial x_j} + \frac{\overline{u_i u_j}}{\partial x_j} = -\frac{\partial \overline{p}}{\partial x_i} + \nu \frac{\partial^2 \overline{U_i}}{\partial x_j \partial x_j} \tag{1b}$$

where $\overline{U_i}$ represents the mean flow velocity components in three directions; x, y, and z, u_i is the fluctuation flow velocity components, $\overline{p}$ is the mean kinematic pressure of the fluid flow, ν is the kinematic viscosity of the fluid, and $-\overline{u_i u_j}$ is the kinematic turbulent Reynolds-stress tensor. Ensuring the solenoidal velocity field is achieved by retrieving the coupling between the continuity and the momentum equations with the Poisson equation, which its averaged equation is expressed as:

$$\frac{\partial^2 \overline{p}}{\partial x_i \partial x_i} = -\frac{\partial \overline{U_i}}{\partial x_j}\frac{\partial \overline{U_j}}{\partial x_i} - \frac{\partial^2 \overline{u_i u_j}}{\partial x_i \partial x_j} \tag{1c}$$

The $k_t - \epsilon_t$ RANS equations are used to model the turbulence in the flow field, and it is depending on the standard model (Launder and Spalding 1983) and the rapid distortion theory compression term (El Tahry 1983). The governing equations of the $k_t - \epsilon_t$ RANS model are expressed as:

$$\frac{\partial k_t}{\partial t} + \overline{U_i}\frac{\partial k_t}{\partial x_j} = -\overline{u_i u_j}\frac{\partial \overline{U_i}}{\partial x_j} - \epsilon + \frac{\partial}{\partial x_j}\left[\left(\nu + \frac{\nu_t}{\sigma_k}\right)\frac{\partial k_t}{\partial x_j}\right] \tag{2a}$$

$$\frac{\partial \epsilon_t}{\partial t} + \overline{U_j}\frac{\partial \epsilon_t}{\partial x_j} = -C_{\epsilon_1}\frac{\epsilon_t}{k_t}\overline{u_i u_j}\frac{\partial \overline{U_i}}{\partial x_j} - C_{\epsilon_2}\frac{\epsilon_t^2}{k_t} + \frac{\partial}{\partial x_j}\left[\left(\nu + \frac{\nu_t}{\sigma_\epsilon}\right)\frac{\partial k_t}{\partial x_j}\right] \tag{2b}$$

where k_t and ϵ_t are the turbulent kinetic energy and the turbulent dissipation rate, respectively. The values of the closure coefficients of the standard $k - \epsilon$ model, C_{ϵ_1}, C_{ϵ_2}, σ_k, and σ_ϵ, are 1.44, 1.92, 1, and 1.3, respectively.

Furthermore, the eddy viscosity, ν_t, and the specific turbulence dissipation rate, ω_t, are defined in terms of the turbulent kinetic energy, k_t, and the turbulence energy dissipation rate, ϵ_t, as:

$$\nu_t = C_\mu \frac{k_t^2}{\epsilon_t} \tag{3a}$$

$$\omega_t = \frac{\epsilon_t}{C_\mu k_t} \tag{3b}$$

where C_μ is the turbulence model constant, and equals 0.09.

Equation Of Motion

A second-order linear ordinary differential equation of time t models the mass-spring-damper system that represents the heave motion of PWEC, and is expressed as:

$$M_b \ddot{z}(t) + C_{pto}\dot{z}(t) + K_{pto}z(t) = F_z(t) - M_b g \tag{4}$$

Where M_b is the mass of the buoy, C_{pto} and K_{pto} are the mechanical damping coefficient and is the elastic stiffness of the PTO. The values of C_{pto} and K_{pto} were kept constant during the simulation, and they were not the optimized values, because the aim of this research was to examine the effect of changing the wave characteristics on the power efficiency of spheroid PWEC. In addition, $F_z(t)$ represents the total hydrodynamic and hydrostatic forces applied on the buoy in the z-direction as a function of time.

A resonance with a natural frequency, f_n, is obtained when an equilibrium between the inertial force, $M_b\ddot{z}(t)$, and the restoring forces, $F_z(t) - M_b g$, happens. Thus, the non-dimensional parameters, the mass ratio, m^*, the natural frequency, f_n, and damping ratio, ζ, are expressed as:

$$m^* = \frac{M_b}{M_d} \tag{5a}$$

$$f_n = \frac{1}{2\pi}\sqrt{\frac{K_{pto} + K_r}{M_b + M_a}} \tag{5b}$$

$$\zeta = \frac{C_{pto}}{2\sqrt{(K_{pto} + K_r)(M_b + M_a)}} \tag{5c}$$

where M_d is the displaced mass, and M_a is the added mass (Sarpkaya 2004). Further, K_r is the restoring spring and is simplified to be $\rho_{water} g A_w$, where A_w is the waterplane area of the buoy (Zhang et al. 2020).

Power And Efficiency Model

The performance of the tested spheroid PWEC in harvesting the energy from the incoming waves was measured using two indicators: the extracted power efficiency, η_P =, and the height efficiency, η_H, (Budar and Falnes 1975). The first one was calculated as the ratio of the average wave energy conversion power, P_c, to the incident wave power, P_w, for a deep-water condition. While the second one was defined as the ratio of the average height of the buoy motion, $\overline{H_b}$, to the incident wave height, H_w.

$$P_c = \frac{C_{pto}}{nT}\int_t^{t+nT} V^2 dt \tag{6a}$$

$$P_w = \frac{\rho_{water} g H_w^2 \omega_w D}{16 n_w} \tag{6b}$$

$$\eta_P = \frac{P_c}{P_w} \tag{6c}$$

$$\eta_H = \frac{\overline{H_b}}{H_w} \tag{6d}$$

where H_w is the incident wave height, T is the wave period, ω_w is the wave frequency in rad/sec ($\omega = 2\pi f_w$ where f_w is the wave frequency in Hertz), n_w is the wave-number, n is the number of the buoy motion period, and V is the buoy velocity.

Computational Domain

The *interDymFOAM* simulates the multi-phase flow using the PIMPLE algorithm, which combines the PISO (Pressure Implicit with Splitting of Operators) and SIMPLE (SemiImplicit Method for pressure-Linked Equations) algorithms Jasak (1996). The buoy is placed at $(x_b, y_b, z_b) = (5, 0, 0)$ in a rectangle computation domain of size $27m \times 10m \times 6m$ in x, y, z, respectively. The inlet wave travels along the x-axis with a given mean velocity,. Furthermore, the wave is dampened $12m$ after the buoy to ensure a smooth exit and negligible reflection. The boundary and initial conditions are shown in Figure 3. A three-dimensional grid was implemented to discretize the domain, with the mesh being finer close to the buoy to improve the accuracy at which the turbulence is captured. The mesh independent analysis, performed by Hamada and Fürth (2021), resulted in the selection of a 11 millions cell mesh. The characteristics of the selected mesh are shown in Table 3.

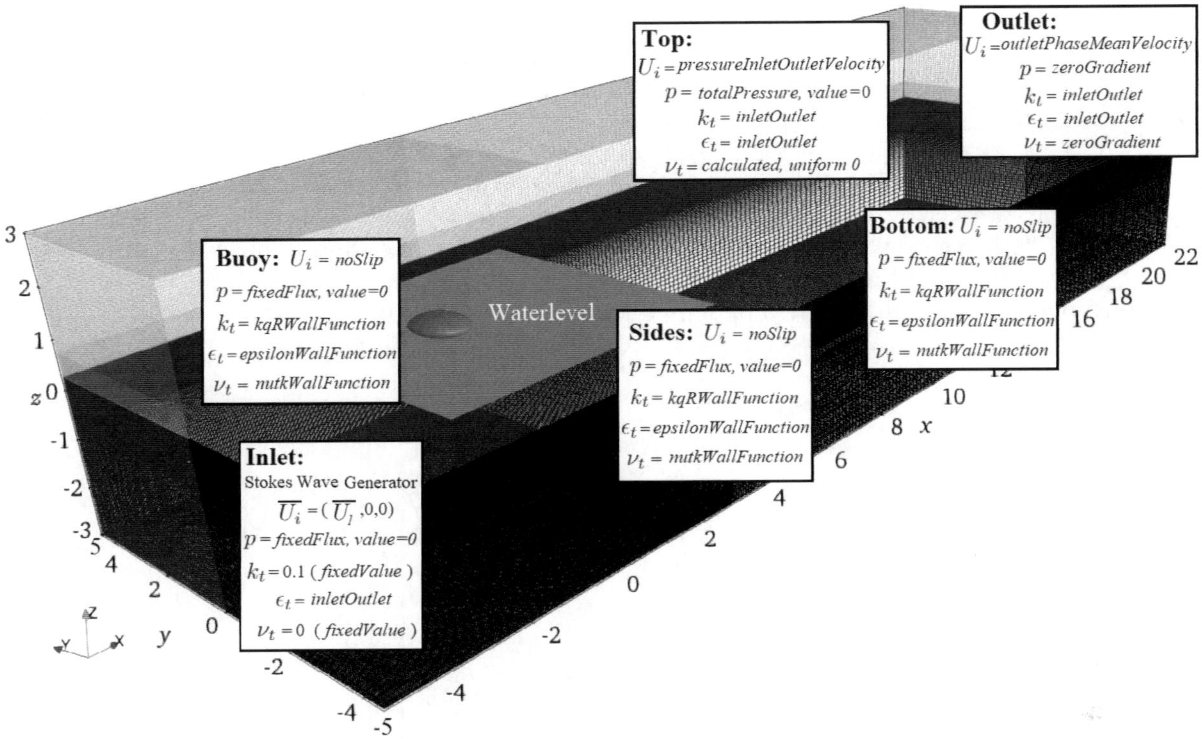

Figure 3: The computational domain, showing the dimension of the domain in terms of the buoy's diameter. The boundary and initial conditions are shown. The grid is clustered in the vicinity of the buoy to capture the turbulence.

Table 3: Characteristics of the selected mesh (Hamada and Fürth 2021).

No. cells	11,026,125
Computational domain (L_x, L_y, L_z)	(27, 10, 6) m
Cell size adjacent to buoy (Δx, Δy, Δz)	(0.024, 0.047, 0.015) m
Dimensionless first wall distance (x_1^+, y_1^+, z_1^+)	(43, 85.6, 27.6) in water & (3.8, 7.6, 2.4) in air
Time step (Δt)	Adjustable
Maximum time step (Δt_{max})	0.1 sec
Maximum Courant–Friedrichs–Lewy (CFL_{max})	1.5

Results And Discussion

The dependence of the hydrodynamic responses (amplitude and frequency ratios) and the power extraction from a single spheroid buoy on the wave characteristics (wave height, wave length, and wave speed) has been investigated. Furthermore, the change in the wave characteristics causes variation in the strength of the applied forces on the submerged part of the buoy. The simulation of each case lasted for 20 seconds to ensure the steady-state solution. After, the last 7.5 seconds of each simulation were post-processed.

Wave Height Dependency on the Extracted Power Efficiency

The steady-state displacement history, z_{cg}/D, for various wave heights of a spheroid buoy is shown in Figure 4 left. It can be observed that the amplitude, A_b, increases when the wave height, H_w, increases, as expected. However, the rate of increase of A_b is less than the rate of increase of H_w. This is reflected in the decrease of height efficiency, η_H of the buoy with the increase of H_w, as shown in Figure 5. Moreover, the frequency ratio, f_b/f_w shows that the resonance phenomenon occurs very close to 1 for smaller wave heights (0.05, 0.1, and 0.2). For the larger wave height, case 3 is very large compared to buoy size with $H_w/D = 0.4$ and $H_w/L = 0.8$, the f_b/f_w decreases and considerably deviates from the resonance frequency, reaching a value of 0.975, as shown in Figure 4 right. The energy harvesting efficiency, η_P, is relatively high when the buoy is exposed to a low H_w. Furthermore, η_P decreases with the increase of H_w, as shown in Figure 5. When the other wave characteristics (wave length, wave frequency, and wave speed) are fixed, the slope of the wave, H_w/L steepens with increasing H_w, causing an impulse-like wave to meet the spheroid buoy shape. This delays the response of the buoy which decreases A_b, pushes f_b/f_w away from 1, and results in sequentially lower η_P. In addition, the gain in η_P is around double when the H_w is decreased by a factor of 4 from 0.2 m.

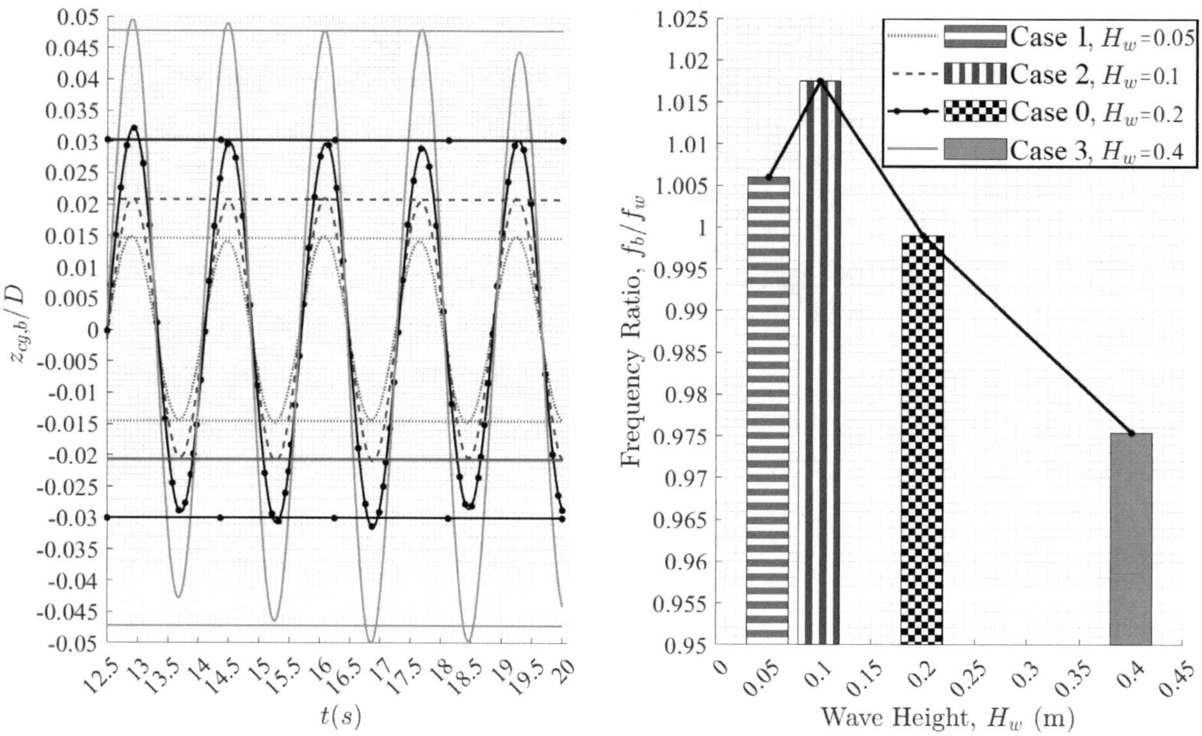

Figure 4: The steady-state displacement history, $z_{cg,b}/D$, and the frequency ratio, f_b/f_w, for PWEC with a spheroid buoy shape with $M_b = 200kg$, $K_{pto} = 1.6 \times 10^6 N/m$, $C_{pto} = 3000 Ns/m$, and different wave heights, $H_w = 0.05, 0.1, 0.2, 0.4$ m, of the Stokes-II wave.

Wave Length Dependency on the Extracted Power Efficiency

The steady-state displacement history, z_{cg}/D, of the buoy increases with increasing wave length, λ_w, as shown in Figure 6 left. Furthermore, increasing λ_w, while fixing the wave height and wave speed at 0.2 m and 0.03 m/s respectively, enables the buoy to trace the wave slope and achieve a higher efficiency height, η_H, as shown in Figure 7. In addition, the variation of the wave number scarcely affects the resonance phenomenon as the frequency ratio, f_b/f_w, is very close to 1 in the range of wave lengths $[3-6]$, as shown in Figure 6 right. Increasing the wave length while fixing the other wave parameters spatially elongates the incoming wave and also increases its wave period, T, as indicated with the dispersion wave equation for deep water. This gives the buoy, which has a relatively large waterplane area, sufficient time to respond to the incoming wave, so that it can follow the wave motion. Thus, the extraction power efficiency, η_P, of the spheroid buoy increases with increasing wave length, as shown in Figure 7. Furthermore, the gain in η_P decreases when λ_w is changed from 3 to 6. Thus, operating the spheroid buoy in waves with $\lambda_w = [4-6]$ provides relatively high η_P around 0.4, compared with low values of λ_w.

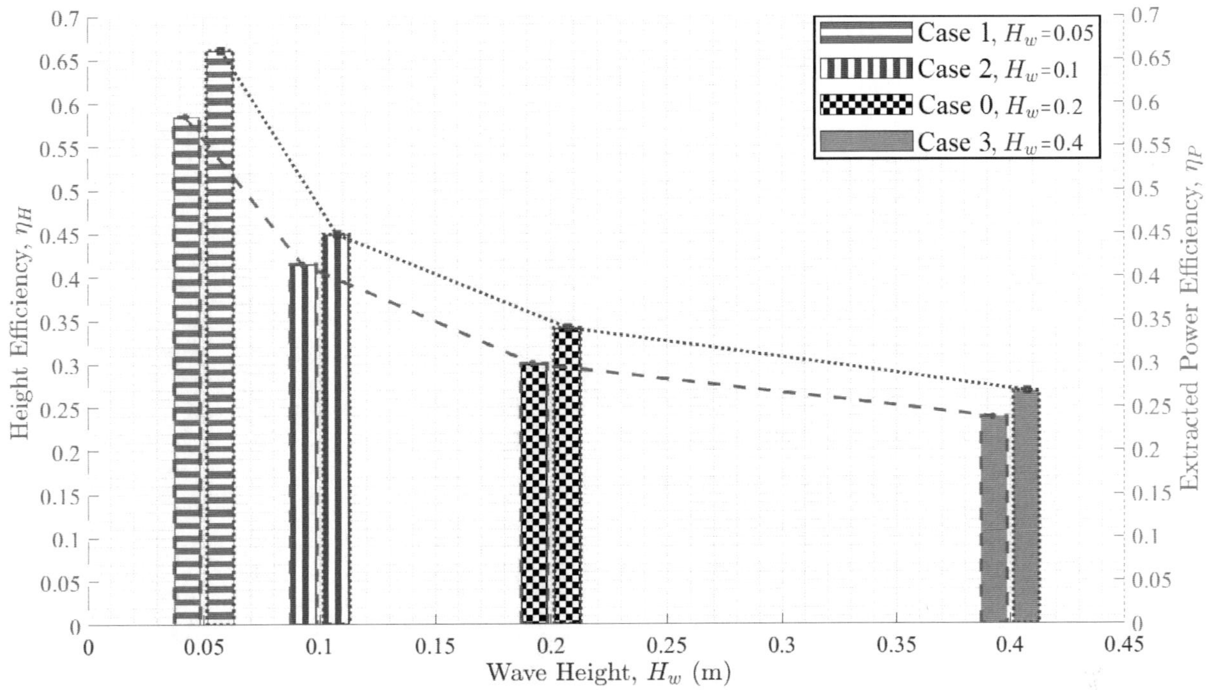

Figure 5: The height efficiency, η_H, and the extracted power efficiency, η_P, for PWEC with a spheroid buoy shape with $M_b = 200kg$, $K_{pto} = 1.6 \times 10^6 N/m$, $C_{pto} = 3000Ns/m$, and different wave heights = 0.05, 0.1, 0.2, 0.4 m, H_w, of the Stokes-II wave.

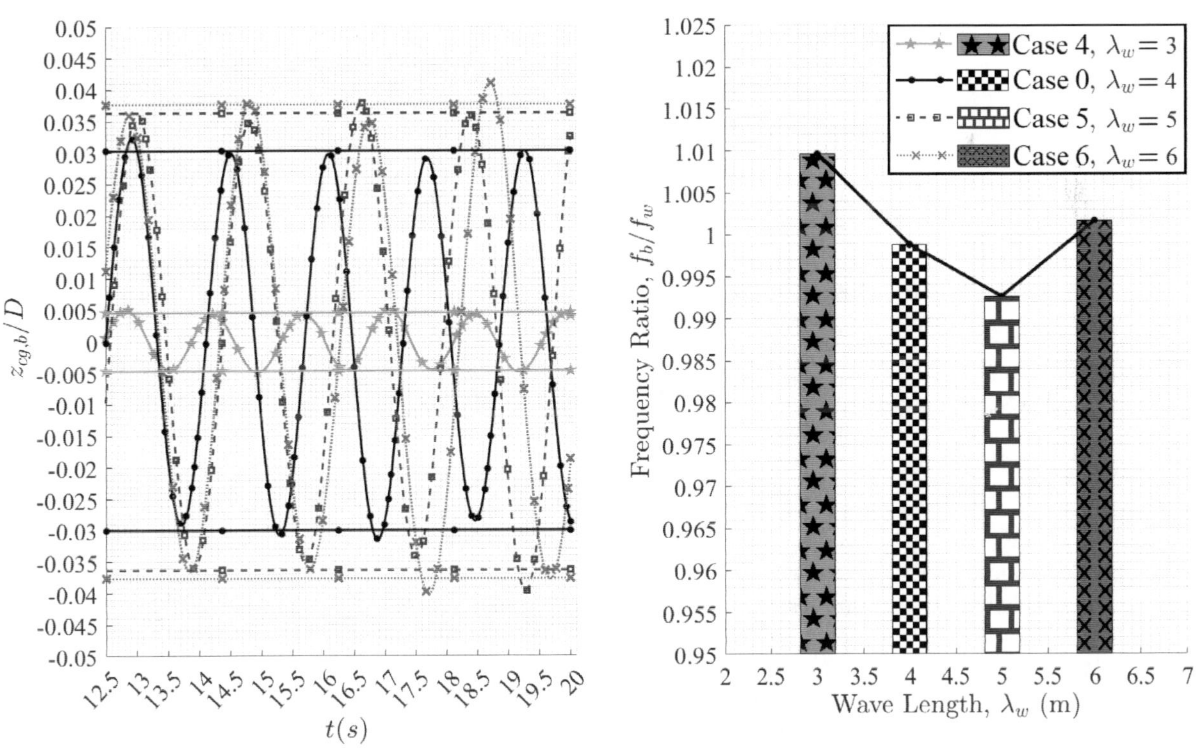

Figure 6: The steady-state displacement history, $z_{cg,b}/D$, and the frequency ratio, f_b/f_w, for PWEC with a spheroid buoy shape with $M_b = 200kg$, $K_{pto} = 1.6 \times 10^6 N/m$, $C_{pto} = 3000Ns/m$, and different wave lengths, $\lambda_w = 3, 4, 5, 6$ m, of the Stokes-II wave.

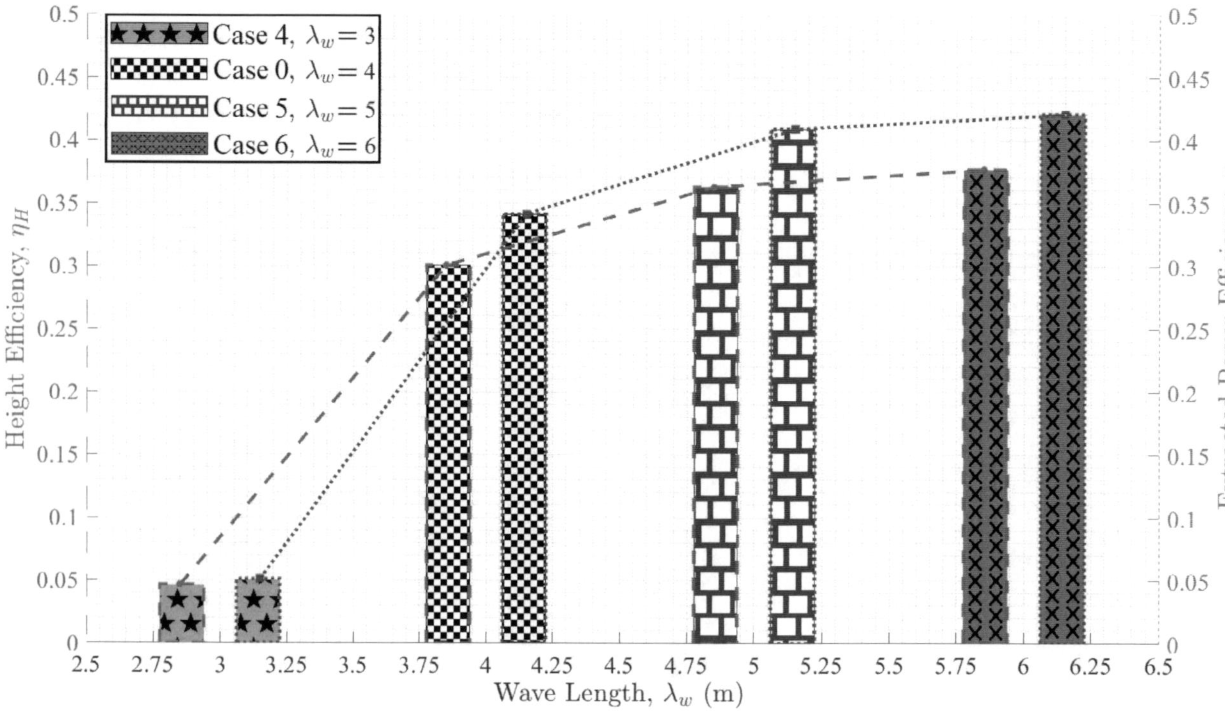

Figure 7: The height efficiency, η_H, and the extracted power efficiency, η_P, for PWEC with a spheroid buoy shape with $M_b = 200kg$, $K_{pto} = 1.6 \times 10^6 N/m$, $C_{pto} = 3000Ns/m$, and different wave lengths = 3, 4, 5, 6 m, λ_w, of the Stokes-II wave.

Wave Speed Dependency on the Extracted Power Efficiency

The steady state displacement, z_{cg}/D, of the buoy increases with an increase in wave speed, $\overline{U_1}$, as shown in Figure 8 left. The wave height, H_w, and wave length, λ_w, were fixed at 0.2m and 4m, respectively. The increase in $\overline{U_1}$ lends momentum to the buoy resulting in a slight increase of η_H as shown in Figure 8 left. The variation of $\overline{U_1}$ also impacts the resonance; when $\overline{U_1}$ is in the range of $[0.03 - 0.07]$ m/s, the frequency ratio remains very close to 1 as shown in Figure 8 right. Then, the frequency ratio continues to increase with further increases in $\overline{U_1}$, leaving the resonance/lock-in range. A spheroid buoy with a high waterplane area will respond quickly to an incoming wave due to its relatively stiff motion geometry (Budar and Falnes 1975). This causes the buoy's extraction efficiency η_P to increase as wave speed increases. The buoy's η_P appears to reach a maximum at a $\overline{U_1}$ of around 0.12 m/s and then begin to decrease gradually, as shown in Figure 9. The same is true of the height efficiency, η_H.

Conclusion

The purpose of this research was to numerically investigate the effect of wave characteristics on energy harvesting from a spheroid Point Wave Energy Converter (PWEC). The selection of the buoy's geometry was determined based on the conclusion of previous work by Hamada and Fürth (2021). They found that the power extraction efficiency of a spheroid buoy increases with the decrease of the geometrical ratio, L/D. The current study was executed by changing each of the following wave characteristics individually to study their direct effect: $H_w = [0.05 - 0.4]$, $\lambda_w = [3 - 6]$, and $\overline{U_1} = [0.03 - 0.15]$. A dynamic two-phase flow solver existing in OpenFOAM was used to conduct three-dimensional turbulent numerical simulations for each case. Finally, the buoy's displacement history, amplitude ratio, frequency ratio, and extracted power efficiency were studied to conclude the following:

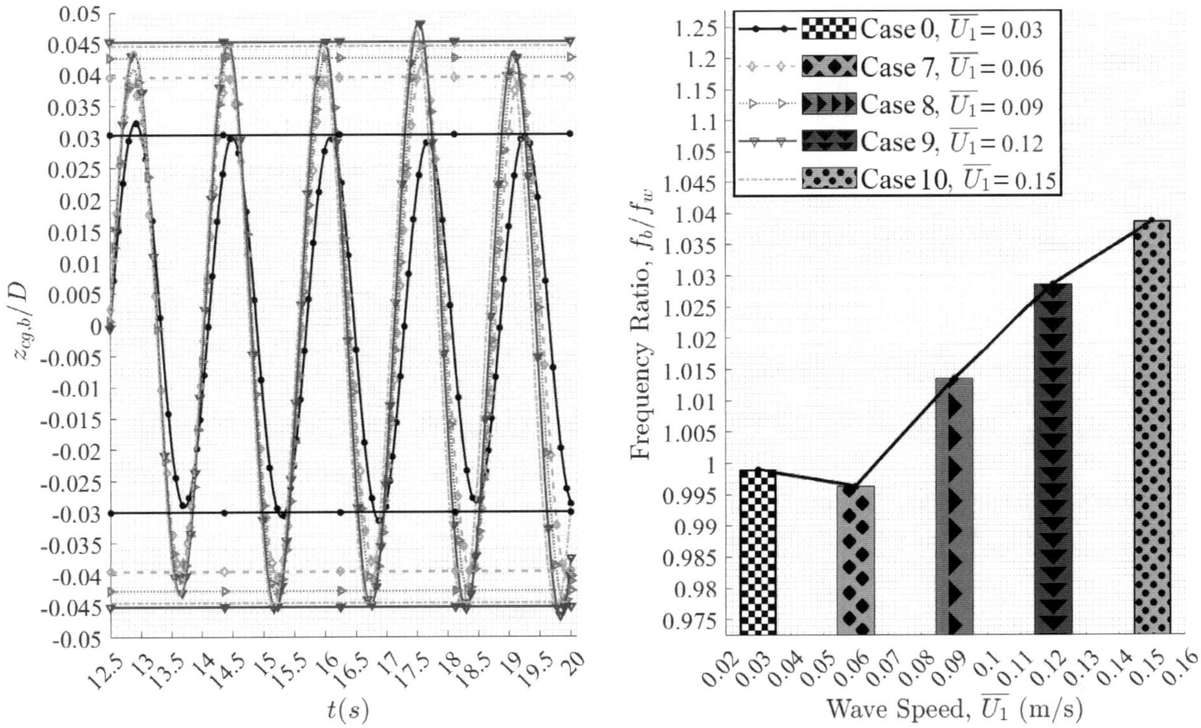

Figure 8: The steady-state displacement history, $z_{cg,b}/D$, and the frequency ratio, f_b/f_w, for PWEC with a spheroid buoy shape with $M_b = 200kg$, $K_{pto} = 1.6 \times 10^6 N/m$, $C_{pto} = 3000Ns/m$, and different wave speeds, $\overline{U_1}$ = 0.03, 0.06, 0.09, 0.12, 0.15 m, of the Stokes-II wave.

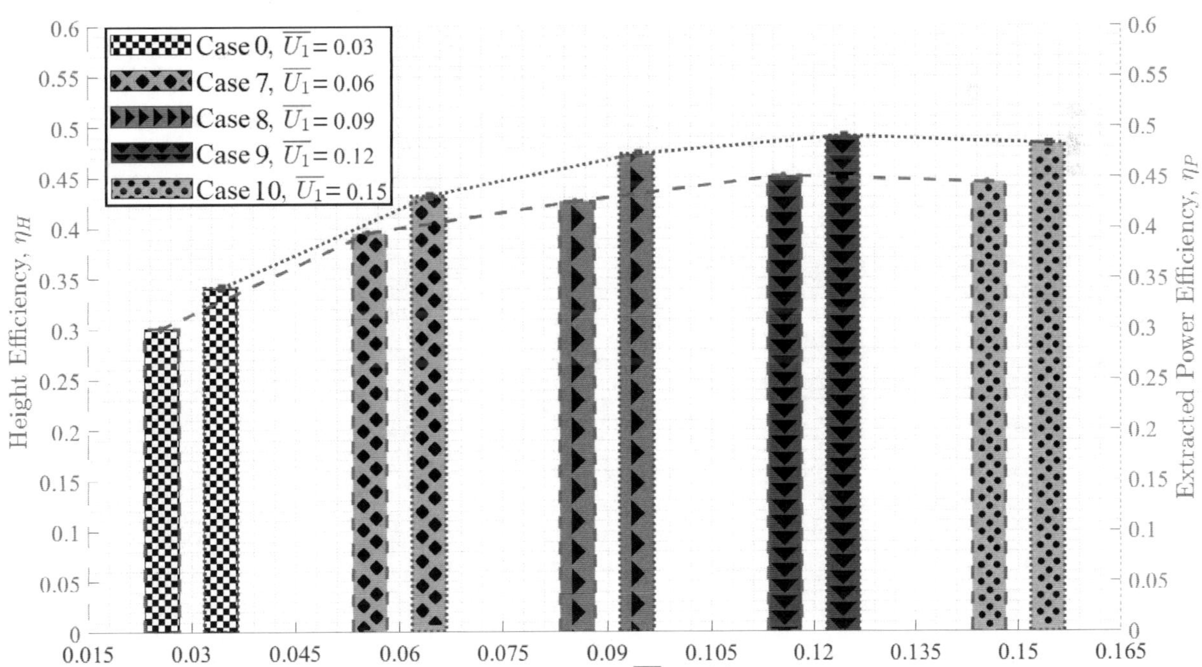

Figure 9: The height efficiency, η_H, and the extracted power efficiency, η_P, for PWEC with a spheroid buoy shape with $M_b = 200kg$, $K_{pto} = 1.6 \times 10^6 N/m$, $C_{pto} = 3000Ns/m$, and different wave speeds = 0.03, 0.06, 0.09, 0.12, 0.15 m, $\overline{U_1}$, of the Stokes-II wave.

- Hamada and Fürth (2021) concluded that a spheroid buoy with low aspect ratio, L/D, had a higher efficiency than the other four evaluated shapes, however the variation in sea state significantly affects its power extraction efficiency. Therefore, to accurately determine the performance of different buoy shapes, they should be evaluated in a range of sea states.
- Increasing both the wave length, λ_w, and the wave speed, $\overline{U_1}$, while decreasing the wave height, H_w, improves the spheroid buoy's energy harvesting efficiency, reaching the highest η_H when operated in waves with $\lambda_w > 4$ m, $\overline{U_1} \in [0.07 - 0.12]$ m/s, and $H_w < 0.15$ m.

In the future, other buoy shapes, such as cylindrical and hemi-ellipsoid, should be tested in a wide range of wave characteristics to recognize the optimal sea states for each buoy shape. This will provide insights when designing a Variable Shape PWEC.

Acknowledgements

The authors would like to acknowledge the Texas A&M University Undergraduate Summer Research Grant program for funding the author Abigail Rolen during the summer of 2021. This made Abigail's contributions to this research possible.

Nomenclature

$-\overline{u_i u_j}$	Kinematic Reynolds-stress tensor.
$\overline{H_b}$	Average height of the buoy motion.
$\overline{p}$	Mean kinematic pressure.
A_w	Wetted surface of the buoy.
C_{pto}	Damping coefficient of PTO.
D	Diameter of buoy.
f_n	Natural frequency.
F_z	Applied force on the buoy in the z-direction.
H_w	Incident wave height.
I_t	Turbulence intensity.
K_r	Restoring force.
k_t	Turbulence kinetic energy.
K_{pto}	Spring stiffness of PTO.
L	Length of buoy.
l_t	Characteristic length scale.
m^*	Mass ratio.
M_A	Added mass.
M_b	Mass of buoy.
M_d	Displaced mass.
N_{cell}	Number of cells in the mesh.
P_c	Average wave energy conversion power.
P_w	Incident wave power.
U_i	Mean flow velocity vector.
u_i	Fluctuation flow velocity vector.
z_{cg}	Center of gravity point.
z_{fp}	Flotation point.
ϵ_t	Turbulence dissipation rate.
η_H	Height efficiency.
η_P	Extracted power efficiency.
ν	Kinematic viscosity.
ν_t	Eddy viscosity.
ω_t	Specific turbulence dissipation rate.
ζ	Damping ratio.
PTO	Power Take Off system.
PWEC	Point Wave Energy Converter.

References

Ossama Abdelkhalik, Shangyan Zou, R Robinett, Giorgio Bacelli, and D Wilson. Estimation of excitation forces for wave energy converters control using pressure measurements. *International Journal of Control*, 90(8):1793–1805, 2017.

Tunde Aderinto and Hua Li. Ocean wave energy converters: Status and challenges. *Energies*, 11(5):1250, 2018.

The World Bank. What is the blue economy? https://www.worldbank.org/en/news/infographic/2017/06/06/blue-economy, 2017. Accessed: 2022-01-10.

Sir Francis Beaufort. Beaufort wind scale. https://www.spc.noaa.gov/faq/tornado/beaufort.html, 1805. Accessed: 2022-01-13.

K Budar and J Falnes. A resonant point absorber of ocean-wave power. *Nature*, 256(5517):478–479, 1975.

SH El Tahry. K-epsilon equation for compressible reciprocating engine flows. *Journal of Energy*, 7:345–353, 1983.

Joel H Ferziger, Milovan Perić, and Robert L Street. *Computational methods for fluid dynamics*, volume 3. Springer, 2002.

Jamie Goggins and William Finnegan. Shape optimisation of floating wave energy converters for a specified wave energy spectrum. *Renewable Energy*, 71:208–220, 2014.

GC Government of Canada. Blue economy strategy. https://www.dfo-mpo.gc.ca/campaign-campagne/bes-seb/index-eng.html, 2021. Accessed: 2022-01-10.

Ahmed A. Hamada and Mirjam Fürth. Numerical simulation of the effect of buoy geometries on pto of wave energy converters. In *SNAME Maritime Convention*. OnePetro, 2021.

Hrvoje Jasak. *Error analysis and estimation for the finite volume method with applications to fluid flows.* PhD thesis, Imperial College London (University of London), 1996.

Danial Khojasteh and Reza Kamali. Evaluation of wave energy absorption by heaving point absorbers at various hot spots in iran seas. *Energy*, 109:629–640, 2016.

Adi Kurniawan and Torgeir Moan. Optimal geometries for wave absorbers oscillating about a fixed axis. *IEEE Journal of Oceanic Engineering*, 38(1):117–130, 2012.

Brian Edward Launder and Dudley Brian Spalding. The numerical computation of turbulent flows. In *Numerical prediction of flow, heat transfer, turbulence and combustion*, pages 96–116. Elsevier, 1983.

M López, F Taveira-Pinto, and P Rosa-Santos. Influence of the power take-off characteristics on the performance of ceco wave energy converter. *Energy*, 120:686–697, 2017.

AP McCabe. Constrained optimization of the shape of a wave energy collector by genetic algorithm. *Renewable energy*, 51:274–284, 2013.

AP McCabe, GA Aggidis, and MB Widden. Optimizing the shape of a surge-and-pitch wave energy collector using a genetic algorithm. *Renewable Energy*, 35(12):2767–2775, 2010.

NOAA. Noaa finalizes strategy to enhance growth of american blue economy. https://www.noaa.gov/stories/noaa-finalizes-strategy-to-enhance-growth-of-american-blue-economy, 2021. Accessed: 2022-01-14.

PNNL Pacific Northwest National Laboratory. The blue economy advancing the power of the ocean. https://www.pnnl.gov/blue-economy, 2022. Accessed: 2022-01-10.

Jeremiah Pastor and Yucheng Liu. Power absorption modeling and optimization of a point absorbing wave energy converter using numerical method. *Journal of Energy Resources Technology*, 136(2), 2014a.

Jeremiah Pastor and Yucheng Liu. Time domain modeling and power output for a heaving point absorber wave energy converter. In *ASME International Mechanical Engineering Congress and Exposition*, volume 46514, page V06AT07A042. American Society of Mechanical Engineers, 2014b.

Turgut Sarpkaya. A critical review of the intrinsic nature of vortex-induced vibrations. *Journal of fluids and structures*, 19(4):389–447, 2004.

Mohamed A. Shabara and Ossama Abdelkhalik. Dynamic modeling of spherical variable-shape wave energy converters, 2022.

Mohamed A Shabara, Shangyan Zou, and Ossama Abdelkhalik. Numerical investigation of a variable-shape buoy wave energy converter. In *International Conference on Offshore Mechanics and Arctic Engineering*, volume 85192, page V009T09A013. American Society of Mechanical Engineers, 2021.

Milad Shadman, Segen F Estefen, Claudio A Rodriguez, and Izabel CM Nogueira. A geometrical optimization method applied to a heaving point absorber wave energy converter. *Renewable energy*, 115: 533–546, 2018.

George Gabriel Stokes. On the theory of oscillatory waves. *Transactions of the Cambridge philosophical society*, 1880.

Yadong Wen, Weijun Wang, Hua Liu, Longbo Mao, Hongju Mi, Wenqiang Wang, and Guoping Zhang. A shape optimization method of a specified point absorber wave energy converter for the south china sea. *Energies*, 11(10):2645, 2018.

Hengming Zhang, Binzhen Zhou, Christopher Vogel, Richard Willden, Jun Zang, and Liang Zhang. Hydrodynamic performance of a floating breakwater as an oscillating-buoy type wave energy converter. *Applied Energy*, 257:113996, 2020.

Shangyan Zou and Ossama Abdelkhalik. Modeling of a variable-geometry wave energy converter. *IEEE Journal of Oceanic Engineering*, 2020.

Shangyan Zou, Ossama Abdelkhalik, and Mohamed Shabara. High-fidelity numerical study of in-line excitation force estimation for wave energy converters. In *The 14th European Wave and Tidal Energy Conference*, 2021.

SNAME
THE INTERNATIONAL COMMUNITY
FOR MARITIME AND OCEAN PROFESSIONALS

PREDICTION OF VESSEL RAOs:
Applications of Deep Learning to Assist in Design

JAMES FRIZZELL
TEXAS A&M UNIVERSITY

DR. MIRJAM FURTH
TEXAS A&M UNIVERSITY

Proceedings of the 27th Offshore Symposium, February 22nd, 2022, Houston, Texas
Texas Section of the Society of Naval Architects and Marine Engineers

ABSTRACT

In an age of high-paced design, a need arises for engineers to quickly estimate the feasibility of their ideas without spending weeks developing a computer model. At the same time, the use of machine learning models, or neural networks, in the maritime industry has grown substantially over the past years. By further extending the use of these predictive models in the design phase, marine engineers and naval architects can expedite their work.

This paper focuses on the creation of a neural network that can estimate the Response Amplitude Operators (RAOs) of a vessel given its characteristic properties such as length, beam, and draft. A dataset was collected through a parametric design analysis of box barges using ANSYS AQWA, and the RAO was simulated for all 6 degrees of freedom. A critically damped spring equation was generated for each degree. A Keras Neural Network Model was trained on the three parameters and the wave heading angle, with the hidden layers and neuron count being adjusted to optimize the loss and maximize R-squared.

To validate the results, a series of box barges with dimensions that were not a part of the training dataset were simulated in ANSYS, while the virtual model with identical characteristics was simulated with the Neural Network. The resulting RAOs were compared to validify the accuracy of the Neural Network.

With this predictive model, engineers can quickly determine a hullform's RAOs, and compare the response with the common sea states along the intended route. Additionally, the model can assist in design iteration. As the hull shape gradually changes, the new RAOs can be estimated to ensure that the design is progressing in an appropriate direction.

Keywords: SNAME, Offshore Symposium, Machine Learning, Deep Neural Network, RAO

INTRODUCTION

Finding the Response Amplitude Operators (RAOs) of a vessel is a numerically expensive process. Industry standard programs such as ANSYS AQWA use complex procedures to solve Equation 1 and return a precise solution.

$$RAO(\omega) = \frac{F_0}{C-(M+A(\omega))*\omega^2+iB(\omega)*\omega} \tag{1}$$

This methodology has benefits and drawbacks. While it is a highly scalable process and can be applied to an object of any shape, the computational time is long. Additionally, the quality of results is directly coupled to the quality of the input geometric model. In an iterative design environment where an engineering is constantly adjusting a hullform to suit a client's needs, the corresponding digital model must be updated as well in order to maintain the integrity of the numerical results. This method, although highly accurate when performed correctly, is time-consuming and by extension, costly.

To eliminate the need for complex mathematical equations, Neural Networks (NN) can be implemented to derive the correlation between inputs and outputs. In this case, a Deep Neural Network (DNN) has been created to predict the RAOs of a vessel with respect to frequency given only the characteristic dimensions of length, beam, draft, and the wave heading angle. With this approach, the calculation of approximate RAOs only requires a few seconds of computational time.

Simply inserting a neural network into a complex problem is never an ex-machina solution, though. To obtain accurate results, a comprehensive dataset must be collected during development to train the NN on. Hundreds or thousands of vessels must have their RAOs solved in the conventional method to provide learning material for the predictive model. Additionally, the model is only able to accurately predict vessel with similar body shapes as those given in the training dataset. To create a robust, accurate model would take a long time to develop.

Although the number of applications of NN in the marine industry are growing rapidly, the data science sector is still rather young. In 2020, a study was performed using neural networks to predict the roll RAO value and the wave frequencies at which they occurred (Jae and Hyo 2020). However, the scope of the previous work was limited in comparison to the objectives of this project. In Gjeraker (2021), a neural network was created to estimate the RAOs of a drillship with differing wave spectrum inputs. The accuracy of this model was highly accurate, but the results are limited to Subsea 7's vessel Cybership Inocean Cat I Drillship.

The purpose of this paper is to detail the process of creating a proof-of-concept neural network that can predict the RAOs of a simple box barge. Future work will be needed to expand the capability to complex hull structures.

METHODOLOGY

Data Collection

The first step in developing any neural network is to collect a dataset. The dataset must include:

1. The model inputs (independent variables)
2. The model outputs (dependent variables)

The inputs are easy to define and mark the bounds of viable prediction. Table 1 outlines the four parameters that were used in this study. The boundary of accuracy is the upper and lower limits of inputs, meaning that this model will work best for box barges with characteristic dimensions that fall within these ranges. The model will still work for barges outside of these values, but then the predicted RAOs are subject to higher inaccuracies. In total, 142 box barges were included in the dataset. The maximum size of the barges was limited by the mesh generation feature in ANSYS AQWA, which limits the number of nodes to 40,000. To streamline the data collection, a parametric study was performed in ANSYS, during which the mesh resolution was fixed. Therefore, the largest model tested fell just within the 40,000-node limit, while the smallest barges maintained a fine enough mesh grid to ensure accurate results.

Table 1: Inputs to Neural Network Model

Input Type	Minimum Value	Maximum Value
Vessel Length (meters)	2	25
Vessel Beam (meters)	1	16
Vessel Draft (meters)	0.15	1.23
Wave Heading Angle (degrees)	-180	180

To collect the model outputs, numerically determined RAO values from ANSYS AQWA were used.

First, a validation study was performed to ensure that the analysis setup would provide accurate results (Lee, Kim and Goo 2012). A barge that had been previously analyzed for pitch RAOs with strip theory was evaluated in ANSYS, and the comparison is shown in Figure 1.

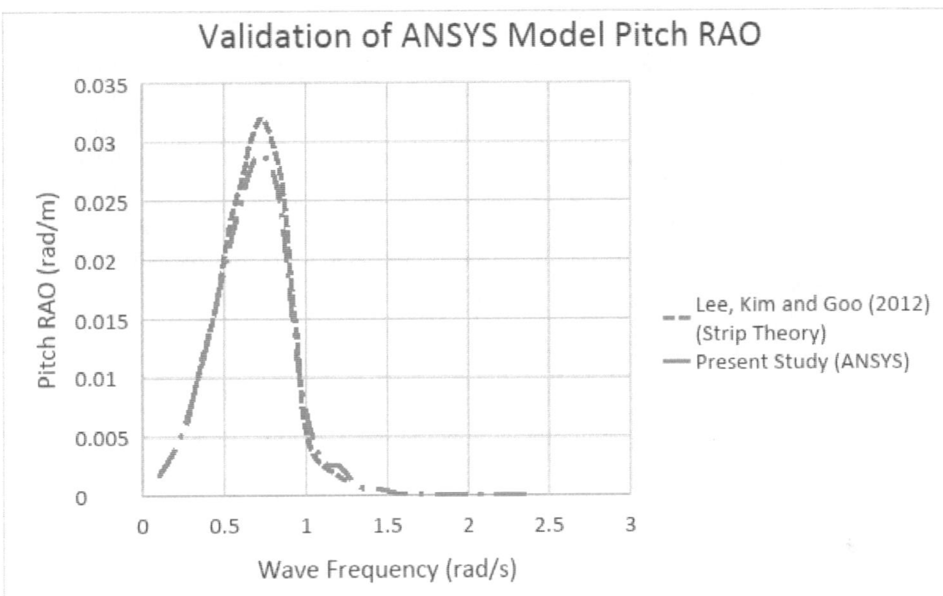

Figure 1: Training data for the NN was simulated in ANSYS AQWA. To ensure accuracy of the ANSYS AQWA model the results were compared with existing strip theory calculations. Overall, good accuracy is seen.

A parametric study was setup in ANSYS DesignModeler, in which a box barge with a length, *L*, beam, *B*, and draft, *T*, would be created and passed into AQWA. In AQWA, a point mass was inserted at the volume centroid to ensure that the weight of the barge would be appropriate to produce the given draft. This weight was determined by Equation 2, where the density of freshwater was used, $\rho = 1000 kg/m^3$.

$$M = LBT\rho \tag{2}$$

The point mass was also given specified moments of inertia based on the dimensions of the box barge. It was assumed that the mass was evenly distributed throughout the vessel, so finding the mass moments of inertia were able to be found with Equations 3, 4 and 5.

$$I_{xx} = \frac{M}{12}(B^2 + D^2) \tag{3}$$

$$I_{yy} = \frac{M}{12}(L^2 + D^2) \tag{4}$$

$$I_{zz} = \frac{M}{12}(L^2 + B^2) \tag{5}$$

The computation time averaged 5 minutes per vessel, for a total of about 17 hours on a 12-Core Dell Precision 5280 with 128 GB RAM.

ANSYS AQWA stores RAO data in a text table format enumerated by frequency, which was saved to be used in the pre-processing. Other information in the text table includes added mass, diffraction, Froude-Krylov Forces, and damping. These could all be used in future studies as alternative predictive models.

Data Pre-Processing

Data preprocessing was conducted in steps. First, the frequency-dependent RAO value for all degrees of freedom of every barge was collected. Then, a Python script fit the frequency-dependent RAO value of each degree of freedom to a curve of a particular form, based on the natural similarity of the response to certain equations. This process was done using the 'curve_fit' function provided by scipy. Equation 6 was used for the surge and sway degrees of freedom, where the response of the barge closely mirrors the pattern of a critically damped spring-mass damper. Equation 7 was for the heave response, where the response at very short frequencies tended to 1, and high frequency waves resulted in 0 response. This closely represents a transformation of the arctangent function, reflected around the x-axis. Finally, the Gaussian distribution shown in Equation 8 was used for the pitch, roll, and yaw responses, where a single peak appeared in mid-range frequencies and the response trended to 0 outside of that range. In Equations 6, 7, and 8, x represents the wave frequency in radians per second, and the result of the function is the RAO value. Other curves were considered, including 3rd, 6th and 9th order polynomials. These were eventually discarded because the curves generated by the neural network were poorly fit to the actual data in all degrees of freedom. Additionally, to produce well-fit polynomials required more coefficient parameters when compared to the selected equations, meaning that the neural network would need to predict more values. This tended to decrease model accuracy and thus was avoided for this project.

Equation	Degrees of Freedom	
$f(x) = Ae^{-Bx} + Cxe^{-Bx}$	Surge, Sway	(6)
$f(x) = A * arctan(B * x + C) + 0.5$	Heave	(7)
$f(x) = A* e^{-\frac{(B-x)^2}{C}}$	Roll, Pitch, Yaw	(8)

The free parameters *A, B* and *C* were then collected and stored. These became the model outputs.

This process was repeated for each degree of freedom for each wave heading of each barge size. The result of the data processing was 1136 datapoints, each with 18 coefficients that described the shape of the RAO curve with respect to frequency – 3 (*A,B,C*) for all 6 degrees of freedom. To check that the curve fitting was accurate, the R-Squared and Mean Average Error (MAE) between the true data and the fit curve was evaluated, and the summary is shown in Table 2. The highest discrepancy was found in sway. However, after comparing the median and average errors, it was found that evaluating the R-Squared metric for the two series which are both very close to zero, results in the R-Squared being equal to zero. This lowered the mean R-squared substantially for surge and sway. Nonetheless, a median R-Squared of 0.955 is very high and suggests that the curve fitting is good.

Table 2: R-Squared Results of Curve Fit to Raw Data

	X (Surge)		Y (Sway)		Z (Heave)		RX (Roll)		RY (Pitch)		RZ (Yaw)	
	R^2	MAE	R^2	MAE	R^2	MAE	R^2	MAE	R^2	MAE	R^2	MAE
Average	0.85	0.038	0.89	0.038	0.94	0.063	0.92	3.734	0.93	5.681	0.96	0.540

Median	0.928	0.044	0.939	0.043	0.989	0.042	0.951	1.276	0.977	0.645	0.980	0.052

Neural Network Architecture

The creation and optimization of neural networks has been made simple with modern packages. Keras and Tensorflow were the two packages used in this study. This paper will not detail the inner workings of machine learning algorithms, as there are better sources to reference for more information on that subject (Tensorflow 2022).

Simple pre-processing methods were employed to prevent any errors – namely dropping any rows with NaN values. These NaN values appeared due to certain barge sizes failing to converge within the default 800 iterations. The number of rows dropped was 178, about 15.7% of the total data. Of these, 163 occurred in the yaw category. It is possible that Equation 8 is a poor representation of the yaw response for certain wave conditions, although the high R-Squared speaks to the overall quality of the Gaussian equation. The input and output data were read into Python and split into a training and test dataset with an 80/20 split, supplying 766 datapoints for the model to train on and 192 for testing.

The neural network itself consists of an input layer, multiple hidden layers, and an output layer, each having a set number of neurons. Between each layer, a dropout layer randomly sets the input units to 0 at a rate of 20% (Keras Team n.d.). Between the input layer and first hidden layer, a batch normalization layer scales the input values. The input and output layers are dictated by the data that will be passed through the model, while the hidden layers are more variable and can be adjusted to optimize the model.

A parametric study was done to select the optimal architecture for the neural network and can be seen in Figure 2 and Figure 3. The number of trainable parameters is defined by Equation 9, where n is the number of hidden layers and A_n is the number of neurons in each hidden layer. 18 is added to the product because of the 18 output parameters that are used in the model. The parametric study consisted of 648 model architectures, each trained for 75 epochs. For each trained model, the R-Squared Score and Akaike Information Criterion (AIC) were evaluated. Figure 2 and Figure 3 summarize the results of the parametric study. Figure 2 also shows that there is a maximum obtainable model accuracy given the data preprocessing, as the R-Squared plateaus at around 0.71.

$$Trainable\ Parameters = \left(\prod_{i=1}^{n} A_n\right) + 18 \qquad (9)$$

Ultimately, a model with 3 hidden layers was selected. Table 3 lists the number of neurons in each layer in the model. This model had the highest R-Squared score and lowest AIC value, implying that it will produce the best results. (Tensorflow, 2022a). A high neuron count additionally increases the ability of the model to fit the input to the outputs due to having more linear combinations, but also requires a larger training time.

The model chosen requires approximately 30 seconds to train for 250 epochs in Python 3.9, on a 12-Core 3.6 GHz system with 32 GB RAM. Training time can be adjusted to optimize the model. A shorter training time tends to avoid the risk of over-fitting, but if the training time is too short the model may not be fully fit at the end of the training. To determine whether the model has been properly fit, a visual inspection of the Loss-Epoch graph shown in Figure 4 was used. If the line is horizontal and mostly unchanging by the final epoch, the model is well-fit (Tensorflow, 2022b). To check that the model is properly predicting the A, B, and C parameters, the true values can be plotted against the predicted values, as seen in Figure 5.

The specific hyperparameters that were used in the model are outlined in Table 4. Due to time constraints, an in-depth hyperparameter tuning analysis was not performed.

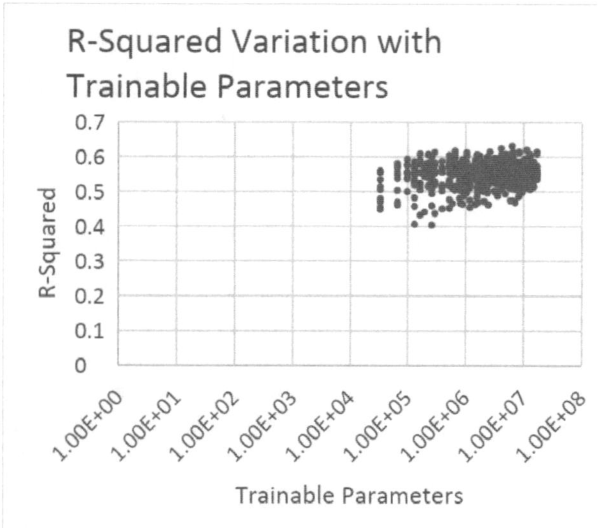

Figure 2: Results of Neural Network Architecture Parametric Study. Increasing the neuron count does not necessarily improve the model accuracy beyond about 10^5 trainable parameters.

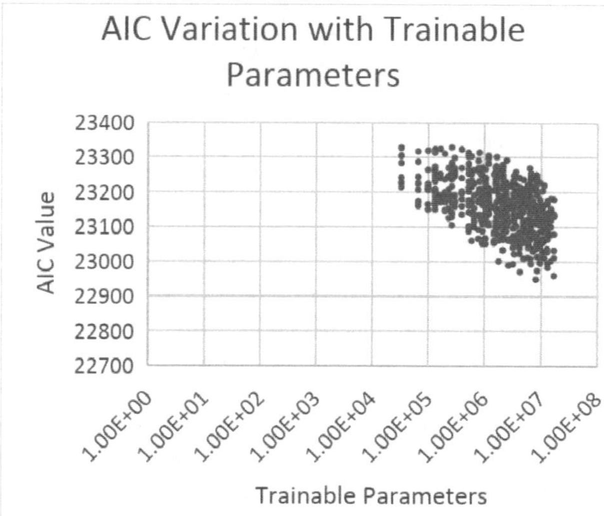

Figure 3: Results of Neural Network Architecture Parametric Study. The Akaike Information Criterion (AIC) is used to compare the relative goodness of fit between different models.

Table 3: Selected Neural Network Architecture. Each layer contains the specified number of neurons.

Input Layer	Hidden Layer 1	Hidden Layer 2	Hidden Layer 3	Output Layer
4	128	256	256	18

Table 4: Hyperparameters of the Neural Network.

Model Optimizer	Adam
Learning Rate	0.001
Activation	ReLu
Loss Type	Mean Absolute Error
Layer Dropout	20% between each hidden layer
Normalization	1 Batch Normalization layer before first hidden layer
Validation Split	0.2
Training Duration	250 Epochs

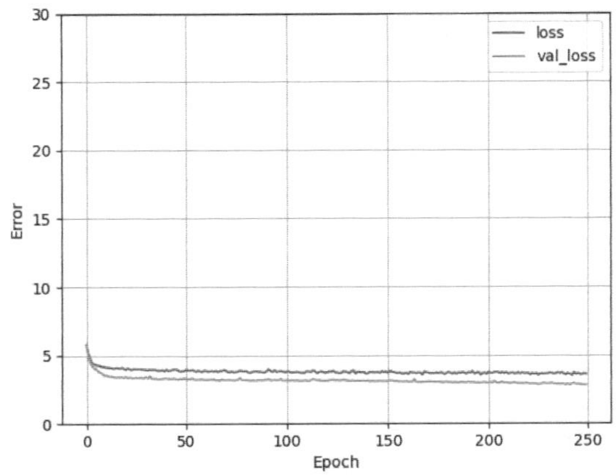

Figure 4: Model Loss During Training Progression. A slope of zero at the final epoch suggests that the model has been appropriately fit.

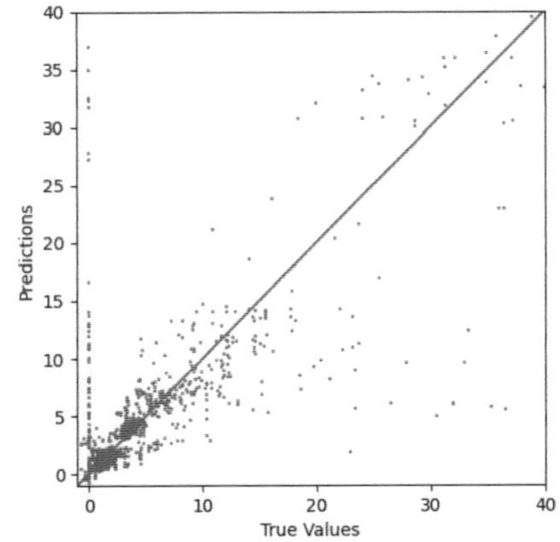

Figure 5: True Values (ANSYS AQWA) Plotted Against Predicted Values (Present Model), R^2 = 0.668. There are many parameters that the model fails to train to be zero, shown by the vertical line of dots on the left side of the plot.

RESULTS

With the model creation and training completed, it can be benchmarked against random datapoints. Figure 7 shows the similarity between the predicted values and true values. Two metrics were used to quantify performance; Relative Percent Difference (RPD) assigns a number between 0 and 200 and explains the difference between the predicted curve and actual curve, as defined in Equation 10. A simple error calculation cannot be used since the true value is often 0 and would result in a division by 0. In the cases where the true value is 0, the RPD is often very high, so the analysis has been supplemented with the MAE. In Figure 7, sway has an RPD of 200 but a MAE of 0.029. So, even though the percent difference is large, the response is only incorrect by about 3 centimeters for every meter of wave height – which is accurate for a barge 15 meters in length.

$$RPD = \frac{|True - Predicted|}{\left(\frac{True + Predicted}{2}\right)} * 100\% \qquad (10)$$

Barge Dimenstions: 15.0 m Length, 10.0 m Beam, 0.867 m Draft, Wave Heading = 180° (Head Seas)

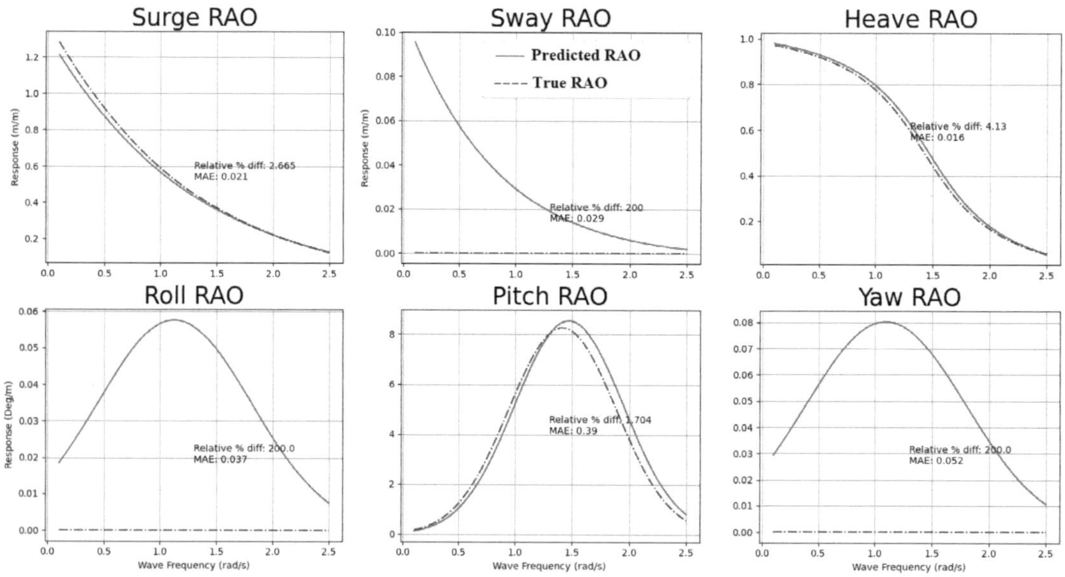

Figure 6: Predicted RAO Values. Surge, sway, and pitch look almost indistinguishable, while the other three degrees of freedom stray from the true values. This is because the true value for these degrees of freedom should be zero. Note the scaling on the y-axis is inflated for these degrees of freedom to better show this phenomenon

Error Analysis

The robustness of this model is highly important. To evaluate the mode performance, 120 barge sizes were randomly sampled from the model predictions and compared to the true RAO values. The RPD and MAE for each datapoint was compared to study the effects of barge size, wave heading, and degree of freedom on the accuracy of the model.

Figure 7 shows the effect of vessel size on the model's prediction accuracy. The average RPD was calculated by averaging the RPD error at each evaluated wave frequency for each degree of freedom of the randomly sampled barges. Admittedly, this tends to smooth over abnormalities, but after seeing that the size and RPD have little correlation, it is assumed that there were few anomalies in the data. To assist in the visualization, the weave heading was differentiated by marker type and color. By considering the wave direction as well, a weak trend suggests that the 45- and 135-degree waves tended to produce generally more accurate results with lower RPD seen in these wave directions. It is important to note that the error distribution for very small barges is more extreme than for larger barges. This implies that the model will give consistent predictions for larger vessels, although there may be some inaccuracy. Note that this does not mean the model will be more accurate for larger barges, but rather that the expected error in the prediction is more uniform. This means that a user can assign error tolerances to an output based on the wave heading input, seeing as certain headings have lower errors.

Figure 8 plots the statistical distribution of the RPD by wave direction, broken down into each degree of freedom. As shown by the box plots, for the 0-, 90-, and 180- degree wave headings, the model had trouble producing accurate parameters for 3 degrees of freedom. These three correspond with the 3 directions in which the actual vessel response should be 0. Because of the way the RPD is calculated, the near-0 in the denominator causes the RPD calculation to become large if the predicted value is not exactly 0 as well. Figure 9 and Figure 10 expand on this with the MAE distributions for the linear and rotational degrees of freedom and show that although the RPD may be high, the actual difference between the true and predicted values are small .

Figure 11, Figure 12, and Figure 13 all show the distribution of error for each degree of freedom. Figure 12 shows that the RPD error for sway, roll, and yaw had the highest relative error. This likely stems from the above-mentioned division issue that arises when determining the RPD. Figure 12 and Figure 13 explain the RPD error, by showing that in general, the difference between the predicted and actual RAOs is small.

Although there are some inaccuracies in the predictive power of this model, it is generally able to produce results that are acceptable. There are a few ways to improve this. First, the model could be split into two models – one for the translational degrees of freedom and one for the rotational degrees of freedom. It is possible that the curve shape of the surge and sway RAOs does not fit well with the exponential equation provided, which would also explain the errors seen in Table 2. If a better general equation was found for these degrees of freedom, a new model could be trained to find parameters and the results may be more accurate.

Additionally, a larger dataset could be collected. With more data, the neural network has more ability to learn the correlations between the input and output parameters, which increases the accuracy. This could possibly increase the R-Squared score from Figure 6.

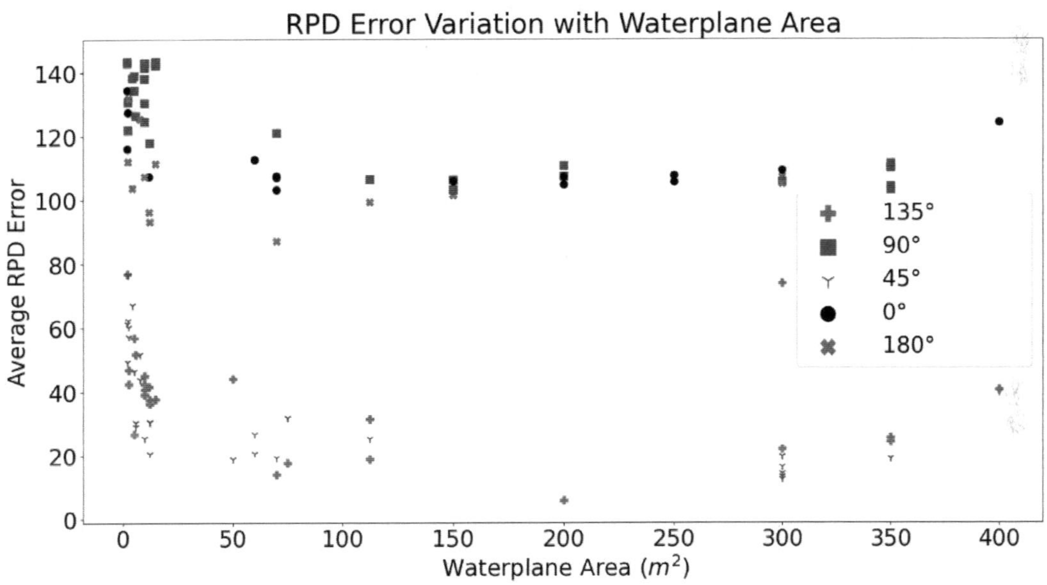

Figure 7: RPD variation with waterplane area. There is no clear correlation between barge size and accuracy of the model. Wave heading is depicted as well, and shows that at 0, 90, and 180 degrees the model has a much higher RPD error.

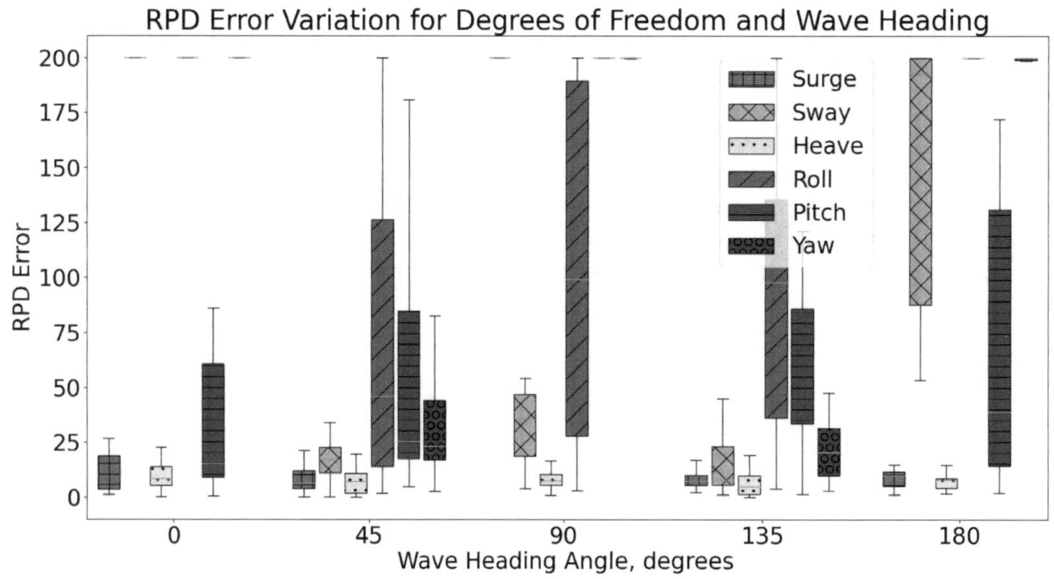

Figure 8: RPD Error Variation for Degrees of Freedom with Wave Heading. Again, at 0, 90, and 180 degrees three of the six degrees of freedom have high error. These three correspond to the directions where the actual response is zero. In these cases, the minimum, median, and maximum RPD is 200.

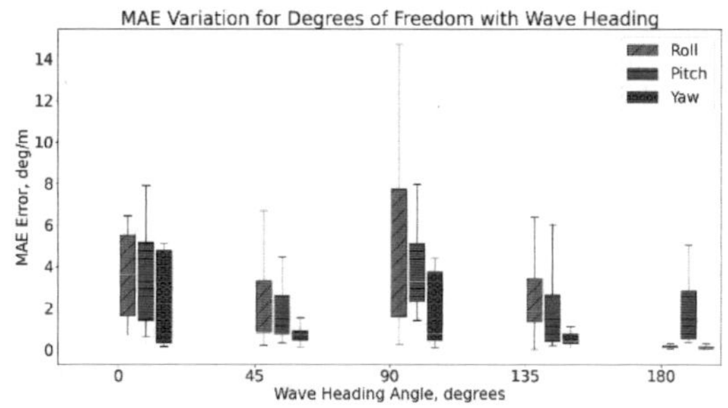

Figure 9: MAE Variation for Rotational Degrees of Freedom with Wave Heading. The degrees of freedom with large MAE generally have low RPD error at the same wave heading, as pictured in Figure 8.

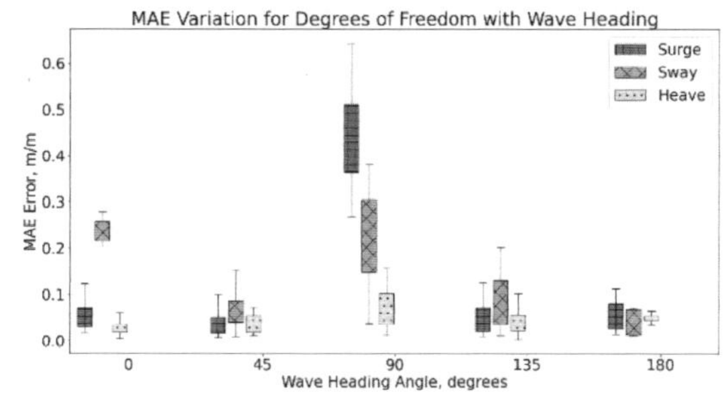

Figure 10: MAE Variation for Linear Degrees of Freedom with Wave Heading. The degrees of freedom with large MAE have low RPD error at the same wave heading, as pictured in Figure 8.

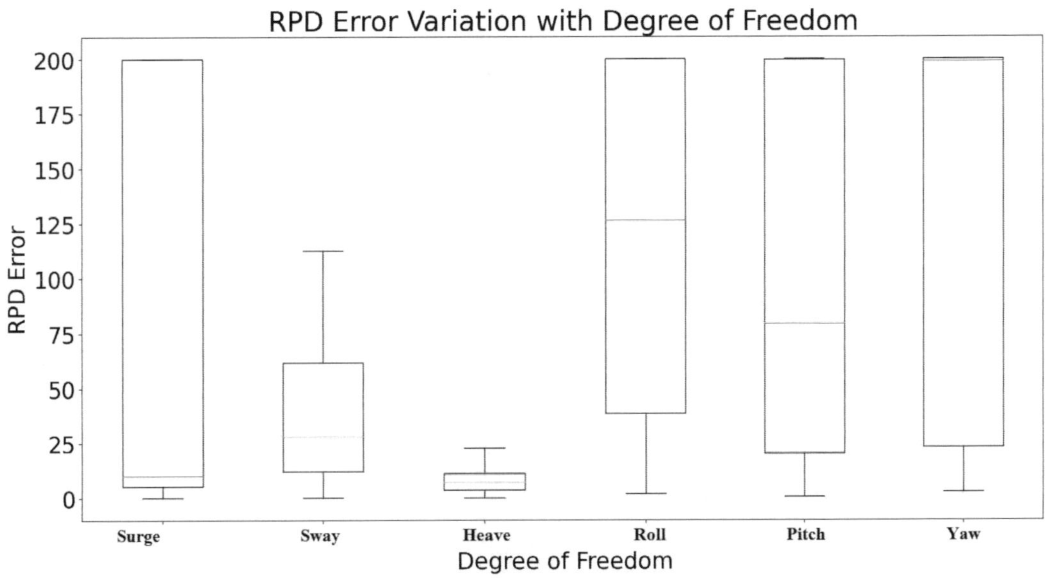

Figure 11: RPD Error Variation with Degree of Freedom. Surge, sway, roll, and yaw have the highest RPD error. Of these, yaw is the most concerning, seeing as the median RPD error is 200, while the lower quartile is around 25. This suggests that the model is failing to predict yaw approximately 50 percent of the time and misses the mark by a large margin in a further 25 percent of cases.

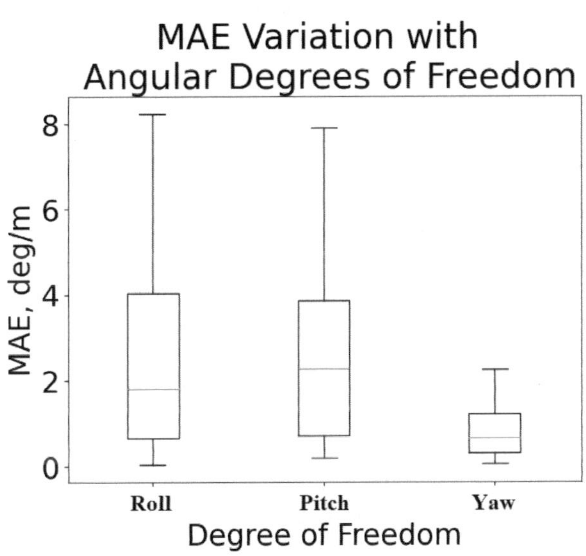

Figure 12: Raw Error Variation with Rotational Degrees of Freedom. Like Figure 9 and Figure 10, the degrees of freedom with high MAE are the same degrees of freedom with low RPD.

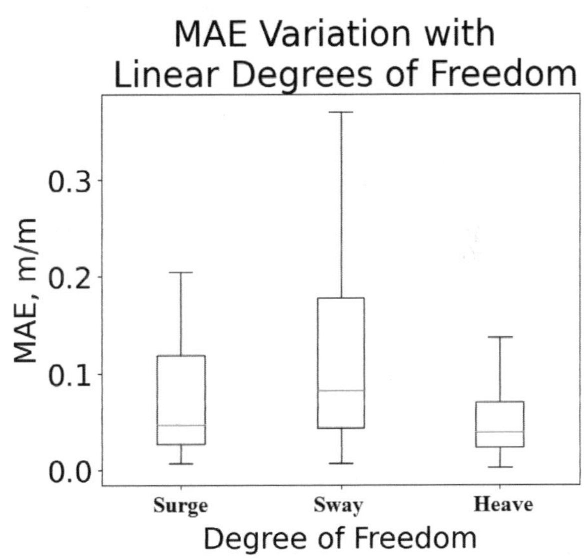

Figure 13: Raw Error Variation with Linear degrees of freedom. The sway degree of freedom has high MAE but low RPD error, while surge has low MAE and high RPD error, seen in Figure 11.

Runtime Analysis

Once the NN was trained and tuned, a software wrapper was created to easily interface with prediction inputs and outputs. Python's built-in libraries 'cProfile' and 'timeit' were used to benchmark the speed of the code. cProfile reported that 437698 function calls were executed within 0.322 seconds. Timeit averaged an execution time of 0.094

seconds across 100 trials. This was performed on a Windows 11 System with 32 GB RAM and a i7-12000KF clocked at 3.6 GHz. The time to specify barge dimensions and run the simulation using the NN averaged less than one minute during testing. Comparatively, creating a CAD model the same barge in ANSYS DesignModeler and simulating the results in AQWA averaged 10 minutes.

When considering the time spent on design modelling, the use of the NN model is much more efficient than the commercial alternatives due to its simplicity. Because the software is based on Python and uses a defined class-structure to process user inputs, it can easily be integrated in other systems in a plug-and-play fashion. However, this comes with downsides – namely a very restrictive modelling capability and non-negligible errors in the RAO predictions. Additionally, this NN cannot predict any hydrodynamic features such as added mass, Froude-Krylov forces, or damping – all of which can be evaluated in ANSYS AQWA in the same amount of time needed to solve for the RAOs. Such capabilities could be added in the future, but the dataset would need to be expanded.

Conclusion

This paper outlines the process by which a predictive model was created and tuned. Bypassing the computationally expensive CAD models and providing another path for vessel design to grow into will assist in the development of digital twins in the modern era. The model creation started with the collection of RAO data from known sources. RAO curves were generated using ANSYS AQWA for a series of box barges. Fitting the raw data to a curve of a general equation allows for the simple expression of the form shape of the responses and allows for the calculation of an RAO at any point in a frequency spectrum. Simplifying the data wherever possible is important when setting up a neural network, as having a lower number of outputs improves accuracy when supplying the same amount of input data. A neural network was set up and adjusted to produce a reasonable estimate of the RAOs for the collected training data. 3 hidden layers and 16.7 million trainable parameters processed inputs such as vessel length, beam, draft, and wave heading, and predicted coefficients for the general curve equation. Error analysis on the model showed that linear degrees of freedom (surge, sway, yaw) are typically predicted to within ±0.25 m/m, and angular degrees of freedom are predicted within ±8 degrees/m. The produced model can produce results around 2500 times faster than ANSYS.

With the model functioning properly for box barges, the first phase of this research project is complete. Given the accuracy and robustness of the model so far, the proof-of-concept has shown the feasibility of using a neural network to supplement the traditional solution process of solving for the RAOs of a vessel. Of course, this model is far from perfect. A 'silver-bullet' model which completely replaces fluid dynamics solvers is unlikely to ever be created. However, this project builds upon past research about the use of neural networks in the maritime industry and will hopefully be useful in the early stages of design where very little is known about a hullform.

Future steps to improve the model are already in motion. The model itself can continually be refined, through changes in the layer architecture or through implementation of different curve fitting functions. Expanding the dataset to include hullforms beyond simple box barges is another such improvement, which will make the predictive power more useful to real-world vessel shapes. Finally, experimental validation of the model predictions should be performed to ensure that the model can obtain the accuracy that is claimed in this project.

References

Gjeraker, Anna Holm. 2021. "Response Amplitude Operator." Norwegian University of Science and Technology, June.

Jae, Hwan Lim, and Jae Jo Hyo. 2020. "Prediction of Barge Ship Roll Response Amplitude Operator Using Machine Learning Techniques." *Journal of Ocean Engineering and Technology* 167-179.

Keras Team. n.d. "Dropout layer." *Keras.* Accessed February 17, 2020. https://keras.io/api/layers/regularization_layers/dropout/.

Lee, S., Y. B. Kim, and J. Goo. 2012. "Analysis of motion response of barge ships in regular waves." *12th International Conference on Control, Automation and Systems.* IEEE. 1920-1922.

Tensorflow. 2022. *Basic regression: Predict fuel efficiency.* 01 19. Accessed 01 21, 2022. https://www.tensorflow.org/tutorials/keras/regression#the_auto_mpg_dataset.

—. 2022. *Overfit and underfit.* 01 19. Accessed 01 21, 2022. https://www.tensorflow.org/tutorials/keras/overfit_and_underfit.

—. 2022. *The Functional API.* 01 10. Accessed 02 25, 2022. https://www.tensorflow.org/guide/keras/functional.

A Critical Assessment of Contracting Strategies and Cycle Times of Post-2014 Sanctioned Floating Platforms in the US Gulf of Mexico

Richard D'souza
President, Richard B Offshore

Shiladitya Basu
Technical Professional Leader, KBR

Proceedings of the 27th Offshore Symposium, February 22nd, 2022, Houston, Texas
Texas Section of the Society of Naval Architects and Marine Engineers

ABSTRACT

Since 1986, about sixty Floating Production Systems (FPSs) have been installed in the US Gulf of Mexico (GoM). They include Spar, Semisubmersible, Tension Leg Platform (TLP) and Floating Production Storage and Offloading (FPSO) platforms. They have been installed in water depths ranging from 500 m to 3,000 m, and production rates ranging from 40,000 boepd to 250,000 boepd. A floating production platform is a major capital investment of a deepwater field development and presents a significant execution challenge. An Operator that opts to install a floating platform for a greenfield development must select a platform type, decide on a contracting strategy and establish credible cycle times for the Appraise, Select, Define and Execute phases of the field development. In their 2021 OTC paper (OTC 31115) the authors undertook a critical assessment of ten major field developments in the GoM, that were sanctioned between 2006 and 2014. They included all four major platform types. A comprehensive database of the field developments including host platform particulars, concept selection, contracting strategies, major engineering, fabrication and installation contractors and cycle times for the four development phases was presented for each project. In the current paper the authors add to the prior body of work, conducting a similar assessment of the seven floating platforms (Appomattox, Argos, Vito, King's Quay, Anchor, Whale and Shenandoah) that were sanctioned between 2015 and 2021. Taken together, the two papers provide the industry with its most comprehensive accounting of the evolution of GoM floating production platforms, development strategies and cycle times. This information will assist Operators and Development Planners plan and benchmark future GoM deepwater developments with a floating platform. Emerging trends in GoM field developments, following the collapse of oil prices in 2014, and the industry pivot to Energy Transition are also addressed.

Keywords: Offshore Field Development, Semisubmersible, semisub, semisub FPS, FPS, Contracting, Cycle Time, Benchmarks

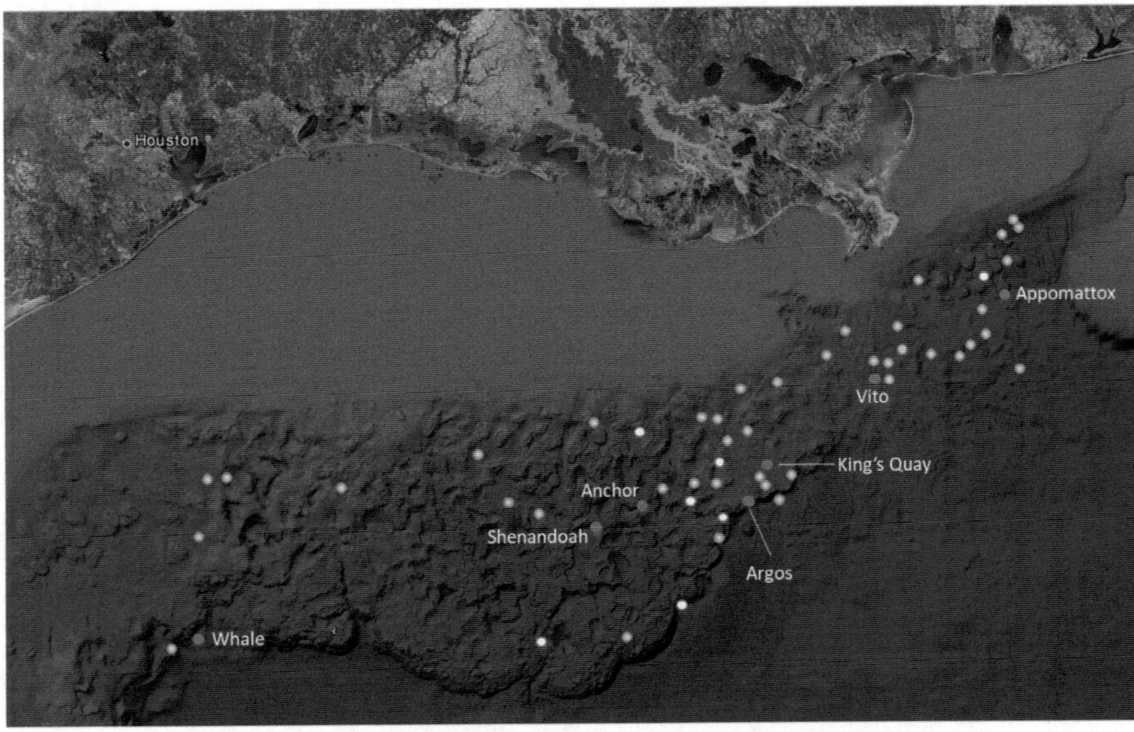

Figure 1: Location of the Seven Post-2014 FPSs sanctioned in US GoM

Introduction

Post-2014, seven greenfield developments with a newbuild floating platform were sanctioned in the US GoM (Figure 1). These developments cover a broad swath of GoM lease areas. Water depths range from 1,200 m (Vito and King's Quay) to 2,600 m (Whale). Nameplate production capacities range from 80,000 boepd (Anchor) to 175,000 boepd (Appomattox). The reservoirs include the Norphlet, Miocene and Lower Tertiary (Paleogene) plays. Estimated recoverable resources range from 250 mmboe (King's Quay) to 650 mmboe (Appomattox). The first post-2014 development to be sanctioned was Appomattox, and the most recent are the 2021 sanctioned Whale and Shenandoah developments. As of December 2021, Appomattox is the only development, among these seven, to come online. The rest are scheduled to begin production between 2022 and 2025.

A brief case history is provided for each development summarizing field development plans, platform particulars, contracting strategies, engineering and fabrication contractors and critical milestone dates (discovery, sanction, and first production). The information is consolidated in a project summary table (Table 1) and a bar graph showing cycle times for major development phases. Information on the Floating Production System (FPS) selection, size and contracting strategy trends, driven primarily by the severe oil and gas price contraction in 2014 and cycle times, exacerbated by the 2020 global Covid pandemic, are presented. The paper concludes with prognostications on technologies and decarbonization initiatives, driven by the energy transition, that are likely to have a significant impact on the future tranche of deepwater projects.

Case Histories

A case history of each of the seven field developments is presented. A list of primary source material is provided in the Reference section.

Appomattox Development

Shell discovered the Appomattox reservoir in Mississippi Canyon (MC) Block 392 in March 2010, about 120 km (80 miles) off the Louisiana coast in 2,200 m of water. The discovery well was drilled to a depth of 7,643 m, followed by several appraisal wells drilled in the next three years. The Vicksburg field was discovered in July 2013 in the adjacent block MC 393. The two fields cumulatively have an estimated ultimate recovery of over 650 mmboe. The reservoirs are in the Norphlet geological formation characterized by high temperatures and pressures and the presence of corrosive gas. These had a significant impact on the size and weight of the topside and platform. Shell made two additional discoveries within 16 km of Appomattox in 2014-2015.

Shell selected a large Semisubmersible hub platform to develop Appomattox and Vicksburg with a production capacity of 175,000 boepd from 15 production wells and 5 water injection wells (Figure 2). A Semisubmersible platform was chosen based on a combination of its ability to support a large topside weight, flexibility for future expansion and Shell's prior experience with the design and execution of large Semisubmersible and TLP projects.

Development engineering was well advanced by 2014, when the price of Brent crude fell by 44% from June 2014 to December 2014 and continued to decline in 2015. In January 2016, the price of oil fell below $30 per barrel, which was less than a third of the price when Appomattox development planning began. To sanction the Appomattox project, the Shell team assessed hundreds of cost-saving ideas, eventually realizing several billions of dollars in capex savings. The biggest saving came from halving the cost of drilling the 20 wells by negotiating a substantial reduction in drilling rig day rates and adopting batch drilling techniques. Shell reduced capital expenditure (capex) by about 20% pre-sanction and another 20% post-sanction by simplifying scope and specifications of the platform and subsea systems. As an example, substantial savings were attained by drawing source water, to cool the massive 150 MW power generation plant, from 600 m below the water line.

Shell sanctioned the Appomattox development in July 2015. The Appomattox Semisubmersible was to be the last of the giant platforms in the GoM, with the hull displacing 125,000 mt, and supporting a 27,000 mt topside operating weight. It was designed for a 40-year service life needed to deplete the known and yet to be discovered reserves that would be tied back to the platform. It was Shell's second Semisubmersible platform in the GoM after Na Kika, that was sanctioned in September 2000. Shell used their time-tested project execution model using handpicked topsides and hull engineering contractors that had worked on most of their previous GoM Semisubmersible and TLP projects. Hull and topside fabrication contracts were negotiated with preferred contractors Samsung Heavy Industries (SHI) and Kiewit Offshore Services (KOS).

Shell awarded the 42,700 mt lightship hull to SHI shortly after sanction. SHI had fabricated the 108,000 mt displacement Olympus TLP hull some five years earlier. On completion, the hull was floated onto one of the largest heavy lift dry transport vessels, COSCO's Xin Guang Hua, outrunning a super typhoon, that was aiming for SHI's Geoje shipyard, by a day. It arrived at the KOS Ingleside yard about two months later, shortly after hurricane Harvey had ravaged the area in October 2017.

KOS fabricated the four major topsides modules (150 MW Power, Process, Utilities, and 180 Person Quarters) and lifted them onto the hull with their 10,000 mt capacity Heavy Lift Device (HLD). Following about ten months of integration and pre-commissioning activities, the platform sailed to the Appomattox site (Figure 3) and was hooked up to the pre-installed 16-point chain-polyester-chain mooring system in August 2018. Export pipelines and production risers to the subsea drill centers were then hooked up to the FPS by Heerema Marine Contractor's (HMC) Balder Semisubmersible Crane Vessel (SSCV). Appomattox achieved First Production in May 2019. Shell is already considering debottlenecking options for future tiebacks.

Figure 2 Appomattox Development Plan (source: Shell website)

Figure 3 Appomattox sail-away (source: Shell website)

Mad Dog Phase 2 Development

BP discovered the Mad Dog field in Green Canyon (GC) Block 780 in 1998. Production began in 2005 from a Spar platform installed in about 1,300 m of water, 304 km (190 mile) south of New Orleans. The platform was designed to produce 80,000 bopd and 60 mmscfd gas. Appraisal drilling in 2009 and 2011 in surrounding leases doubled the resource estimate of the greater Mad Dog field to more than 4 billion boe in place, justifying the need for a second platform in the field.

In 2010 BP began work on a second hub facility, a large wet tree Spar platform dubbed Big Dog, to develop Mad Dog Phase 2. BP recycled the Mad Dog Phase 2 development in 2013, as the field development plan they were progressing was uneconomic even before the 2014 oil price collapse. This time BP selected a Semisubmersible host platform over a TLP and Spar, based in part on the excellent execution and operational performance of the BP operated Atlantis Semisubmersible platform that began producing in 2007 in the Green Canyon area. A key to the selection of the Semisubmersible FPS was the acceptance of Steel Lazy Wave Risers (SLWR) by BP riser engineers.

BP achieved a 60% capex reduction from the scrubbed development plan while producing the same resources. Much of this was achieved by simplifying and standardizing the host platform, subsea systems and well construction. Reservoir recovery factors were greatly improved by implementing ocean bottom nodes and advanced enhanced recovery technologies (Low Salinity or LoSal water injection) while reducing well count. Significant savings were also realized from industry collaboration and standardization.

Figure 4 Mad Dog Phase 2 Development Plan (source: BP website)

The revised Mad Dog Phase 2 development plan (Figure 4) includes a Kellogg Brown and Root (KBR) designed box-deck Semisubmersible hull with the capacity to produce 140,000 bopd from 14 production wells and inject 140,000 bwpd LoSal water into 5 water injection wells. The platform, named Argos, a reference to Odysseus' loyal dog from the Odyssey, is moored about 10 km southwest of the original Mad Dog truss spar. The project was sanctioned in December 2016 at an estimated break-even cost of $40/boe. Planned first oil at sanction was the fourth quarter of 2021.

BP deviated from the prevailing segmented execution model for large host platforms at the time, by awarding SHI a turnkey Engineer Procure Construct (EPC) contract to fabricate and integrate the FPS (topsides and hull) in their Geoje yard. The award followed several rounds of competitive bidding involving multiple contractors. Construction began in March 2018. The completed platform was loaded onto the BOKA Vanguard heavy transport vessel, setting sail for the KOS yard on February 8, 2021. After a 26,000 km voyage from South Korea, the Argos FPS was offloaded at the KOS yard on April 13, 2021. It underwent final preparatory work and regulatory inspections before being wet towed (Figure 5) to the site and hooked up to its pre-installed moorings in November 2021, after battling a strong, persistent loop current. It is currently undergoing final hook-up and commissioning to the subsea wells and export pipelines. First oil is anticipated in the second quarter of 2022, about six months later than planned. Much of the delay is attributable to the global Covid pandemic.

Figure 5 Argos Platform Sail-away (September 2021)

King's Quay Development

The Khaleesi and Mormont (Figure 6) reservoirs were discovered by LLOG in 2017 in GC blocks 390 and 478 in about 1,200 m of water. They are Miocene reservoirs 9,300 m below the seabed. Murphy Oil assumed operatorship of the Khaleesi/Mormont field development, following their acquisition of LLOG's Green Canyon assets in May 2019. The reservoirs are estimated to contain a combined 165 mmboe of recoverable reserves. Murphy discovered the Samurai field in 1,060 m of water in GC 432 with recoverable resources estimated at 90 mmboe.

LLOG initiated the King's Quay project as a seven well subsea tieback from Khaleesi/Mormont to a Semisubmersible FPS host located on GC block 433. Following the acquisition of LLOGs assets, Murphy added Samurai as a four well, 10 km subsea tieback. Murphy formally sanctioned the King's Quay development in August 2019. In January 2020, Murphy entered into a memorandum of understanding with energy investment firm Arclight Capital Partners to sell its 50% ownership in the FPS, replicating a strategy employed by LLOG for the Delta House development.

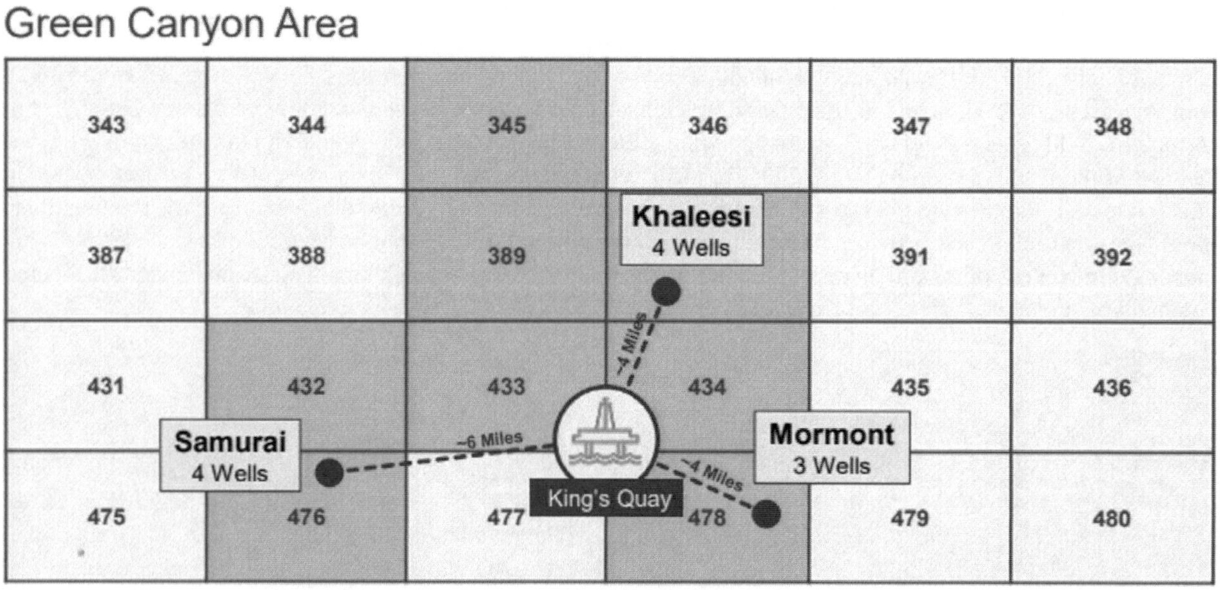

Figure 6 Schematic of the King's Quay Field Development (source: Offshore Magazine)

The FPS is an Exmar standardized OPTI-11000$^{(TM)}$ design (Figure 7), much like the Delta House FPS operated by LLOG that began production in 2015. The topside nameplate capacity is 80,000 bopd and 100 mmscfd gas, with peak capacity of 100,000 bopd and 240 mmscfd gas. It has the flexibility to host future subsea tiebacks. The hull is a four column, ring pontoon configuration with a truss deck connecting the column tops. The deck offers a flat area that facilitates installation, hook-up and pre-commissioning topsides modules. The deck and topsides are fabricated and integrated at ground level and installed onto the hull in a single heavy lift.

Figure 7 King's Quay Dry Transport to KOS (source: see note below)[1]

[1] The project is being executed by Murphy Exploration and Production. Owners of the semisubmersible facility include entities managed by Ridgewood Energy Corp., ILX III Holdings, LLC and Third Coast Super Holdings, LLC. The facility will be operated by Murphy.

Exmar Offshore provided engineering and construction supervision services for the hull, while Audubon Companies was awarded a contract for topsides engineering and procurement services. LLOG negotiated a turnkey EPC contract with Hyundai Heavy Industries (HHI) to fabricate and integrate the topsides and hull, which they later assigned to Murphy. The deck and topsides were placed on the hull in a dry dock in a single lift. This was a departure from the Delta House project execution, where HHI fabricated the hull in Korea and KOS fabricated and integrated the topsides and deck with their HLD in a single lift at their Ingleside facility in Texas. The completed FPS was dry transported on the COSCO heavy lift vessel Xiang An Kou (Figure 7) and delivered to the KOS shore base in September 2021 for final commissioning prior to sail away. First oil is expected in the second quarter 2022, following hook-up of the moorings and risers and final commissioning.

Vito Development

The Vito discovery well in the subsalt Miocene sands was drilled by Anadarko in July 2009 at a depth of 9,750 m below the mudline. Shell assumed operatorship in September 2009. After a four-year appraisal program Shell estimated that the field held recoverable resources of 300 mmboe, sufficient to support a stand-alone host platform.

Shell intended to sanction Vito in 2015, shortly after Appomattox, but the 2014 oil price collapse caused Shell to put the brakes on. A roughly three-year redesign exercise from the subsurface up followed. Shell succeeded in reducing the capex estimate by more than 70% from the original concept, which was a Semisubmersible FPS comparable in size to the Appomattox FPS. This was achieved in several ways: reduced well count, improvements in drilling efficiency, a smaller, fit for purpose FPS with limited spare capacity, and competitively bid fabrication. The leaner FPS design incorporated automation technology that reduced manning significantly. With this development plan, Shell estimated that they could develop Vito for a forward-looking breakeven price less than $35 per boe. Shell announced the final investment decision on April 24, 2018.

Figure 8 A 3D rendering of Shell's Vito platform (source: Shell website)

The Vito FPS is shown in Figure 8. The topside is supported by a structurally efficient modular support frame (MSF) with knee-braces. The approximately 10,500 mt topsides can process 100,000 boepd setting a benchmark for topside design efficiency. The lightship weight of the four-column, ring pontoon hull, including the MSF, is about 13,000 mt. Hull displacement at a 27.5 m operating draft is 38,000 mt, like the OPTI-11000$^{(TM)}$.

Shell broke from their traditional split construction contracting practice for prior GoM floating platforms by tendering to a broad selection of EPC fabrication yards. Shell awarded a turnkey EPC contract to Sembcorp Marine (Sembcorp) to deliver the Vito FPS. The yard installed the topsides onto the hull in a single lift in a dry dock with twin gantry cranes (Figure 9). Shell awarded the contract in May 2018. Sembcorp handed over the FPS to Shell in the final weeks of 2021. Dry transport of the Vito FPS onboard Boskalis' Mighty Servant I, from Singapore to Texas began in the first week of 2022. The delivery was delayed by several months because of the Covid pandemic. The FPS is expected to arrive at KOS in March 2022.

Figure 9 Vito in fabrication yard (source: Internet)

Shell has advised that first oil from Vito is expected in the second half of 2022.

Anchor Development

Chevron discovered the Anchor field in late 2014 on block GC 807 in 1,400 m of water (Figure 10). It lies about 225 km off the Louisiana coast and due North of their giant Jack and St. Malo fields, that began production in December 2014 from one of the world's largest Semisubmersible FPSs. The reservoir is in the Lower Tertiary Wilcox formation, about 10,300 m below the mudline. At an operating pressure of 20,000 psi, it is one of the first ultra-high-pressure projects in the world. Industry had begun the design and qualification of 20,000 psi subsea and drilling equipment several years earlier in anticipation of producing several of these massive, high-pressure reservoirs.

The timing of the discovery was not propitious as it coincided with the 2014 oil price collapse. Chevron, strongly committed to growing its already material production in the GoM, drilled two appraisal wells in the next two years to confirm reservoir productivity and recoverable reserves. From the appraisal program they estimated that the Anchor field had at least 440 mmboe of total recoverable resources, sufficient for them to begin Select phase activities for the Anchor development with a floating production platform in 2016.

Anchor is a staged development with an initial seven subsea production wells tied back to a Semisubmersible FPS. Chevron estimated it would take 250 days to drill and complete each deep, ultra-high pressure well. The FPS nameplate production capacity is 75,000 bopd and 28 mmscfd gas.

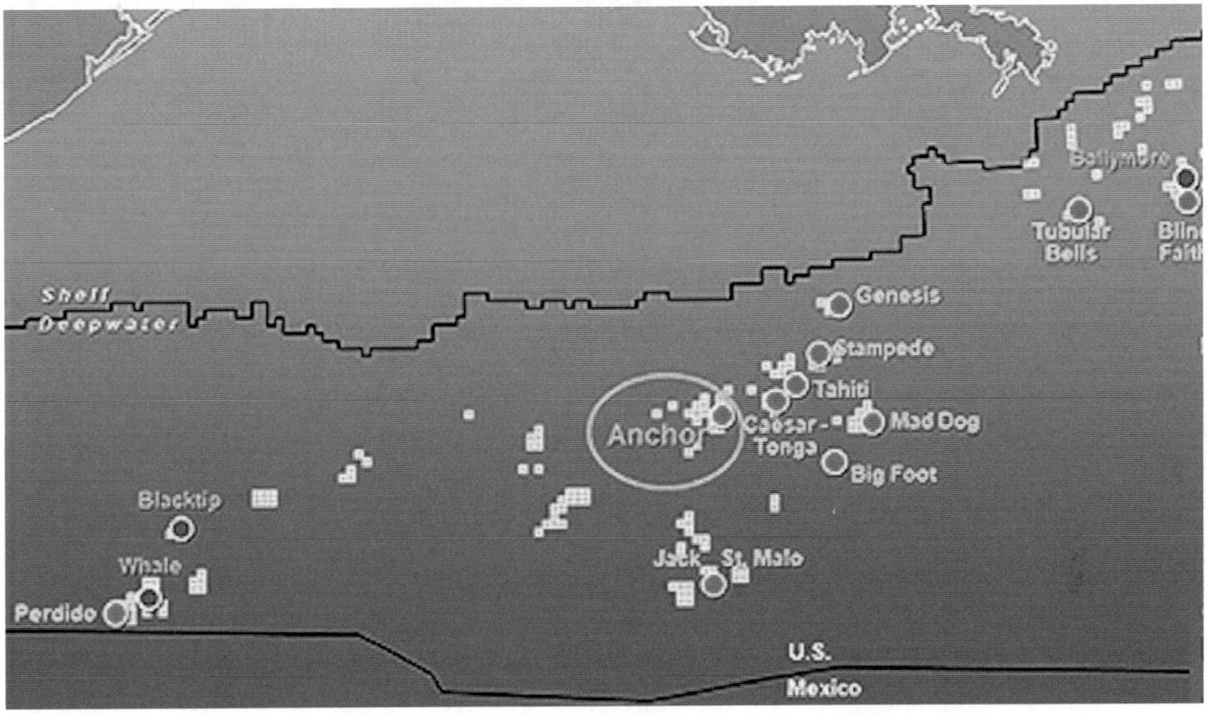

Figure 10 Anchor Project location in Green Canyon Block 807

The Define phase kicked off in early 2017 with Wood designing the topsides and integrated deck and the hull designed by KBR (Figure 11). These were the same contractors that designed the successful Jack St. Malo Semisubmersible. Topsides operating weight, including the integrated truss deck, is 17,000 mt. The hull lightship weight is about 20,000 mt, displacing 69,500 mt at a 35 m operating draft. Chevron greenlighted the $5.7 billion development in December 2019 having reduced development costs for new projects in the GoM by nearly a third, compared to their last generation of greenfield deepwater investments. Anchor is a phased development with the FPS designed to accommodate future subsea tiebacks. The initial phase targets the southern fault block of the reservoir. The second phase targets the northern fault block with a four well drill center. The performance of the southern wells will inform the decision to move forward with the next phase.

Chevron awarded Transocean a contract to drill the Anchor wells just before project sanction. Transocean will upgrade a deepwater drillship under construction to accommodate 20,000 psi blowout preventers, a net hook load capacity of 3 million pounds and an enhanced dynamic positioning system.

Chevron opted for a traditional segmented project execution approach for the FPS i.e., hull built in the Far East and topsides built and integrated in the GoM. In December 2019 Daewoo Shipbuilding and Marine Engineering company (DSME) of South Korea was awarded a contract to build the hull. The award ended a five-year drought in new offshore platform orders for the yard. At the same time, KOS was awarded the topsides fabrication and integration contract. First oil date at project sanction is the second quarter of 2024. As of this writing, it appears that the project is on schedule.

Figure 11 Anchor platform 3D model

Whale Development

Shell, the Operator of the Whale development, made the discovery in 2017 in Alaminos Canyon (AC) Block 773. The discovery, in 2,600 m of water, is approximately 16 km from their Perdido development, 320 km southeast of Houston. Perdido began production in March 2010 from a drilling-production Spar platform. Whale is a Paleogene reservoir some 9,000 m below the mudline. Shell labelled the find as one of its largest in the GoM in the past decade. Following several years of appraisal activities, Shell estimated a recoverable resource volume of 490 mmboe, sufficient to support a standalone greenfield platform. Well productivity, one of the biggest uncertainties in a Paleogene development, was largely derisked by their ten-year operating experience at the Perdido and Silvertip developments. In April 2020 Shell decided to postpone sanctioning Whale because of supply chain constraints and economic uncertainty created by the pandemic. However, they continued to advance a development plan that closely resembled the Vito development, sanctioned some two years earlier. The hull is a 99% replication and the topsides an 80% replication of the Vito platform (Figure 12).

The development plan features fifteen subsea production wells tied back to the Semisubmersible FPS. Shell describes the Whale FPS as a "simplified, cost-efficient host design" like the Vito FPS, as it is configurationally and dimensionally identical to Vito (Figure 13). Like Vito, the nameplate production capacity is 80,000 bopd and 100 mmscfd of gas, the topside operating weight is 10,500 mt and the displacement is 38,000 mt. To lower greenhouse gas intensity, Shell is utilizing energy-efficient gas turbines and compression systems.

Anticipating a final investment decision (FID), and to accelerate first production, Shell contracted Sembcorp to build and integrate the Whale FPS topsides and hull in November 2019. Shell had earlier contracted Sembcorp for the same scope on the Vito FPS (EPC of the FPS and deliver to a shore base in the GoM). By taking advantage of the synergies from the Vito FPS supply chain and construction processes Shell estimated that they could achieve first oil 7.5 years after discovery. Shell sanctioned the Whale development on July 26, 2021, at a forward-looking breakeven cost below $35/boe. Whale will be Shell's 12th deepwater platform in the GoM and is currently scheduled to begin production in 2024.

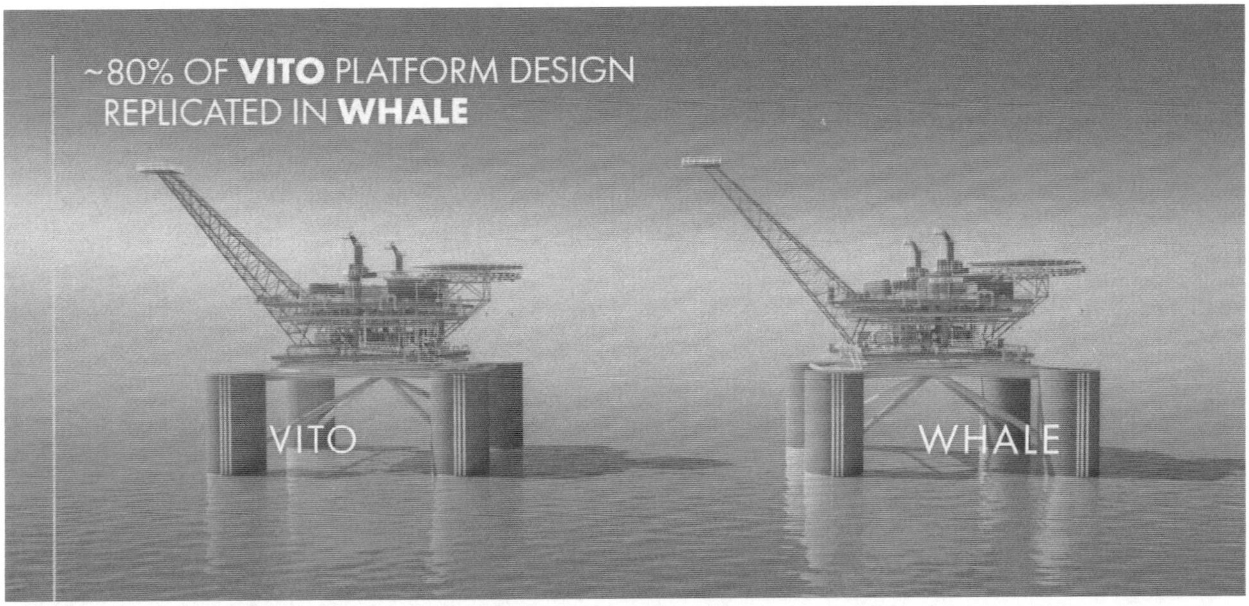

Figure 12 Whale replicates Vito platform design (source: OGJ.com)

Figure 13 Schematic of Shell's Whale platform (source: Offshore Magazine)

Shenandoah Development

The Shenandoah field is located on Walker Ridge (WR) Blocks 51, 52 and 53 (Figure 14). The discovery well was drilled by Anadarko in February 2009 in about 1,700 m of water. The reservoirs are in the Upper and Lower Wilcox sands about 9,500 m below the mudline. Like Anchor, the reservoir is ultra-high pressure, requiring 20,000 psi rated equipment.

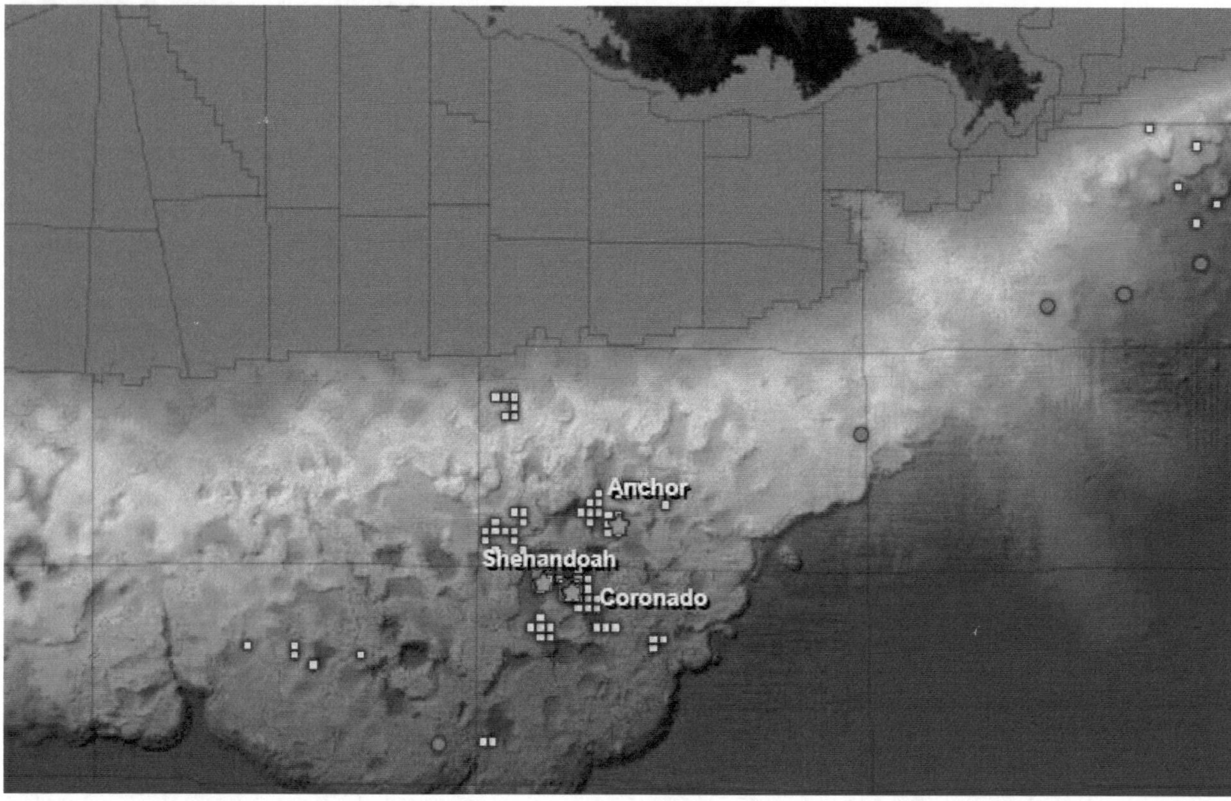

Figure 14 Shenandoah project location in Walker Ridge Blocks 51, 52, 53

There was considerable excitement surrounding the discovery, as the initial appraisal wells suggested the potential for upwards of 1 billion boe in place. After drilling five appraisal wells from 2012 to 2017, Anadarko concluded that the reservoir was highly compartmentalized, with recoverable reserves considerably lower than originally anticipated. Concurrent with appraisal, Anadarko had initiated a competitive FEED for a Semisubmersible and Spar FPS as the host facility and selected a Semisubmersible. Anadarko relinquished the field in 2017 after spending a reported $1 billion from 2009 to 2017 on exploration and appraisal.

The project was revived in 2018 when LLOG became Operator after Anadarko withdrew from the project and partner Cobalt Energy divested their interest in the project. Based on the results of the extensive appraisal program, LLOG concluded that recoverable resource from Shenandoah was about 280 mmboe. They quickly settled on a phased development plan, with the first phase consisting of tying back four producing subsea wells to a Semisubmersible FPS. The development plan drew heavily from LLOG's experience with Who Dat and Delta House developments. The second phase would add four more producers, informed by the performance of the four initial wells. LLOG contracted TechnipFMC to provide the 20K subsea trees in October 2019.

On 28 July 2020, Beacon Offshore (Beacon) acquired LLOG's stakes in Shenandoah and Yucatan, the latter being a discovery just southwest of Shenandoah that will eventually be included in the Shenandoah development. Beacon was formed in 2016 by private equity group Blackstone with the goal of building a significant, high-quality portfolio in the deepwater GoM.

Beacon sanctioned the Shenandoah development, as envisioned by LLOG, on August 27, 2021. The Semisubmersible FPS is an Exmar OPTI-11000(TM) (Figure 7), similar to LLOG's Delta House FPS and Murphy's King's Quay FPS , with a nameplate production capacity of 100,000 boepd. Shortly thereafter, they awarded HHI an EPC contract for delivery of the FPS following a competitive bid. HHI had delivered the King's Quay FPS in July 2021. HHI plans to start constructing the hull at its Ulsan yard in the third quarter of 2022. Work on the FPS is scheduled to be completed by September 2024, with installation in the field expected in fourth quarter of 2024.

Beacon awarded Transocean a contract to provide a drillship to drill and complete the first phase four well drilling and completion program. The eighth generation Deepwater Atlas drillship will begin drilling with a 15,000-psi rated BOP, and the program is expected to last 255 days. A 20,000 psi BOP will then be installed for the completion program, which is expected to last 275 days. Beacon anticipates first production from Shenandoah in the first quarter of 2025.

Critical Assessment of Post-2014 Sanctioned GOM FPSs

Table 1 provides summary information of the field development, FPS particulars, platform contracting strategies and contractors and significant project milestones (Discovery, Sanction, and First Production) of the seven post-2014 sanctioned field developments in the deepwater US GoM.

Table 1 Summary Data of the seven post-2014 sanctioned FPSs in the GoM

	omattox	s	Vito	King's Quay	Anchor	Whale	Shenandoah
Block	392	80	MC 939	GC 433	GC 807	AC 772	WR 52
Operator	l		Shell	Murphy	Chevron	Shell	Beacon Off
Water Depth (m)	2265		1200	1190	1580	2600	1700
Reservoir Depth (m)	)		5500	9300	10300	9000+	9500
Reservoir Rock	hlet	ene	Miocene	Miocene	Wilcox	Paleogene	Paleogene
EUR (mmboe)	+		300	250	440	490	280
Host Platform	*Shell Semi*	*KBR Semi*	*Shell Semi*	*Opti-11000(TM) Semi*	*KBR Semi*	*Shell Semi*	*Opti-11000(TM) Semi*
Peak Oil Rate (mbopd)			80	80	75	80	80
Peak Gas Rate (mmscfd)			100	100	30	100	100
Enhanced Recovery		.oSal)	Gas lift	-	SSB (Future)	-	-
Well Count	, 5 I	8 I	8P	11P	7P (staged)	15P	8P (staged)
Topsides Op Wt (mt)	00	0	10,500	11,000	17,000	10,000	11,000
Hull L'ship Wt (mt)	00		13,000	11,000	19,300	15,000	11,000
Displacement (mt)	200	00	38,000	39,000	69,500	38,000	39,000
Hull Type	Shell MSF with Knee-braces	GVA Box-deck	Shell MSF with Knee-braces	Exmar Truss-deck with MSF	GVA Integrated Truss Deck	Shell MSF with Knee-braces	Exmar Truss-deck with MSF
FPS Contracting	Segmented Fab, Int & Inst. Contracts	Turnkey EPC Contract for FPS fabrication	Turnkey EPC Contract for FPS fabrication	Turnkey EPC Contract for FPS fabrication	Segmented Fab Contracts	Turnkey EPC Contract for FPS fabrication	Turnkey EPC Contract for FPS fabrication
Top. Eng. Co	bs		Jacobs	Audubon	Wood	Jacobs	-
Hull Eng. Co.	l		Shell	Exmar	KBR	Shell	Exmar
Top. Fab.	s		Sembcorp	HHI	KOS	Sembcorp	HHI
Hull Fab.			Sembcorp	HHI	DSME	Sembcorp	HHI
Int. Cont.	s		Sembcorp	HHI	KOS	Sembcorp	HHI
Int. Method	yside Lift	Floating Crane	Single Lift DD	Single Lift DD	Quayside Lift	Single Lift DD	Single Lift DD
Inst. Cont.	C	ea 7	Jumbo Offshore	Subsea 7	HMC	-	Subsea 7
Discovery	2010		Jul 2009	2017	Q4, 2014	Jan 2018	Feb 2009
Sanction	2015	016	Apr 2018	Aug 2019	Dec 2019	Jul 2021	Aug 2021
Hull Delivered	2017	021	Mar 2022 (exp.)	Sep 2021	-	-	3Q 2024 (exp)
FPS Mooring Hook-up Completed	2018	2021	-	Jan 2022	-	-	-
First Production	2019	022	Q3, 2022	Q2, 2022	Q1, 2024	2024	Q1, 2025

FPS Selection and Displacement Trends

Figure 15 is a bubble chart showing the type, displacement, and water depth versus sanction date of ten deepwater FPSs sanctioned before 2014 and the seven deepwater FPSs sanctioned from July 2015 to August 2021. The trends are very revealing.

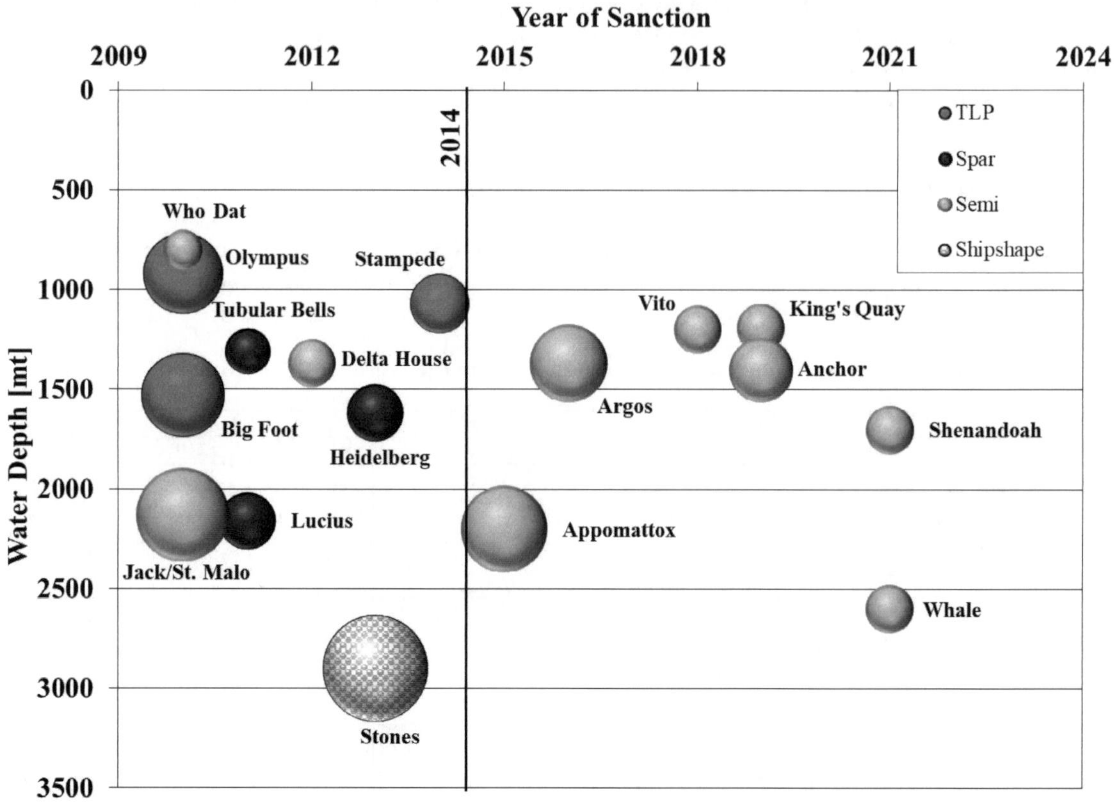

Figure 15 US GoM FPS Type and Displacement Trends since 2010. Bubble size corresponds to platform displacement

Of the ten pre-2014 sanctioned projects, there were three TLPs, three Semisubmersibles, three Spars and one FPSO. The average displacement of the four projects executed by major Operators (Jack St. Malo, Olympus, Big Foot, Stones) was about 140,000 mt and average water depth about 1,900 m. The average displacement of the six projects executed by independent Operators (Who Dat, Tubular Bells, Lucius, Delta House, Heidelberg, Stampede) was about 45,000 mt and average water depth about 1,500 m.

Just seven deepwater greenfield projects were sanctioned in the seven years since Hess sanctioned the Stampede TLP in October 2014. Significantly, all seven platforms are Semisubmersible FPSs. Of these, five (Appomattox, Argos, Vito, Anchor, Whale) were sanctioned by three major Operators (Shell, Chevron, BP) with average displacement about 75,000 mt and average water depth about 1,800 m. Average displacement of the two Semisubmersible FPSs (King's Quay, Shenandoah) sanctioned by independent Operators (Murphy, Beacon) was about 38,500 mt and average water depth about 1,350 m.

The selection of a Semisubmersible FPS for all post-2014 sanctioned platforms is not coincidental. In 2018, the authors presented the case for a Semisubmersible FPS as a universal host for future GoM platforms (OTC 28970). It is informative to revisit the rationale for this selection as it has become a reality. In the run-up to the dramatic oil price drop in late 2014, deepwater projects witnessed a spectacular growth in investment, production and capex inflation. Much of the latter was attributable to inflation in drilling, equipment and labor cost, and lack of capital discipline. Oil prices in 2012-2013 averaged over $100/bbl and were expected to continue rising. Breakeven costs of deepwater projects in the GoM often exceeded $70/boe. Most deepwater projects were late and over budget.

In the wake of the 2014 oil price collapse, the industry was confronted with a stark choice: to enable sanction of future deepwater projects, break-even costs had to be reduced to $40/boe or less. By 2015, breakeven costs had fallen to between $50-$60/boe, driven largely by commercial gains extracted from the supply chain. Shell reduced capex of the Appomattox project, sanctioned in July 2015, by 40% without altering the development plan. They did this principally by halving the cost of drilling and completing the twenty wells.

To realize an additional $10-$15/boe reduction in breakeven costs, the industry had to do more. Significant break-even cost compression was achieved by simplifying and standardizing designs and specifications of everything from drilling and completions, subsea systems, and the floating platforms. Operators realized that by picking a universal FPS building block they could reap the benefits of greater project delivery certainty, accelerated cycle time to first oil and reduced capex, thereby achieving the harder to obtain cost reductions. The question was which of the four major platform types deployed in the GoM deepwater best fit the requirements of a universal host, i.e., water depth and payload bandwidth, flexibility for phased development, ease of integration and installation and contracting flexibility. The Semisubmersible FPS checked all the boxes.

BP sanctioned the Mad Dog Phase 2 development in December 2016 at an estimated breakeven cost of $40/boe. They achieved a 60% reduction from the original development plan while maintaining the same production rates by utilizing many of the cost reduction levers at their disposal. One of the levers was to replace the Spar with a Semisubmersible FPS, building upon their successful execution of and lessons learned from the Atlantis FPS.

Shell rewrote their deepwater GoM development playbook with the Vito development. They delayed a presumptive 2015 sanction and spent three years on a revised plan that could be developed at a break-even cost of about $35/boe. A key was scaling back the topside nameplate capacity, simplifying the topsides processing facilities and reducing manning with automation. This combination led to an enhanced production rate to topsides weight ratio and a semisubmersible that displaced about a third that of the Appomattox FPS. The tradeoff came at the expense of limiting flexibility for future expansion. Buoyed by the success of Vito, Shell sanctioned the Whale development about three years later with a near replica of the Vito semisubmersible design and project execution.

King's Quay and Shenandoah are being developed by independent Operators Murphy and Beacon. They are following the successful template developed by LLOG (a partner on both projects) for their Delta House project by utilizing the lean, standardized OPTI-11000(TM) semisubmersible design. The efficient topsides throughput (100,000 boepd) to topside operating weight (11,000 mt) ratio results in a compact and efficient hull (39,000 mt displacement). By standardizing and repeating the FPS design and using the same engineering and fabrication contractors (design one build two), the Operators were able to accelerate cycle times and realize significant cost efficiencies, much as Shell hope to do for their Vito and Whale projects.

Shenandoah, however, is a frontier development that falls into the ultra-high-pressure category (20,000 psi), requiring high specification subsea and drilling equipment. To manage the uncertainty in well productivity, Beacon has opted for a phased development where the second phase will be predicated on the performance of the first phase wells.

Chevron also opted for a phased approach for their ultra-high-pressure Anchor development that faces many of the challenges as the Shenandoah development. This provides the dual advantages of a) optimizing the second phase well designs based on information from the first phase and b) specifying a lower nameplate production (80,000 boepd) resulting in a leaner platform. The 17,000 mt topsides is about 50% heavier than the Shell Vito and OPTI-11000(TM) semisubmersibles partly because of the inclusion of subsea boosting to enhance ultimate recovery.

Figure 16 shows how FPS topsides operating weights are trending with time. Topside operating weight of Argos and Appomattox FPSs averaged about 26,000 mt. Topside operating weights of the later sanctioned Vito, King's Quay, Anchor, Whale and Shenandoah FPSs average about 12,500 mt, reflecting increased operator focus on capital efficiency.

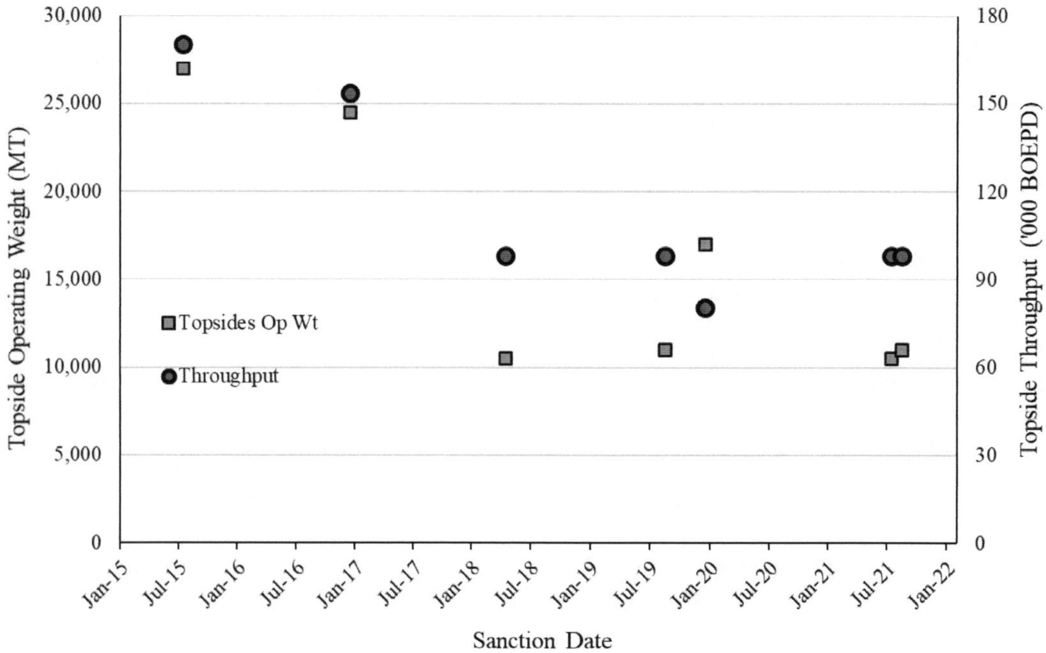

Figure 16 Post-2014 Sanctioned FPS Topside Operating Weight and Throughput Trends

Discovery to First Production Cycle Times

The Figure 17 bar graph shows discovery to first oil cycle times of the seven post-2014 sanctioned deepwater projects.

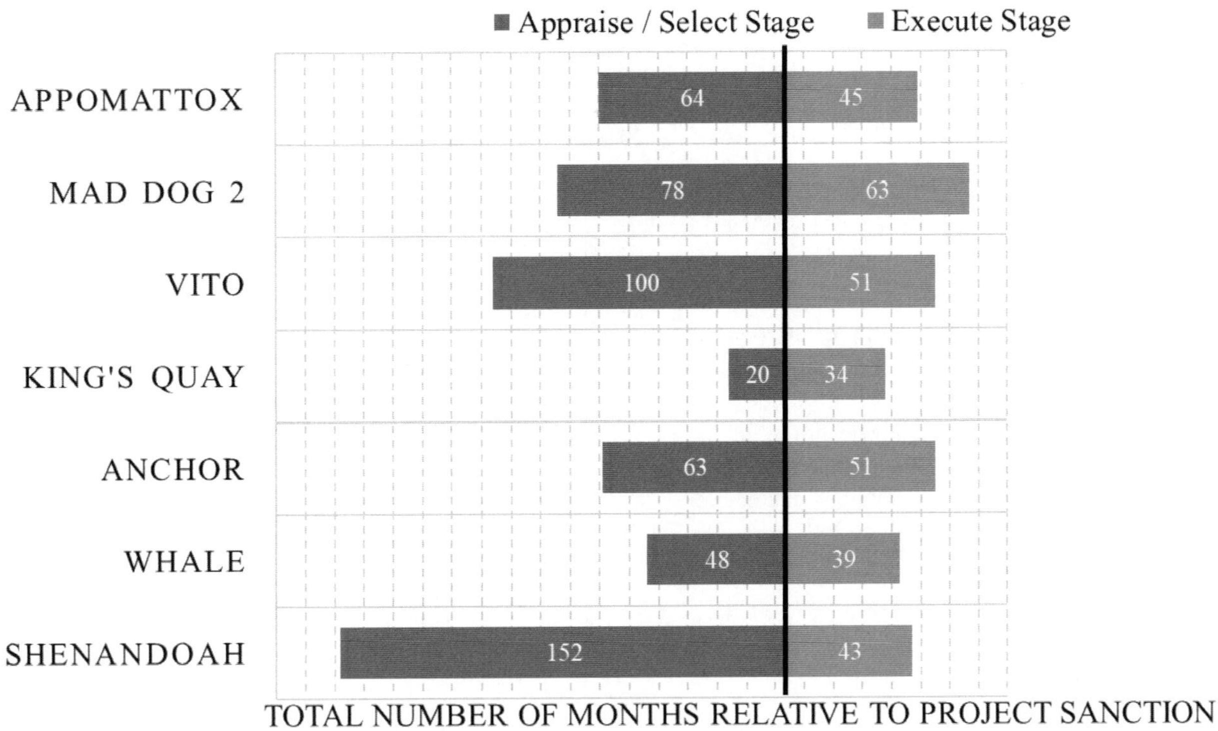

Figure 17 Discovery to First Production cycle time trends for Post-2014 GoM Projects

The average discovery to sanction cycle time is 75 months, sanction to first production, 46.6 months and thus discovery to first production, 121.6 months, or about 10 years and 2 months. First oil dates for six of the seven projects are current best estimates as the projects are still in execution. Contrast these with the ten pre-2014 sanctioned deepwater project corresponding cycle times, which are 46.4, 34, and 80.4 months respectively. The average topside operating weight of the post-2014 sanctioned FPSs is about 16,000 mt, whereas the average topside operating weight of the pre-2014 sanctioned FPSs is about 16,500 mt. It is instructive to assess why the cycle times for the post-2014 cycle times are so much longer than the pre-2014 cycle times, even though the average topside operating weights are similar.

Discovery to Sanction Cycle Times

Generally, the discovery to sanction cycle time is dictated by the appraisal programs. Deeper, high-pressure, Lower Tertiary reservoirs require more extensive appraisal drilling to derisk the reservoir and increase confidence in the expected recoverable resource. Four of the ten pre-2014 sanctioned developments were in the Lower Tertiary with an average 78-month cycle time. The five developments in the Miocene reservoirs averaged 28 months. Tubular Bells was an outlier as there was a change in Operatorship seven years after the discovery.

Four of the seven post-2014 developments were high or ultra-high-pressure and/or high temperature reservoirs (Appomattox, Anchor, Whale and Shenandoah) requiring either extensive appraisal drilling or technology qualification of subsea and drilling equipment. Shenandoah (152-month cycle time) is an outlier because Anadarko spent eight years in appraisal before relinquishing their stake to LLOG, who two years later sold it to Beacon. The project also had to deal with the 2014 price collapse. Considerable time for technology qualification of ultra-high-pressure (20,000 psi) equipment was required. Average discovery to sanction cycle time for the other three high pressure developments is about 58 months, about 20 months faster than the corresponding pre-2014 sanctioned high-pressure projects.

Average discovery to sanction cycle time for the three Miocene discoveries (Mad Dog Phase 2, Vito, King's Quay) was 64 months, almost three years longer than the corresponding pre-2014 sanctioned projects. The Mad Dog Phase

2 development plan was recycled three years into Select/Define activities, as the capex had ballooned to $20 billion. The recycled plan was executed in 36 months, despite being affected by the mid-2014 oil price collapse. Shell queued up the Vito project for sanction shortly after sanctioning the Appomattox development, about six years after the 2009 discovery. They then spent an additional three years redesigning the development from the subsurface up to drive the break-even cost below $35/boe because of the 2014 oil price collapse.

The King's Quay discovery to sanction cycle time of 20 months was consistent with Operator Murphy's aggressive approach to developing projects.

Sanction to First Production Cycle Times

Figure 18 compares pre-2014 and post-2014 sanction to first oil cycle times versus topside operating weight. The pre-2014 cycle time curve is the best fit of the ten projects assessed by the authors in their earlier paper. There is no discernable trend for the post-2014 sanctioned projects. Appomattox and King's Quay follow the pre-2014 sanctioned projects trend line. Cycle times for the other five projects are nine to twenty months greater than the pre-2014 trend line. Each was affected to varying degrees by the global Covid pandemic that created labor shortages, work stoppages and supply chain disruptions. In the last two years there has also been discernable escalation in equipment and bulk costs, labor rates and drilling rig day rates. These first oil delays and cost inflation may adversely impact sanctioned project rate of return, depending on the realized cost of the product (oil and gas) at the time of sale.

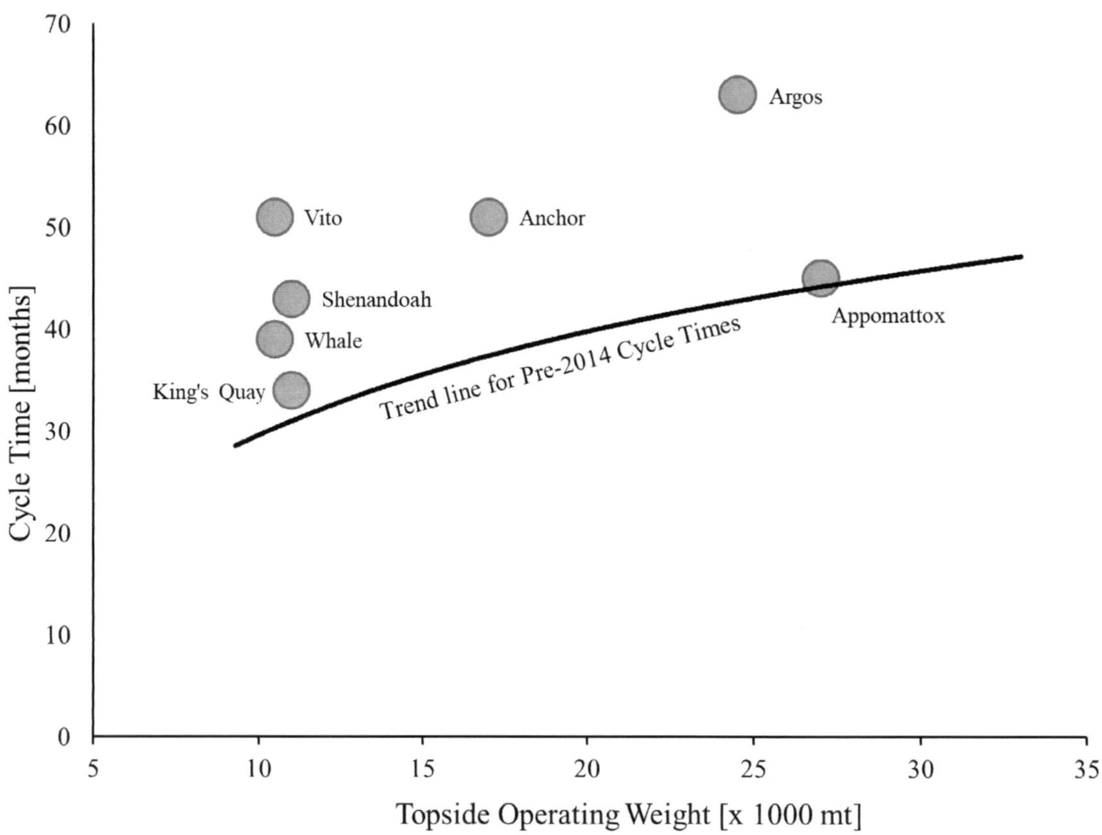

Figure 18 Pre- and Post-2014 FPS Sanction to First Oil Cycle Time Trends vs Topside Weight

FPS Contracting Strategies

Most FPS projects sanctioned from 2005 to 2014 followed a familiar template: build the hull overseas and fabricate the topside in a Gulf Coast yard. Topside integration to TLP and Semisubmersible hulls was mostly done quayside by KOS with their HLD. Topsides integration on Spar hulls was done on site with SSCVs. With a few exceptions, contracts for topside, hull fabrication and integration and installation were awarded separately. Shell, Anadarko and BP installed a majority of the FPSs during this period. They typically negotiated engineering, fabrication, integration and installation contracts with preferred contractors. There was the occasional EPC contract (Red Hawk Spar) or a lease-operate contract (Gomez and Thunderhawk Semisubmersibles, Stones FPSO).

Post-2014, Operators were under relentless pressure to substantially reduce capex. The big breakthrough was to design leaner, standardized, repeatable FPSs. As the number of sanctioned projects fell, competition from Korean and Singapore fabricators became more intense, creating an opportunity for further cost reductions. By then many yards had established track records for EPC delivery of large integrated FPSs or sixth and seventh generation semisubmersible drilling rigs. In addition, the capabilities of heavy lift dry transport vessels kept growing, with the BOKA Vanguard (introduced in 2013 with Chevron's Jack St. Malo FPS as its first cargo) capable of transporting a large FPS with a dry weight exceeding 100,000 mt. This combination of events led to a rethink of contracting strategies of post-2014 sanctioned FPSs.

When Shell sanctioned Appomattox in July 2015, they had already committed to their traditional contracting approach of awarding segmented fabrication contracts for the hull (SHI), topsides (KOS) and integration (KOS). Their radical rework of the Vito development included the FPS contracting strategy. They awarded a turnkey EPC contract for the Vito FPS to Sembcorp following a round of competitive bidding. They repeated the contract strategy for the Whale FPS delivery but negotiated the contract with Sembcorp to take advantage of the efficiencies derived from the Vito FPS learning curve.

BP was the first major Operator to depart from the traditional contracting approach for a major FPS, when they sanctioned the Argos FPS. After several bidding rounds with multiple contractors, they awarded a turnkey EPC contract for the FPS delivery to SHI. The 70,000 mt FPS was dry towed from South Korea to the GoM on the BOKA Vanguard.

Chevron, however, elected to stay with the traditional segmented contracting strategy for the Anchor FPS: hull fabricated in South Korea by DSME, and the topsides fabricated and integrated by KOS in Texas.

Murphy (King's Quay) and Beacon (Shenandoah) chose a contracting strategy similar to that adopted by Shell for Vito and Whale, to take advantage of the efficiencies of building two similar hulls back-to-back. HHI was awarded an EPC contract for turnkey delivery of both FPSs.

The jury is still out as to whether a turnkey EPC contracting approach yields a superior commercial/operational result than the traditional segmented approach. The issue has been complicated by the pandemic that has caused unforeseen delays and distorted sanction to first production cycle times.

Future Trends

In their 2021 OTC paper (OTC-31155) the authors identified three trends for the post-2014 tranche of FPS projects in the GoM. The first was the selection of the semisubmersible FPS as a universal host platform. The second was specifying leaner platforms, standardized specifications, and phased developments. The third was turnkey EPC contracting for FPS delivery. Taken together, these have enabled the more recent projects to be sanctioned at break-even costs in the $35 to $40/boe range.

The next tranche of GoM floating platforms will respond to two principal drivers:

- Continued focus on capex and opex reduction, and
- Reducing greenhouse gas emissions from operations

As we see it, these will be addressed by digitization, reduced manning, and decarbonization.

Digitization and Reduced Manning

Manning on an FPS is a significant contributor to capex, operating costs, safety incidents and greenhouse gas emissions. A substantial reduction in manning will reduce all the above. Reduced manning has been a particular focus of Operators in the North Sea and Western Australia for many years because of platform remoteness, harsh metocean conditions and high labor costs. By borrowing from rapid advances in automation, robotics, and digitization in other industries, they have reduced manning on several complex fixed and floating platforms. Several normally unmanned fixed platforms have been operating successfully for many years and some upcoming floating platforms will also be normally unmanned. An enabler has been industry collaboration with government and classification agencies who are also vested in improving safety, reducing emissions and driving new production.

The design of a minimally manned platform must use high reliability equipment to minimize inspection and maintenance, remote diagnostics for critical equipment, full control from a remote (preferably shore based) location, digital twins and special purpose walk to work vessels. Much of this technology is transferrable to/from offshore wind farms, that are normally unmanned.

As an example, Chevron, with partners Exxon and Shell recently (July 2021) sanctioned the Jansz-Io compression project offshore Western Australia. It includes the construction and installation of a 27,000 mt displacement normally unattended Semisubmersible platform in 1,350 m of water, powered by a 135 km long high voltage cable from shore. It houses electric drives, transformers and breakers for the subsea compression station on the seabed below, that will ship gas to the Gorgon LNG facility on Barrow Island.

Decarbonization

Energy transition is a reality as the world rebalances energy supplied by fossil fuels and energy supplied by renewables. The Oil and Gas industry is being pressed by governments, institutional investors, and other stakeholders to reduce its carbon footprint. Most Operators and NOC's have expressed their intent to do so with some major Operators publicly declared their goals to achieve carbon neutrality by 2050. Fossil fuels will continue to be an important contributor to the energy mix, but greenhouse gas emissions related to oil and gas production will have to be substantially reduced.

Focusing strictly on an operational FPS, its largest source of greenhouse gas emissions is the power generation facility, followed by transport and resupply. Power generated on future FPSs will reduce emissions by some combination of

- Increasing engine efficiency
- Switching to cleaner burning fuels, and
- Using electrical power supplied from renewable sources

The implementation of these cleaner power options will mean added upfront costs but there will be value accretive savings from improved operational efficiencies, greater uptime, lower maintenance and carbon taxes. Many of these initiatives are underway, particularly in the North Sea, where electrification of offshore platforms with power generated from onshore renewable energy is being implemented. With rapid advances in floating offshore wind, where electricity generated from a single turbine could provide sufficient energy to power a medium sized FPS, and with costs becoming competitive with conventional power generation, it is quite possible that future deepwater FPSs in the GoM could have base power supplied by a floating wind turbine moored alongside.

Carbon emissions related to transportation and resupply will be substantially reduced by transitioning to minimally or normally unattended installations. Reducing manning and carbon emissions will be accretive to capex and opex reductions.

Conclusions

The dramatic oil and gas price drop in mid-2014 had a profound impact on deepwater developments in the GoM. Anticipating persistent low oil prices, the industry dialed back the number of sanctioned projects to about one a year. The focus shifted to greater capital discipline and shorter cycle times. The bar for project sanction was set at a breakeven cost of $40 per boe. Much of the early reductions came from capex deflation of the supply chain. A step change came from vastly improving drilling efficiencies and simplifying and standardizing designs and specifications of subsea and floating production systems. Further reductions were achieved by designing smaller, more efficient, repeatable platforms. By doing so, Operators gave up flexibility for future expansion and enhanced recovery. The Semisubmersible platform was selected for all seven of the post-2014 sanctioned developments.

Contracting strategy for FPS fabrication shifted from the traditional segmented approach to a turnkey EPC approach, in part to take advantage of Far East fabrication yards desperation for work. Discovery to sanction cycle times suffered as Operators reset and revised development plans to lower prices. An additional factor, in some cases, was developing high pressure Paleogene and Norphlet reservoirs that required qualifying drilling, completion and subsea equipment for these ratings and extensive appraisal programs. Unfortunately, sanction to first production cycle times increased relative to pre-2014 sanctioned projects because of the global pandemic that affected six of the seven projects causing delays up to twelve months.

The next capex and opex breakthrough will be from reducing or eliminating manning. Most operators have committed to or will be required to reduce carbon emissions for their next tranche of projects. Over the years, the offshore industry has shown great resilience in responding to oil price and external shocks by being flexible, adaptable and innovative. These qualities have been amply demonstrated by the post-2014 FPS projects sanctioned in the deepwater Gulf of Mexico.

Acknowledgements

The authors acknowledge the support of Richard B Offshore and KBR Consulting in preparing this manuscript. Thanks also to Bruce Beaubouef, at Offshore magazine; Jordan Barnes and Colin White at Rystad Energy and Rick Fowler for their input, relevant publications and feedback. Industry magazines (Offshore Engineer, Offshore Technology and Upstream) and conferences such as this SNAME Symposium and the Offshore Technology Conference, that provide industry with current and relevant information on field developments and technologies have been a wonderful source of material for this paper.

References

Shell Advances Deepwater Projects with Capital Efficiency, Re-Engineering; Offshore Magazine, November 2017

Khurana, Sandeep, D'Souza Richard et al, "Bringing Upstream Projects to Final Investment Decision", OTC 27949, Houston 2017

Aimin Wang et al, "Case Studies of Industrial Standardizing and Cost-Efficient Offshore Concept", OTC 27813, Houston, 2017

D'Souza Richard, Basu Shiladitya, "A Universal Field Development Approach for Advancing Stalled Gulf of Mexico Projects", OTC 28970, Houston, 2018

D'Souza Richard, Basu Shiladitya. "A Critical Assessment of Deepwater Gulf of Mexico Floating Platform EPCI contracting Strategies and Cycle Times", OTC 31115, Houston 2021

Turner William, "Floating Production Systems: What's next?", OTC 30715, Houston 2020

D'Souza Richard, Basu Shiladitya, "The Semisubmersible Floating Production System: From Argyll to Appomattox", OTC 30907, Houston 2020

Lagrange, Elgonda, "New Concepts for Normally Unattended Installation NUI-Design, Operation and Digitization", OTC 29367, Houston 2019

Francis Thomas, "Appomattox: The Energy Project that Defied the Odds", May 2019

Shenandoah gets the go ahead in the Deepwater GoM, Offshore Magazine, August 2021

Shenandoah Deepwater Oil Field, GoM, Offshore Technology, August 2021

Shell Sanctions Deepwater Whale Project in the GoM, Offshore Magazine, July 2021

Topsides and Hull Work moves Forward on Anchor, Whale, Offshore Magazine, April 2020

Chevron Hits Go on Anchor, Offshore Engineer, December 2019

Vito Deepwater Oilfield Development, Offshore Technology, April 2018

King's Quay Arrives in Texas, Offshore Magazine, September 2021

King's Quay FPS Facility, Offshore Technology, January 2020

Re-Engineered Mad Dog 2 Gets Greenlight, Offshore Magazine, January 2018

Mad Dog 2 FPU arrives in Texas, Offshore Magazine, April 2021

Shell Advances Deepwater Projects with Capital Efficiency, Re-Engineering, Offshore Magazine, November 2017

NOMENCLATURE

20K	20,000 psi system
AC	Alaminos Canyon
boepd	barrels of oil equivalent per day
bopd	barrels of oil per day
bwpd	barrels of water per day
COSCO	China Ocean Shipping (Group) Company
DD	Dry dock
DSME	Daewoo Shipbuilding and Marine Engineering
EPC	Engineer-Procure-Construct
FID	Final Investment Decision
FPS	Floating Production System
FPSO	Floating Production Storage and Offloading
FPU	Floating Production Unit
GC	Green Canyon
GoM	Gulf of Mexico (US Exclusive Economic Zone)
HCM	Heerema Marine Contractor
HHI	Hyundai Heavy Industries
HLD	Heavy Lift Device, shore-based fixed 10,000 ton lifting system at KOS' Ingleside yard
I	Injection well
KOS	Kiewit Offshore Services
mmboe	million barrels of oil equivalent (m = millennium; 1,000)
mmscfd	million standard cubic feet per day
MSF	Modular Support Frame

mt	metric ton
MW	megawatt
NOC	National Oil Companies
OTC	Offshore Technology Conference
P	Production well
psi	pound per square inch
SHI	Samsung Heavy Industries
SLWR	Steel Lazy-Wave Risers
SSB	Subsea Boosting
SSCV	Semisubmersible Crane Vessel
TLP	Tension Leg Platform
WI	Water Injection
WR	Walker Ridge

HIGH-FIDELITY TURBULENCE MODELING FOR NUMERICAL WAVE TANK APPLICATIONS: EVALUATION OF THE SCALE-RESOLVING PANS APPROACH

AUTHOR (HANEESHA IPHINENI)
AFFILIATION (DEPARTMENT OF OCEAN ENGINEERING, TEXAS A&M UNIVERSITY)

AUTHOR 2 (BJÖRN WINDÉN)
AFFILIATION (DEPARTMENT OF OCEAN ENGINEERING, TEXAS A&M UNIVERSITY)

AUTHOR 3 (SHARATH S. GIRIMAJI)
AFFILIATION (DEPARTMENT OF AEROSPACE ENGINEERING, TEXAS A&M UNIVERSITY)
AFFILIATION (DEPARTMENT OF OCEAN ENGINEERING, TEXAS A&M UNIVERSITY)

Proceedings of the 27th Offshore Symposium, February 22nd, 2022, Houston, Texas
Texas Section of the Society of Naval Architects and Marine Engineers

Abstract

For analyzing vessel seakeeping, platform wave loads and similar tasks, potential flow-based methods are still the workhorse Numerical Wave Tank (NWT) used by industry. Recently, NWTs have employed more advanced viscous flow simulation methods. These are commonly based on the Reynolds-Averaged Navier-Stokes (RANS) approach. RANS is often preferred by industry due to it's lower computational cost in relation to better scale resolving methods. The RANS approach however, has limits in NWT applications. Firstly, it is not suited for studying unsteady turbulence phenomena at different scales of wave and fluid motion. Secondly, RANS turbulence models assume that the energy transport in turbulent flows can be described as a dissipative process. In a NWT where large parts of the flow is unsteady, but not dominated by turbulent variations, this could lead to non-physical dissipation of waves.

Focusing on the second issue, the applicability of the Partially-Averaged Navier-Stokes (PANS) approach in NWTs is investigated in this paper. PANS is a suite of turbulence models of varying modeled to resolved scale ratios ranging from RANS to Direct Numerical Simulation (DNS). PANS modeling allows for more direct control of how energy dissipation is treated. It is hypothesized that this will be beneficial in NWTs by reducing non-physical dissipation of waves.

Initial results for a regular wave tank are promising. About 40% of the wave amplitude was lost downstream with the RANS approach versus less than 3% with the PANS approach; using the same computational grid and with only a slight increase in the computational effort. Similar results were obtained for a self-contained "dam break" simulation.

Keywords: Numerical Wave Tank, PANS, dam break, wave elevation

1 Introduction

Numerical wave tanks (NWTs) have been used for many years as a computational tool that supplements physical wave basins in a various fields of engineering. NWTs can capture pertinent hydrodynamic non-linearities, including turbulent phenomena, providing accurate results [1]. The accessibility of open-source software is one of the driving factors behind the increased use of NWTs. Numerical methods used in NWTs range from potential flow to fully viscous CFD [1]. The advantage of potential flow methods is their low computational cost and ability to accurately represent propagating waves. However, they are unable to accurately represent wave/object interaction in many situations where non-linear and viscous effects are prominent. Viscous methods, such as those based on Reynolds Averaged Navier Stokes (RANS) equations offer the ability to study such effects in more detail for many engineering applications. However, current higher order methods may only be applicable to studying off-design and highly specific conditions due to their associated computational effort [2].

One issue with using RANS-based methods to study wave-object propagation is their dissipative nature. RANS turbulence closure models assume that the energy transport in turbulent flows can be described as a dissipative process. In a fluid domain where large parts of the flow is unsteady, but not dominated by turbulent variations, applying such a dissipative process could artificially dampen the flow. This is the case with propagating gravity waves in a NWT. Other factors, such as the choice of numerical schemes to model convection and the interface capturing scheme, also affect the accuracy of the predicted wave propagation. This often means that, in order to ensure that waves are not artificially dampened by the RANS model, a finer discretization in required near the free surface. This in turn leads to increased computational effort.

In this study, we investigate an alternative to RANS known as Partially Averaged Navier Stokes (PANS). The PANS method is an 'accuracy on demand' turbulence simulation approach that is a bridging method developed by [3]. PANS is essentially a suite of turbulence closure models of varying modeled to resolved scale ratios ranging from RANS to Direct Numerical Simulation (DNS) of the Navier-Stokes (NS) equations. The degree of turbulence resolution is controlled by two parameters - unresolved-to-total ratios of kinetic energy and dissipation: $f_k = \frac{k_u}{k}$, $f_\varepsilon = \frac{\varepsilon_u}{\varepsilon}$. At one extreme, PANS reduces to the classical RANS when $f_k = f_\varepsilon = 1$. For high Reynolds number flows, f_ε can be taken to be unity as most of the dissipation will occur in the unresolved scales. With decreasing f_k more and more scales of turbulent motion are resolved. As $f_k \to 0$, PANS asymptotes to DNS. The closure model coefficients are functions of f_k such that the required degree of eddy viscosity is achieved [4]. The numerical resolution (grid) required will increase with reducing f_k [5]. Over the years, PANS has been tested in a variety of wake flows and wall-bounded flows [6–8].

In contrast to RANS, PANS modeling allows for more direct control of how energy dissipation is treated. It is hypothesized that this will be beneficial in NWTs. The turbulence modeling accuracy and computational cost increases by decreasing the f_k and the optimal choice for f_k for a particular simulation is determined by a balance between computational resources available and desired accuracy of computed physics.

In the present study, a PANS method based on a standard two equation $k - \varepsilon$ formulation is used with four cutoff filter (f_k) widths, $f_k = 1.0$(RANS), $f_k = 0.6$, $f_k = 0.3$ and $f_k = 0.2$. $f_\varepsilon = 1.0$ for all different f_k values. In this paper, the PANS formulation will be employed to study a transient "dam break" case with an obstacle. The dam break case is interesting because of its closed and transient nature which generates its own turbulence, eliminating potential bias to the result from factors such as inlet turbulence, wave generation/absorption and so on; allowing for a direct analysis of the energy dissipation. In addition to the dam break, initial results for the application of the PANS method to a regular wave tank are presented. This shows the promise of the method for fulfilling the ultimate goal of developing an efficient Numerical Wave Tank for studying industry-relevant problems of wave/object interaction.

OpenFOAM version v2106 was used to perform all simulations. OpenFOAM [9] is an open source CFD toolbox which allows for sharing of the developed NWT methodology to a wide audience of both industry and academia.

This introduction is followed by a brief description of the governing equations, numerical simulation set-up, results and finally, conclusions are drawn from the analysis.

2 Governing Equations

Partially-Averaged Navier Stokes equations are based on the filtering approach proposed by [10]. In partial filtering of the Navier Stokes equations, the Reynolds rule of the mean is not valid and the classical analysis of turbulence in terms of means and fluctuations becomes complicated. [10] presented a new operational approach to study the turbulence based on the algebraic properties of the filtered representations of a turbulence field at different levels; unlike the statistical approach.

The PANS formulation reduces to DNS at the extreme value of f_k of 0 and to RANS at a f_k value of 1.0, thereby effectively acting as a bridging method between the two ends of the turbulence modeling spectrum. PANS model development is based on filtering the parent Navier Stokes equations using an arbitrary filter. This partial decomposition of velocity V(x,t) given in Equation 1 consists of a filtered velocity $U(x,t) = \langle V(x,t) \rangle$ and residual velocity u(x,t) where $\langle \rangle$ denotes arbitrary (explicit or implicit) filter.

$$V_i(x,t) = \langle V_i(x,t) \rangle + u_i(x,t) = U_i(x,t) + u_i(x,t); \langle u_i(x,t) \rangle \neq 0 \tag{1}$$

The parent instantaneous incompressible Navier Stokes equations, given in Equations 2 and 3 are partially filtered.

$$\frac{\partial V_i}{\partial t} + V_j \frac{\partial V_i}{\partial x_j} = -\frac{\partial p}{\partial x_i} + \nu \frac{\partial^2 V_i}{\partial x_j \partial x_j} \tag{2}$$

$$\frac{\partial^2 p}{\partial x_i \partial x_i} = -\frac{\partial V_i}{\partial x_j} \frac{\partial V_j}{\partial x_i} \tag{3}$$

If the angular brackets in the partially averaged flow variables, $-U_i = \langle V_i \rangle$; $p_U = \langle p \rangle$ denotes a filter which is constant preserving and commutes with spatial and temporal differentiation, then the governing equations for the resolved velocity field U_i are given by Equations 4 and 5 as suggested by [3].

$$\frac{\partial U_i}{\partial t} + U_j \frac{\partial U_i}{\partial x_j} + \frac{\partial \tau(V_i, V_j)}{\partial x_j} = -\frac{\partial p_U}{\partial x_i} + \nu \frac{\partial^2 U_i}{\partial x_j \partial x_j} \tag{4}$$

$$-\frac{\partial^2 p_U}{\partial x_i \partial x_i} = \frac{\partial U_i}{\partial x_j} \frac{\partial U_j}{\partial x_i} + \frac{\partial^2 \tau(V_i, V_j)}{\partial x_i \partial x_j} \tag{5}$$

In Equation 5, $\tau(V_i, V_j)$ is the generalized central second moment. Generalized central second and third moments of random variables A, B, and C are defined as

$$\tau(A, B) = \langle AB \rangle - \langle A \rangle \langle B \rangle \tag{6}$$

$$\tau(A, B, C) = \langle ABC \rangle - \langle A \rangle \tau(B, C) - \langle B \rangle \tau(C, A) - \langle C \rangle \tau(A, B) - \langle A \rangle \langle B \rangle \langle C \rangle \tag{7}$$

The evolution equation can be written as

$$\frac{\partial \tau(V_i, V_j}{\partial t} + U_k \frac{\partial \tau(V_i, V_j)}{\partial x_k} = P_{ij} + \phi_{ij} - D_{ij} + T_{ij} \tag{8}$$

In Equation 8, the various terms on the right-hand side can be classified as production (P_{ij}), pressure-correlation(ϕ_{ij}), dissipation(D_{ij}) and transport(T_{ij}) of Sub Filter Scale (SFS) stress:

$$P_{ij} = -\tau(V_i, V_j) \frac{\partial U_j}{\partial x_k} - \tau(V_j, V_k) \frac{\partial U_i}{\partial x_k}; \tag{9}$$

$$\phi_{ij} = 2\tau(p, S_{ij}); \tag{10}$$

$$D_{ij} = 2\nu\tau \left(\frac{\partial U_i}{\partial x_k}, \frac{\partial U_j}{\partial x_k} \right); \tag{11}$$

$$T_{ij} = -\frac{\partial}{\partial x_k} \left(\tau(V_i, V_j, V_k) + \tau(p, V_i)\delta_{ik} - \nu \frac{\partial \tau(V_i, V_j)}{\partial x_k} \right) \tag{12}$$

The unclosed term $\tau(V_i, V_j)$, known as the SFS stress term, needs to be modelled to obtain fully closed system of equations. To model the SFS term, the Boussinesq approximation is invoked as follows

$$\tau(V_i, V_j) = -\nu_u S_{ij} \tag{13}$$

where ν_u is the eddy viscosity of unresolved scales and is given as

$$\nu_u = C_\mu \frac{k_u^2}{\varepsilon_u} \tag{14}$$

The subscript u in Equation 14 indicates PANS unresolved statistics and k_u and ε_u are unresolved turbulent kinetic energy and unresolved dissipation respectively.

The resolved stress tensor $\overline{S_{ij}}$ is given in terms of the resolved velocity field as

$$\overline{S_{ij}} = \frac{1}{2}\left(\frac{\partial U_i}{\partial x_j} + \frac{\partial U_i}{\partial x_i}\right) \tag{15}$$

To close the above system of equations, unresolved turbulent kinetic energy and unresolved turbulent specific dissipation have to be modelled using partial averaging. For this purpose, a two equation linear eddy viscosity $k-\varepsilon$ model is used and the following section presents a modification of the model to be used with the PANS equations.

2.1 Modification of $k-\varepsilon$ turbulence model for PANS application

From the form of the evolution equation of the SFS and $\tau(V_i, V_j)$ for the closure SFS obtained in the above section, it can be seen that the equations closely resemble their respective RANS counterparts. Therefore, it can be deduced that the structure of the filtered equations is invariant to the particular averaging operation. This is called the averaging invariance property. This property can be utilized to broadly describe the structure of the evolution equations for the filtered $k-\varepsilon$ turbulence model which is similar in essence to that in the equivalent RANS turbulence model.

The extension of the RANS $k-\varepsilon$ turbulence model to PANS is presented in the following manner.

The unresolved turbulent kinetic energy, k_u and unresolved dissipation, ε_u can be defined as

$$k_u = \frac{1}{2}\tau(V_i, V_j) \tag{16}$$

$$\varepsilon_u = \nu\tau\left(\frac{\partial V_i}{\partial x_j}, \frac{\partial V_i}{\partial x_j}\right) \tag{17}$$

The ratios of the unresolved to total kinetic energy and unresolved to total dissipation are given as

$$f_k = \frac{k_u}{k}; f_\varepsilon = \frac{\varepsilon_u}{\varepsilon}; \tag{18}$$

To derive the k_u and ε_u evolution equations, their RANS counterparts (k and ε) are assumed as a starting point. The RANS $k-\varepsilon$ turbulence model is obtained from [11]. If the filter width f_k is assumed constant then the evolution of kinetic energy in PANS can be directly related to RANS. Only the final form of the k_u and ε_u equations are presented below. The unresolved kinetic energy transport equation is given by:

$$\frac{\partial k_u}{\partial t} = P_u - \varepsilon_u + \frac{\partial}{\partial x_j}\left((\frac{\nu_u}{\sigma_{ku}} + \nu)\frac{\partial k_u}{\partial x_j}\right) \tag{19}$$

and the unresolved dissipation transport equation is given by:

$$\frac{\partial \varepsilon_u}{\partial t} = f_k\left(C_{\varepsilon 1}\frac{P_u \varepsilon_u}{k_u} - C^*_{\varepsilon 2}\frac{\varepsilon_u^2}{K_u}\right) + \frac{\partial}{\partial x_j}\left((\frac{\nu_u}{\sigma_{\varepsilon u}} + \nu)\frac{\partial \varepsilon_u}{\partial x_j}\right) \tag{20}$$

The modified model coefficients are given by:

$$C^*_{\varepsilon 2} \equiv C_{\varepsilon 1} + \frac{f_k}{f_\varepsilon}(C_{\varepsilon 2} - C_{\varepsilon 1}); \sigma_{ku} \equiv \sigma_k \frac{f_k^2}{f_\varepsilon}; \sigma_{\varepsilon u} \equiv \sigma_\varepsilon \frac{f_k^2}{f_\varepsilon} \tag{21}$$

where $C_{\varepsilon 1} = 1.44$; $C_{\varepsilon 2} = 1.92$; $\sigma_k = 1.0$; $\sigma_\varepsilon = 1.3$.

The unresolved production term, P_u is due to special transfer of kinetic energy from resolved to unresolved scales, and not due to direct interaction with the mean flow. A zero transport model

(ZTM) is used in the above turbulence model development which is apt for high Reynolds number flows and assumes that the resolved fluctuations do not contribute to the SFS energy transport. The transport model for PANS application is clearly stated in [5].

3 Numerical Simulation Setup for DamBreak

The free surface flow, in 3D, following the release of a fluid column (dam break) towards an obstacle is modelled with the PANS approach. After the fluid column is released, a standing wave is formed in the tank. The hypothesis is that the dissipative nature of RANS vs. PANS will be seen clearly in how quickly the amplitude of the standing wave decays. The computational domain used in the simulations is shown in Fig. 1. This domain is not meant to be representative of any existing experimental facility or engineering application. It is merely designed to test the above mentioned hypothesis. The obstacle generates turbulence upon impact by the fluid column, thereby eliminating inlet turbulence as a potential bias in the conclusions.

The OpenFOAM solver interFoam was used which is a large time-step transient solver for two incompressible, isothermal immiscible fluids using a Volume Of Fluid (VOF) method to capture the water/air interface and the PIMPLE(merged PISO-SIMPLE) algorithm for pressure-velocity coupling. This standard solver is used in conjunction with the modified $k - \varepsilon$ model from Section 2.1. The flow is assumed to be unsteady, incompressible and an adjustable time step is specified to maintain a maximum Courant number of 1. Euler and bounded Gauss upwind schemes were used to solve time derivatives, gradients, divergence- and Laplacian terms ([12]).

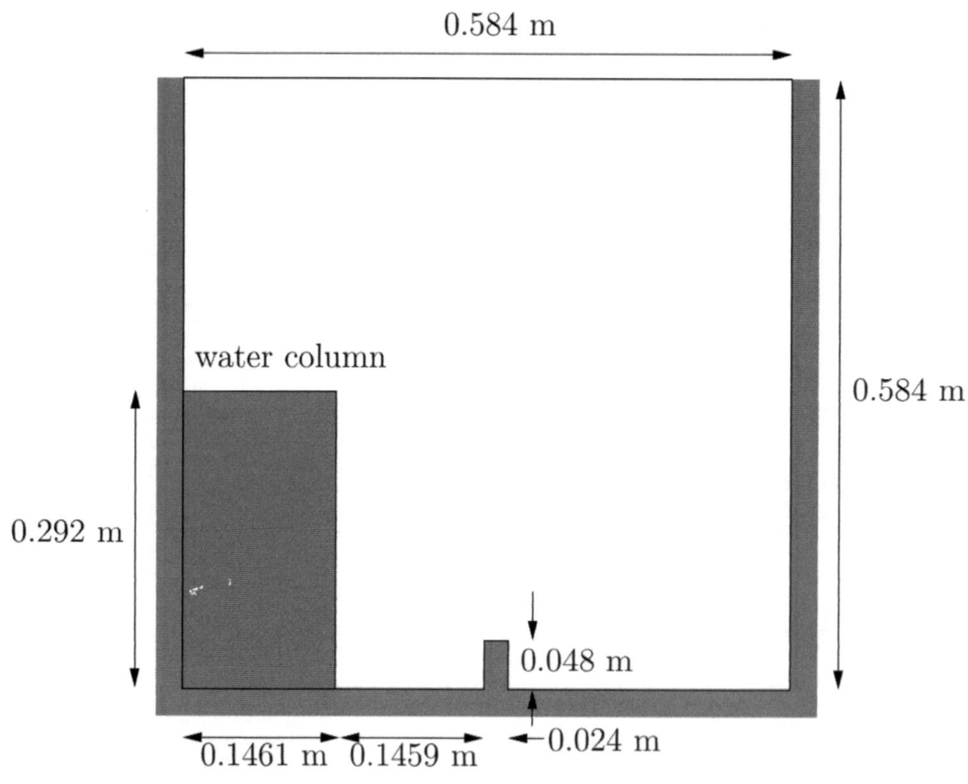

Figure 1: Overview of the computational domain

3.1 Boundary- and Initial Conditions

A no-slip boundary condition is imposed on the left and right walls as well as the tank bottom. A cyclic boundary condition at the sides of the domain. At the top, an adaptive boundary condition that changes between inlet and outlet depending on the flow gradient is imposed to emulate a boundary open to the atmosphere. For the turbulent properties, an initial kinetic energy of 0.1 was assumed throughout the domain and wall functions were employed on all wall boundaries. Gravity is considered to be -9.81 m/s^2 in the y-direction. The physical properties for water is defined as $\nu = 10^{-06} m^2/s$ and $\rho = 1000 kg/m^3$ and for air as $\nu = 1.48^{-05} m^2/s$ and $\rho = 1 kg/m^3$.

3.2 Mesh description

The computational domain is a square domain with a side of 0.584 m. The water column is of dimensions 0.1461 m x 0.292 m and is located at the lower left of the domain. The domain also features an obstacle close to the center. To discretize this domain, a block structured mesh, created with the blockMesh utility in OpenFOAM is employed as shown in Fig. 2. A side view of the resulting mesh is shown in Fig. 3.

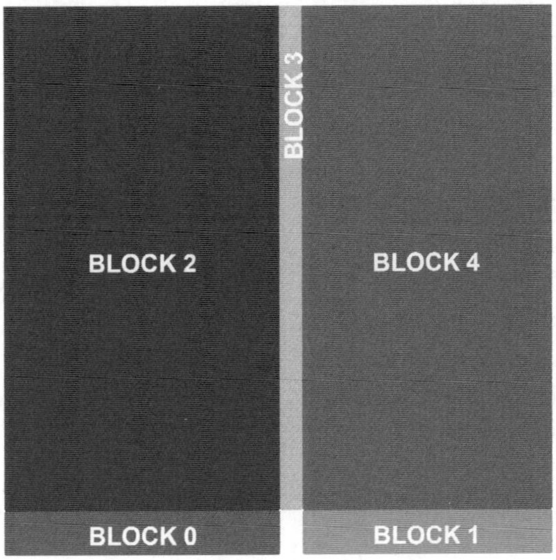

Figure 2: Schematic view of block structured mesh in the dam break case

Figure 3: Final mesh structure for the dam break case

4 Results

The data collected from the dam break simulations is used to study the evolution of the fluid column upon release. This is done by tracking the interface (VOF volume fraction = 0.5) at certain locations (probes) in the tank.

Fig. 4 shows the evolution of the water column, where alpha.water is the VOF volume fraction, as it hits the obstacle for the $f_k = 0.2$ case. A standing wave can be seen to have formed at 5.7 s.

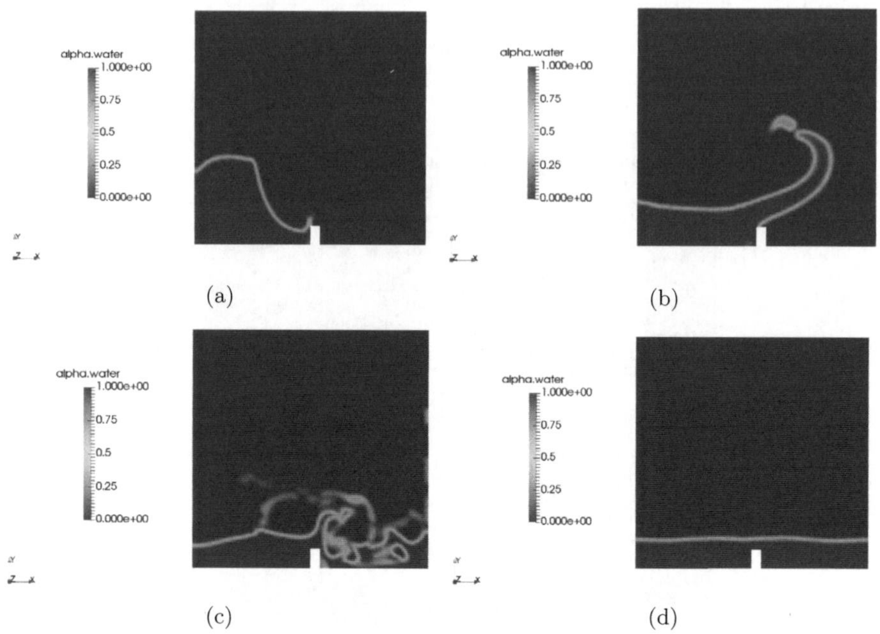

Figure 4: Evolution of the water column for $f_k = 0.2$; shown at (a) $t = 0.2s$, (b) $t = 0.4s$, (c) $t = 0.8s$ and (d) $t = 5.7s$.

Two probes are set at different locations along the center of the tank ($z = 0$) to study the developing waves. Probe 1 is located at $x = 0.292m$ and probe 2 at $x = 0.316m$ where x is the horizontal distance from the left wall. Fig. 5 shows the wave elevation at these locations from the CFD simulations with f_k of 0.2, 0.3, 0.6 and 1 (RANS). The presented wave elevation is relative to the height of the obstacle ($y = 0.048m$).

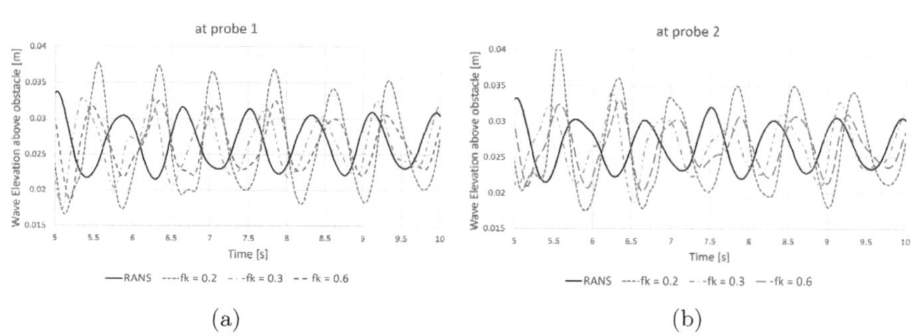

Figure 5: Wave elevation plots for different f_k values at (a) probe 1 and (b) probe 2. The wave elevation is relative to the height of the obstacle ($y = 0.048m$).

Table 1 shows the total computational time (normalized with the computational time of RANS) for different values of f_k.

f_k	*Normalized computational time*
1 (RANS)	1
0.6	1.45
0.3	1.56
0.2	1.64

Table 1: Normalized computational time required for Dam Break

5 Initial results for application of PANS in a Numerical Wave Tank

The free surface flow in a rectangular basin, 20 m x 4 m x 1.5 m, with Stokes 2nd order waves being generated at the inlet is modeled. The computational domain is shown in Fig. 6. The simulation setup is similar to that in the dam break case with some differences highlighted below.

Figure 6: Geometry of numerical wave tank

The boundary conditions and physical properties are the same as for the dam break case with the exception of the inlet and the outlet boundary conditions. The inlet is given a volume fraction and velocity profile as to generate a 2nd order Stokes wave propagating downstream. The outlet is set to absorb any incoming wave. Turbulence was treated in the same way as in the dam break case but with the inlet turbulence being modified for different values of f_k to be consistent with the PANS formulation. The generated waves have a wave height = 0.1 m and wave period = 3 s.

The instantaneous wave elevation ([13]) at time = t is given as:

$$\eta = \frac{H}{2}\cos\left(kx - \omega t + \phi\right) + k\frac{H^2}{4}\frac{3-\sigma^2}{4\sigma^3}\cos\left(2(kx - \omega t + \phi)\right) \tag{22}$$

where H = wave height, k = wave number, ω = angular frequency, σ = radian wave frequency, ϕ = phase shift and t = time.

The mesh is refined near the free surface with 20 cells per wave height in the region where the waves propagate (from z = 0.6 m to z = 0.8 m) ([14]).

Wave probes are located at a distance of 15 m and 18 m from the inlet of the tank. The presented wave elevation is relative to the still water level of the tank.

Fig. 7 shows the wave elevation for f_k of 0.2, 0.3, 0.6 and 1 (RANS) at different locations. It can be seen that the value of f_k has a large impact on the downstream wave amplitude with $f_k = 1$ (RANS) showing the largest loss of energy.

6 Summary and Conclusions

Simulations of the flow after a dam break and wave generation in a rectangular wave tank using the Partially Averaged Navier Stokes (PANS) equations with modified $k-\varepsilon$ model have been carried out

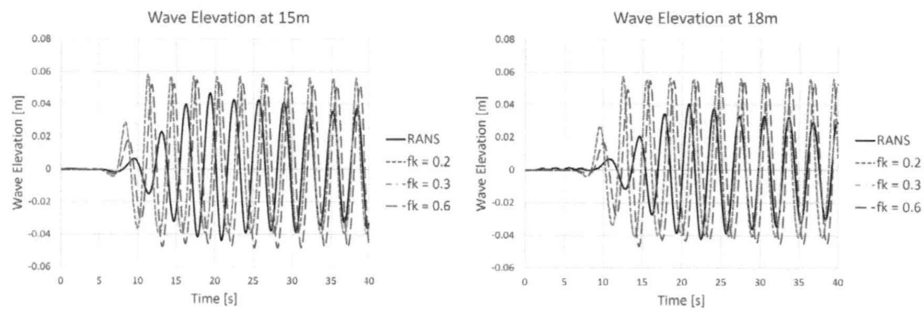

(a) (b)

Figure 7: Wave elevation (relative to the tank still water level) at (a) 15 m and (b) 18 m from the inlet in the regular wave tank

with the open-source CFD toolbox OpenFOAM. The hypothesis behind this study is that PANS will provide a benefit over RANS in these simulations in that it would be less dissipative and thus allow wave propagation with less loss of energy (wave amplitude.)

The time histories of wave elevation have been plotted for both the dam break case and the regular wave tank. In the dam break case, the amplitude of the standing wave is increased by 18% for PANS with $f_k = 0.2$ compared to RANS. This is achieved with only a small increase in the computational cost. While there is no experimental data to compare with, it is expected that a higher amplitude wave is closer to reality. Similarly, in the regular wave tank, the RANS simulation loses 40.72% of the input wave amplitude at 18 m downstream whereas PANS with $f_k = 0.2$ only loses about 2.34%. Here, it is expected that the amplitude loss would be very small in a physical wave tank meaning that PANS comes closer to reality.

More research into the intricacies of the flow in both the dam break and the regular wave tank is required. Furthermore, comparisons with experimental data need to be done in order to verify the conclusions drawn here in terms of wave amplitude loss when compared to a physical wave tank. It should also be noted that this study has focused on wave propagation and energy dissipation, an area where potential flow methods would yield superior accuracy at much lower costs. However, when considering practical applications of NWTs to industry-relevant problems of viscous wave-object interactions, it is expected that the PANS approach will prove to be a very useful tool.

These findings show great promise for the PANS method in future wave modeling and NWTs, with enhanced accuracy and minimal additional computational work. Future work will involve studying in detail what specific aspects of the flow regime give rise to the difference between PANS and RANS models in terms of energy dissipation. An acceptable trade-off in terms of accuracy vs. computational effort also needs to be found in terms of the choice of f_k value. Ultimately, the choice of f_k value will depend on the needs of the specific application and NWT user. Increased accuracy at the same computational effort or maintained accuracy with reduced computational effort could both be achieved.

References

[1] Christian Windt, Josh Davidson, and John V Ringwood. "High-fidelity numerical modelling of ocean wave energy systems: A review of computational fluid dynamics-based numerical wave tanks". In: *Renewable and Sustainable Energy Reviews* 93 (2018), pp. 610–630.

[2] Josh Davidson and Ronan Costello. "Efficient nonlinear hydrodynamic models for wave energy converter design—A scoping study". In: *Journal of Marine Science and Engineering* 8.1 (2020), p. 35.

[3] Sharath S Girimaji. "Partially-averaged Navier-Stokes model for turbulence: A Reynolds-averaged Navier-Stokes to direct numerical simulation bridging method". In: (2006).

[4] Sharath S Girimaji, Eunhwan Jeong, and Ravi Srinivasan. "Partially averaged Navier-Stokes method for turbulence: Fixed point analysis and comparison with unsteady partially averaged Navier-Stokes". In: (2006).

[5] Sharath Girimaji and Khaled Abdol-Hamid. "Partially-averaged Navier Stokes model for turbulence: Implementation and validation". In: *43rd AIAA Aerospace Sciences Meeting and Exhibit.* 2005, p. 502.

[6] Sunil Lakshmipathy and Sharath S Girimaji. "Partially averaged Navier–Stokes (PANS) method for turbulence simulations: flow past a circular cylinder". In: *Journal of Fluids Engineering* 132.12 (2010).

[7] Branislav Basara et al. "Near-wall formulation of the Partially Averaged Navier Stokes turbulence model". In: *AIAA journal* 49.12 (2011), pp. 2627–2636.

[8] Branislav Basara, Sinisa Krajnovic, and Sharath Girimaji. "PANS vs. LES for computations of the flow around a 3D bluff body". In: *Proc. of ERCOFTAC 7th Int. Symp.-ETMM7, Lymassol, Cyprus.* Vol. 2. 2008, p. 3.

[9] *OpenFOAM.* `https://www.openfoam.com/documentation/guides/latest/doc/index.html`.

[10] Massimo Germano. "Turbulence: the filtering approach". In: *Journal of Fluid Mechanics* 238 (1992), pp. 325–336.

[11] David C Wilcox et al. *Turbulence modeling for CFD.* Vol. 2. DCW industries La Canada, CA, 1998.

[12] *Dambreak.* `https://cfd.direct/openfoam/user-guide/v6-dambreak/`.

[13] *Dambreak.* `https://www.openfoam.com/documentation/guides/latest/doc/guide-wave-regular-stokesii.html`.

[14] Christian Windt et al. "On the assessment of numerical wave makers in CFD simulations". In: *Journal of Marine Science and Engineering* 7.2 (2019), p. 47.

VIScous Vorticity Equation (VISVE) Model Applied to oscillatory Turbulent Flow around a Cylinder

Rui You
Ph.D. Candidate, Ocean Engineering Group, CAEE, The University of Texas at Austin

Akhilesh S. Tiwari
Graduate Student, Ocean Engineering Group, CAEE, The University of Texas at Austin

Spyros A. Kinnas
Professor, Ocean Engineering Group, CAEE, The University of Texas at Austin

Proceedings of the 27th Offshore Symposium, February 22nd, 2022, Houston, Texas
Texas Section of the Society of Naval Architects and Marine Engineers

ABSTRACT

The VIScous Vorticity Equation (VISVE) method has previously been applied to the laminar oscillatory flow around a circular cylinder (Li and Kinnas (2017); Wu and Kinnas (2021a)). In this work, the VISVE method is applied to the oscillatory turbulent flow around a cylinder. To handle the flow within the turbulent regime, the $k-\omega$ SST turbulence model is employed for calculating the turbulent viscosity. The VISVE and the $k-\omega$ SST 2-equation turbulence model are both discretized based on the Finite Volume Method (FVM), with Quadratic Upstream Interpolation for Convective Kinematics (QUICK) method dealing with the convective terms, and the Alternating Direction Implicit (ADI) method calculating the unsteady terms. The method is applied to various Reynolds and Keulegan-Carpenter numbers, and the results, including vorticities and velocities are compared with those from a Reynolds-Averaged Navier-Stokes (RANS) solver. The results of turbulent VISVE agree well with those from RANS for KC =2, but deviate from RANS at a higher KC number.

Keywords: Viscous Vorticity Equation (VISVE), Oscillatory Turbulent Flow, $k-\omega$ SST Turbulence Model, Keulegan-Carpenter Number, Reynold Number

Introduction

Oscillating flows represent an important aspect of fluid dynamics and appear in various applications in ocean engineering. Oscillating flow means that the fluid motion possesses a dominant frequency, which can be maintained either by an oscillating boundary condition, by an oscillating external force, or by self-oscillations of a flow. A classic example is the generation of Vortex-Induced Vibrations (VIV). Vortex shedding generates a continuously changing pressure, resulting in fluctuations of forces on the cylinder, which is an important source of fatigue. Comprehensive reviews have been delivered for Vortex-Induced Vibrations problems by Williamson and Govardhan

(2008). When applying numerical approaches, many offshore structures can be simplified as cylinders, such as the legs and risers of offshore platforms.

Vortex methods can be characterized into two regimes: the grid-based Eulerian approach and the Lagrangian discretization. The Lagrangian approach, as described by Stock (2006), can be classified as particles and segments in 2-D, and correspond to filaments and sheets in fully 3-D models. A comparison of singular and regularized vortex particle methods, and their vorticity updating equations are given by Winckelmans and Leonard (1993). An example of a modern approach is presented by Ploumhans, Winckelmans, Salmon, Leonard and Warren (2002). A primary advantage of vortex particle methods is that they do not rely on structural properties of the tracked vorticity elements. Vortex filaments in a 3D vortex method are employed by Leonard (1980, 1975). Chorin (1980) uses simpler, but still filament-connected method. Leonard (1975) proposes topology change for vortex filaments in very close proximity. Agishtein and Migdal (1989) demonstrate a three-dimensional vortex method based on vorticity discretization into sheets. Knio and Ghoniem (1992, 1991) employ triangular and quadrilateral elements to discretize the vorticity and scalar gradient in their simulations of vortex sheets. Discretization of vorticity can be based on volumes, which has one higher dimension. Huyer and Grant (2000) demonstrate two-dimensional volume discretizations using Delaunay triangulations. Marshall, Grant, Gossler and Huyer (2000) apply tetrahedra to discretize the vorticity. The combination of the above methods is used by Airiau and Giovanni (1999). Other methods that can be applied to solve vortex are Level Set Method (Osher and Sethian (1988); Chang, Hou, Merriman and Osher (1996)), semi-Lagrangian particles (Wiin-Nielsen (1959); Robert (1981)) and impulse elements method (Buttke (1992); Cortez (1998)). For the realm of Eulerian approach, Napolitano and Catalano (1991) apply a multigrid solver; Tryggvason, Abdollahi-Alibeik, Willmarth and Hirsa (1992) use a finite di˙erence method to solve a collision of a viscous, two-dimensional vortex pair; Weinan and Liu (1997) apply finite difference methods to 3-D unsteady viscous incompressible flows on non-staggered grids.

In the previous work, Vorticity Equation (VISVE) for laminar flow was solved by using a finite volume method (Tian and Kinnas (2015); Tian (2014)). Tian (2014) developed the VISVE solver for laminar flows. Laminar VISVE has been applied on 2-D and 3-D hydrofoils, circular cylinders, and propeller blades (see Tian and Kinnas (2015), Wu, Kinnas, Li and Wu (2019), Wu and Kinnas (2019) and Wu and Kinnas (2021a)). Comprehensive parallelization using MPI and OpenMP hybrid program is accomplished by Wu and Kinnas (2021b). Xing (2018) and Iliopoulos and Kinnas (2021) consider the compressibility and implement the laminar cavitating flow for VISVE. Kinnas (2020) derives the vorticity equation for turbulent compressible flows with variable viscosity. Yao and Kinnas (2019) and Wu and Kinnas (2020) develop a coupling method by incorporating VISVE and a turbulence model in OpenFOAM, to solve turbulent flow. However, this coupling method does not solve the turbulence in VISVE and it needs to read the turbulent viscosity from OpenFOAM at every time step, which means VISVE solver cannot solve turbulent flow independently. The VIScous Vorticity Equation (VISVE) method has previously been applied to the laminar oscillatory flow around a circular cylinder (Li and Kinnas (2017)).

In this paper, the turbulent oscillating flow around a cylinder will be investigated by the turbulent VIScous Vorticity Equation (VISVE) Model. The turbulent $k-\omega$ sst model is discretized and solved in the VISVE code, to enable VISVE solver to solve turbulent flow independently. A different gridding system compared with Li and Kinnas (2017) was employed to consider the effects of turbulence. The predicted vorticities and velocities are compared to those from a commercial CFD software (ANSYS Fluent).

METHODOLOGY

1. Turbulent VIScous Vorticity Equation

The viscous vorticity equation for turbulent flows with variable density and viscosity is derived by Kinnas (2020). By taking the curl of the Navier-Stokes equation, it gives the vorticity transport equation. It only depends on vorticity, velocity, and viscosity. This transport equation for vorticity has several simplifications, which are essentially 2-dimensional, incompressible, constant-temperature, single-phase turbulent flows. Because of $\omega = \nabla \times q$, we have $\nabla \cdot \omega = 0$. There is no continuity equation needed for ω in 2D.

$$\frac{\partial \omega}{\partial t} + \nabla \cdot (\omega q) = \nabla^2\left(\left(v_m + v_\tau\right)\omega\right) \quad (1)$$

Where **q** is the velocity vector; ω is the vorticity; ν_m is the kinematic molecular viscosity; ν_τ is the kinematic turbulent viscosity of the fluid.

2. Vorticity-Velocity solver

The relation of stream function ψ and velocity q was defined by:

$$\nabla \times \psi = q \tag{2}$$

Take curl of both sides, with $\omega = \nabla \times q$, we can get:

$$\nabla^2 \psi = -\omega \tag{3}$$

The stream function ψ can be calculated from vorticity ω using three Poisson's equations via the Green's function in free space:

$$\psi(x_f) = -\int_V \omega(x) G(x, x_f) dV + \psi_b \tag{4}$$

Where $G(x, x_f) = -1/(4\pi|x - x_f|)$ is the Green's function for the Laplacian in 3D. x_f is the field point where ψ is calculated and x is the dummy variable running over the computational domain. $\psi_b = (0, xU_{\infty,z} - zU_{\infty,x}, 0)$ is the stream function corresponding to the background flow.

3. Turbulent model

$k - \omega$ SST model was implemented in the code to take the turbulence into account. This model is proposed by Menter (1994) and it belongs to a two-equation eddy viscosity model. It is composed of two transport equations for turbulent kinetic energy k and specific dissipation rate ω.

$$\frac{\partial k}{\partial t} + U_j \frac{\partial k}{\partial x_j} = P - \beta^* k\omega' + \frac{\partial}{\partial x_j}\left[\left(\nu + \sigma_k \nu_T\right)\frac{\partial k}{\partial x_j}\right] \tag{5}$$

$$\frac{\partial \omega'}{\partial t} + U_j \frac{\partial \omega'}{\partial x_j} = \frac{\gamma}{\nu_T} P - \beta \omega'^2 + \frac{\partial}{\partial x_j}\left[\left(\nu + \sigma_\omega \nu_T\right)\frac{\partial \omega'}{\partial x_j}\right] + 2\left(1 - F_1\right)\sigma_{\omega 2}\frac{1}{\omega'}\frac{\partial k}{\partial x_i}\frac{\partial \omega'}{\partial x_i} \tag{6}$$

Where k is the turbulent kinetic energy and ω' is the specific dissipation rate. (Note: This ω' is the specific dissipation rate from $k - \omega$ SST turbulence model, and it needs to be distinguished from the vorticity ω in the previous VISVE model.) P is the production for turbulent kinetic energy or dissipation rate.

4. Oscillating flow

The oscillating flow was realized by

$$U = -U_m \cos\sigma t \tag{7}$$

Where U_m is the amplitude of the velocity, t is the time, and σ is the frequency.

The Keulegan–Carpenter (KC) number can give information on the relative importance of the drag forces over inertia forces and is defined as:

$$K_C = \frac{UmT}{D} \tag{8}$$

Where T=2 π/σ is the period of the oscillation, and D is the diameter of the cylinder.

NUMERICAL METHODS

The VISVE model and the $k-\omega$ SST model are both solved based on the Finite Volume Method (FVM). Detailed descriptions on numerical methods for VISVE model can be found in Wu and Kinnas (2020). Since the same numerical schemes are applied on k equation and ω equation, only the manipulations of k equation will be presented. Take integral over each control volume, and apply Gauss divergence theorem to get the discretized form:

$$\frac{\partial k}{\partial t}\Big|_P \Omega = -\sum_f \vec{q}_f \cdot \vec{n}_f k_f s_f + \sum_f \left(v + \frac{v_T}{\sigma_k}\right)_f \frac{\partial k}{\partial n}\Big|_f s_f + \bar{S}\big|_P \Omega \tag{9}$$

As shown in Figure 1, $\vec{q}_f$ is the velocity vector in the middle of the face; $\vec{n}_f$ is the unit vector perpendicular to the face, pointing outwards of the cell, while Ω and S are the volume of the cell and the area of the surface, respectively.

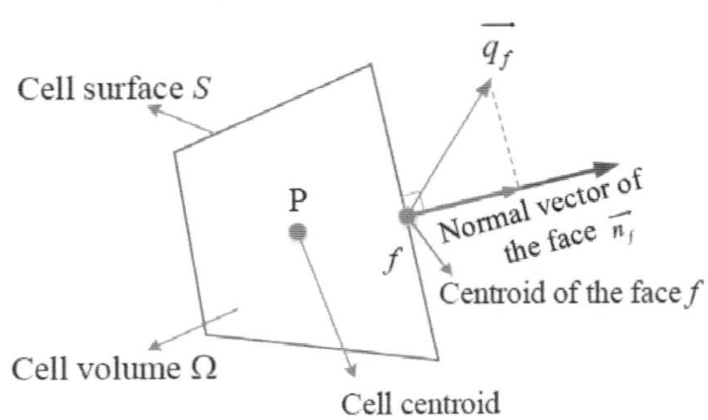

Figure 1. Finite Volume Method over a cell

The convective terms are treated by Quadratic Upwind Interpolation (QUICK) scheme. The diffusion term is discretized by Central Differential Scheme (CDS). The unsteady term is solved by using the Alternating Direction Implicit (ADI) method.

APPLICATION TO TURBULENT OSCILLATING FLOW AROUND A CYLINDER

Schematic view of the turbulent oscillating flow around a cylinder was shown in Figure 2. The velocity is changing its direction around the cylinder.

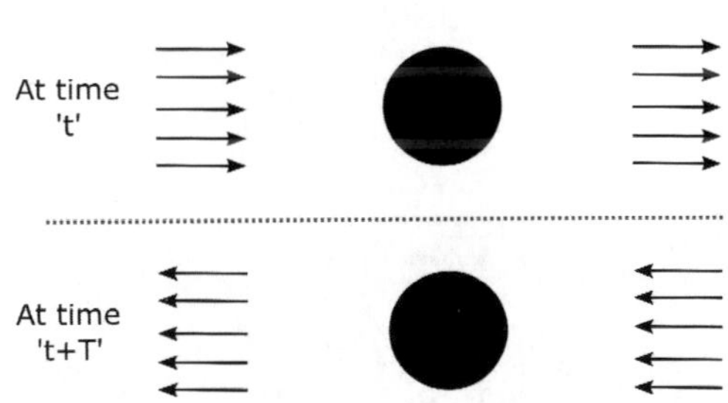

Figure 2: Schematic view of the turbulent oscillating flow around a cylinder

1. VISVE

A Coons is a bi-cubic parametric surface formed by four corner points as shown in Figure 3. Coons patch was applied to form the unstructured mesh in block 2. The sketch drawn in Figure 3 shows the two blocks of the mesh. Block 1 is placed on the left side of $d_1(t)$ curve, including the polar grids and the curved surface composed of $c_0(s)$, $c_1(s)$, $d_0(t)$, $d_1(t)$. The right-hand side of $d_1(t)$ locates block 2. A bi-linearly blended Coons patch is the surface:

$$C(s,t) = L_c(s,t) + L_d(s,t) - B(s,t) \quad (10)$$

Where, $L_c(s, t)$ is the linear interpolation between c_0 and c_1; $L_d(s, t)$ is the linear interpolation between d_0 and d_1; and these two producing two ruled surfaces. $B(s, t)$ is the bi-linear interpolation on the four corners.

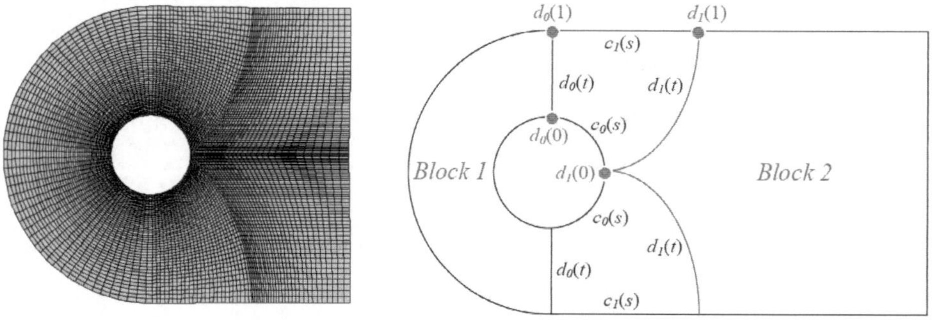

a) Mesh applied in VISVE b) Coons patch

Figure 3. Computational domain used in VISVE

2. Fluent

We have used the commercial package ANSYS-Fluent which is based on unsteady RANS modeling to compare our results. In the given Fluent scheme, RANS turbulent flow is computed using the k-omega SST model. The SIMPLEC scheme is used for the pressure-velocity coupling. Further, discretization schemes for pressure and momentum are done using "standard" and "second order upwind" schemes respectively.

Figure 4 shows the direct comparison of the computational domain. The VISVE method requires a significantly smaller computational domain than the N-S solver, because the vorticity vanishes much closer to the wall boundaries compared with the velocity. The diameter of the outer boundary of the domain used in Fluent is usually 30 times the diameter of the cylinder to exclude the influence from the boundary. Velocity inlet was set at the left-hand side of the boundary and pressure outlet was utilized at the right-hand side of the boundary. Wall boundary condition was assigned on the surface of the cylinder.

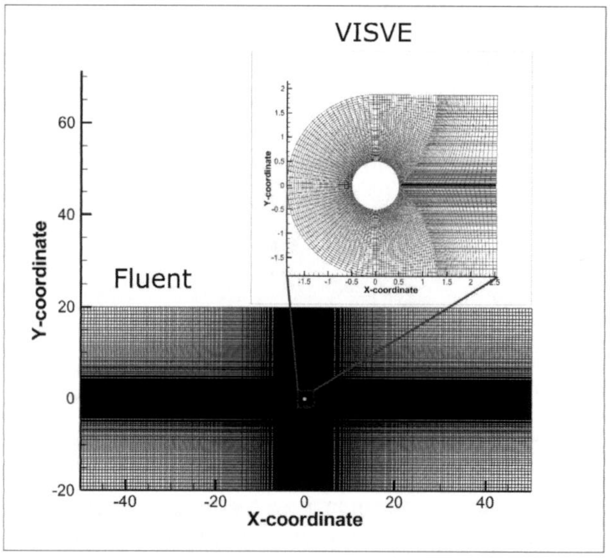

Figure 4. Comparison of computational domain in VISVE and RANS

RESULTS AND DISCUSSIONS

1. Comparisons of VISVE and RANS

Figure 5 and Figure 6 show the comparisons of vorticity contour among laminar VISVE, turbulent VISVE and RANS solvers, with Reynolds number being 1,000 and 40,000, KC=2. The snapshots were taken at 0.5T, 1T, 1.5T and 2T. Agreements can be observed. The conclusion can be made that the laminar results were validated with RANS and the turbulence model does not take effects for a small Reynolds number. Figures 7 and 8 show the comparisons of velocity among laminar VISVE, turbulent VISVE and RANS solvers. Figure 9 shows the comparison of the turbulent kinetic energy between VISVE and RANS. There are differences between them. VISVE has more concentration of the turbulent kinetic energy near the cylinder. The reason for the differences were not known yet and will be investigated further in the future.

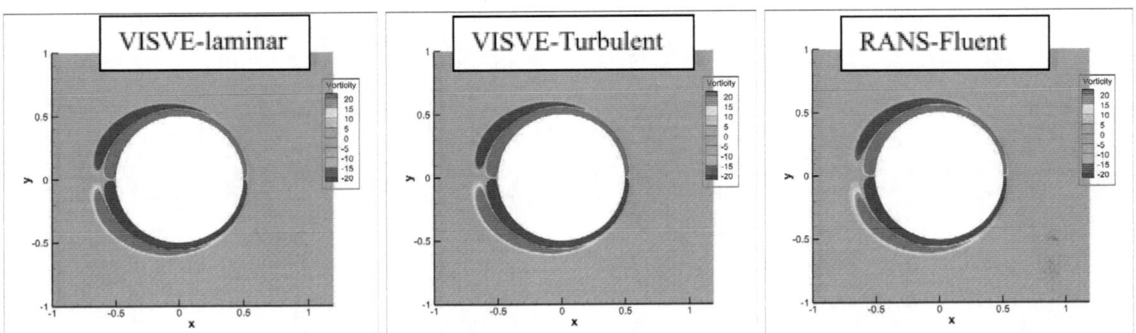

a) t/T=0.5

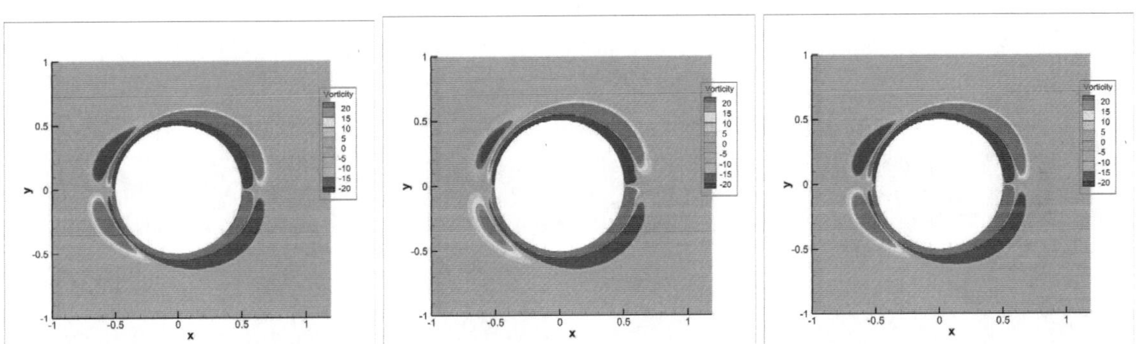

b) t/T=1

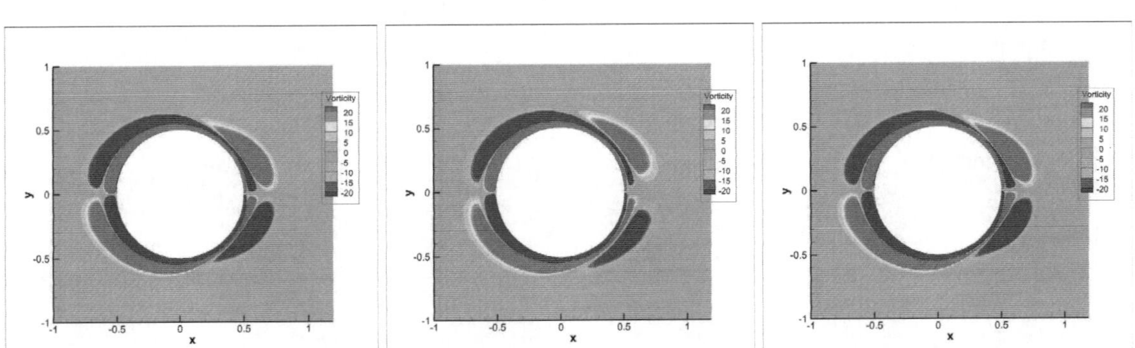

c) t/T=1.5

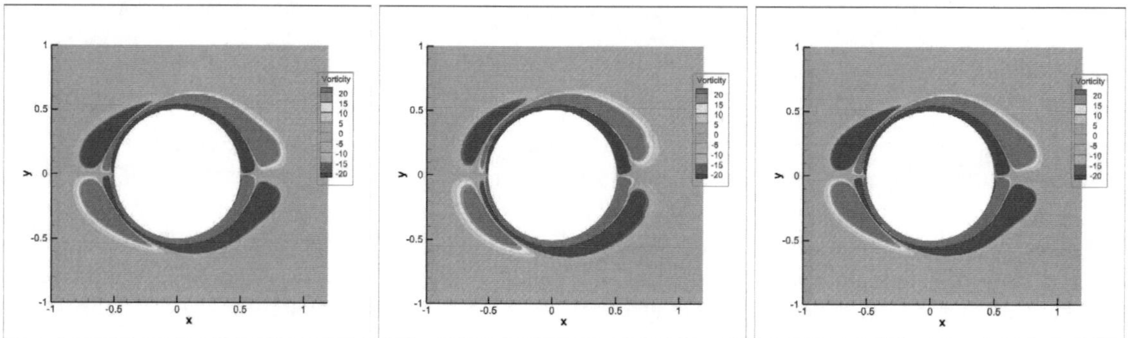

d) t/T=2

Figure 5. Vortices evolving with the different solvers. Left: VISVE –laminar; Middle: VISVE –Turbulent; Right: RANS-Fluent; Re=1,000, KC=2

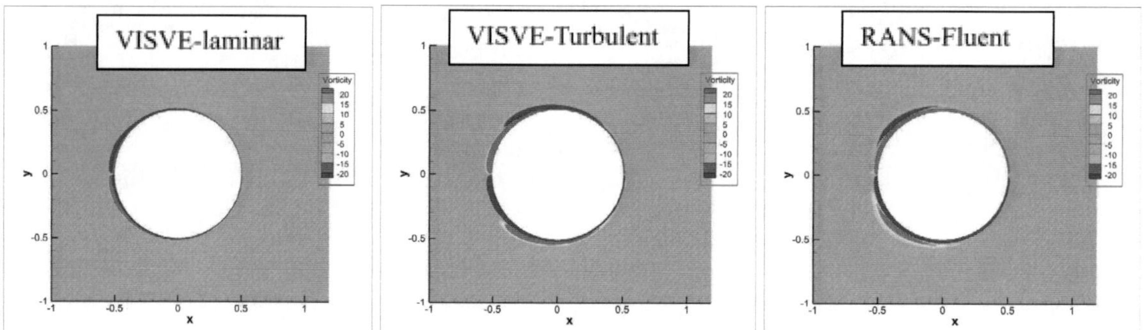

a) t/T=0.5

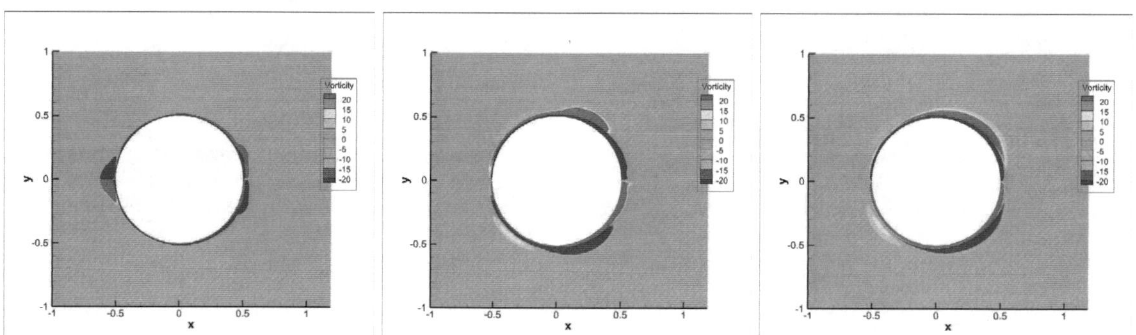

b) t/T=1

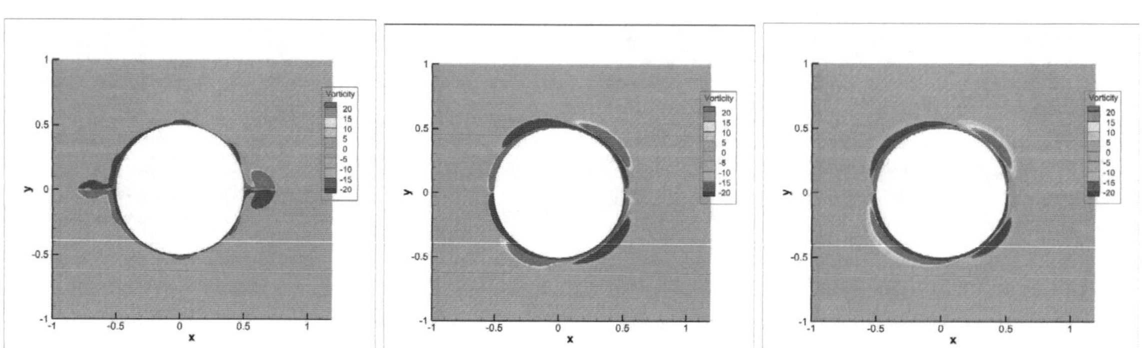

c) t/T=1.5

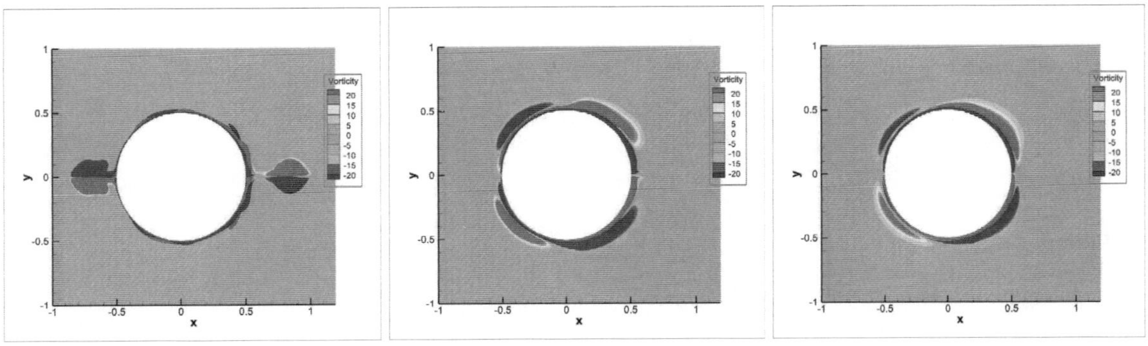

d) t/T=2

Figure 6. Vortices evolving with the different solvers. Left: VISVE –laminar; Middle: VISVE –Turbulent; Right: RANS-Fluent; Re=40,000, KC=2

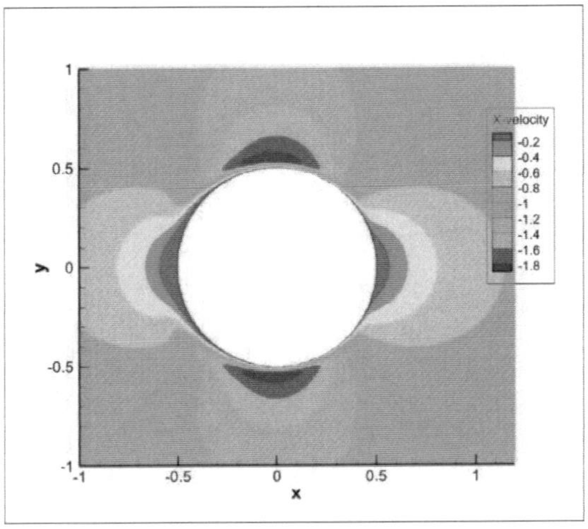

a) VISVE-laminar

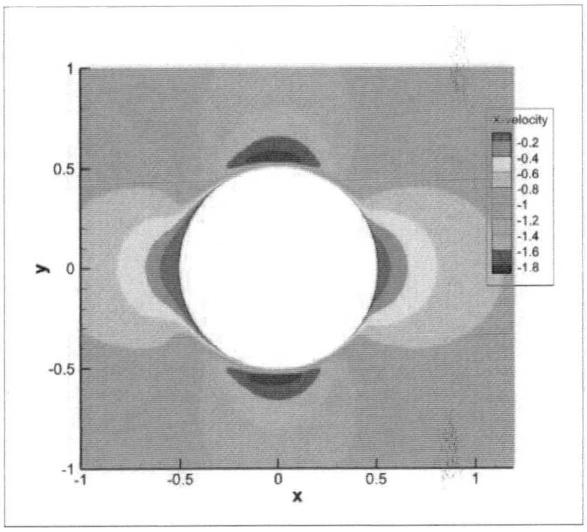

b) VISVE-turbulent

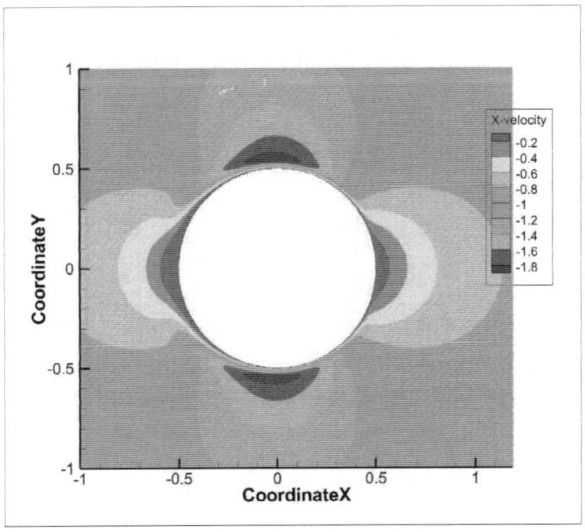

c) RANS

Figure 7. Velocity contours: Re=1000, KC=2; t/T=0.5

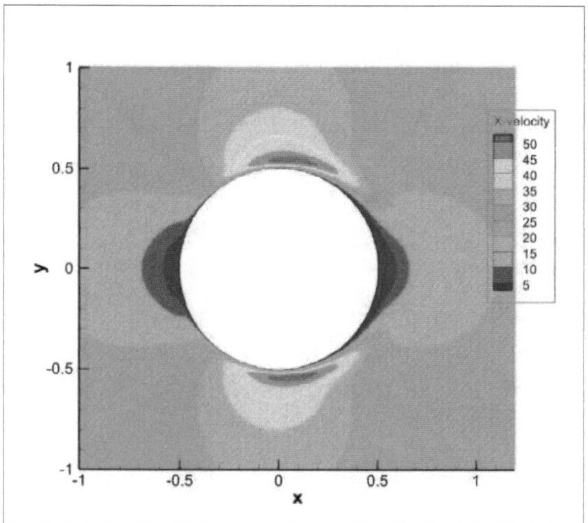

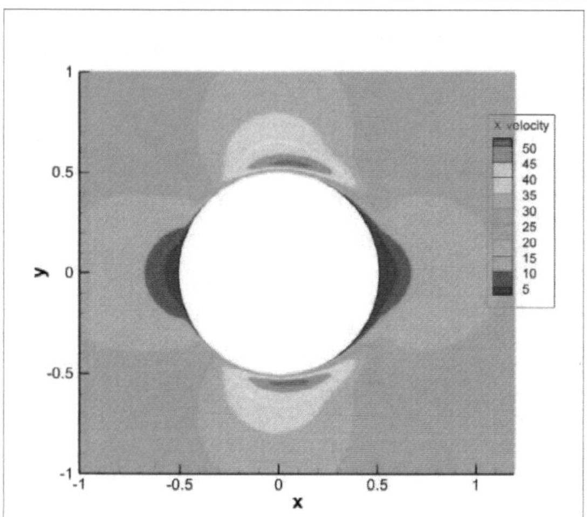

a) VISVE-laminar

b) VISVE-turbulent

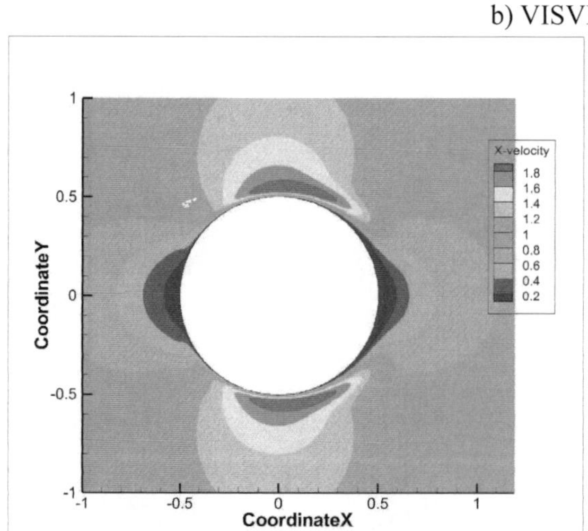

c) RANS

Figure 8. Velocity contours: Re=1000, KC=2; t/T=1

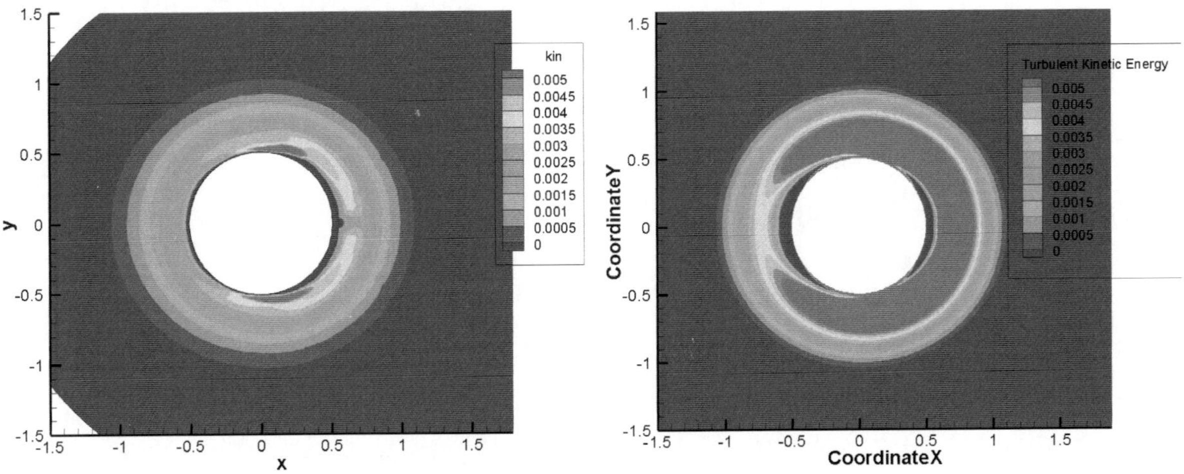

a) VISVE-turbulent   b) Fluent

Figure 9. Turbulent kinetic energy for Re=40000, KC=2, t/T= 1

2. Effects of different Re numbers

The effects of different Re numbers were investigated in this section. Two Reynolds numbers (Re=2,985 and Re=40,000) were examined for KC=2. Figures 10 and 11 show the case with Reynolds number Re=2985, at t/T=0.5, and Figures 12 and 13 show results of Re =2985 at t/T=1. Figure 14 and 15 show that the results of turbulent VISVE agree better with RANS solver than laminar VISVE. At Reynolds number Re=40,000, the inclusion of the turbulence model will contribute to the improvements of the results. Figure 16 shows the comparison of the boundary layers for different Reynolds numbers. The profiles were extracted from the cross-section at x =0, on the top of the cylinder. With the increasing of Re, the boundary layer became thinner.

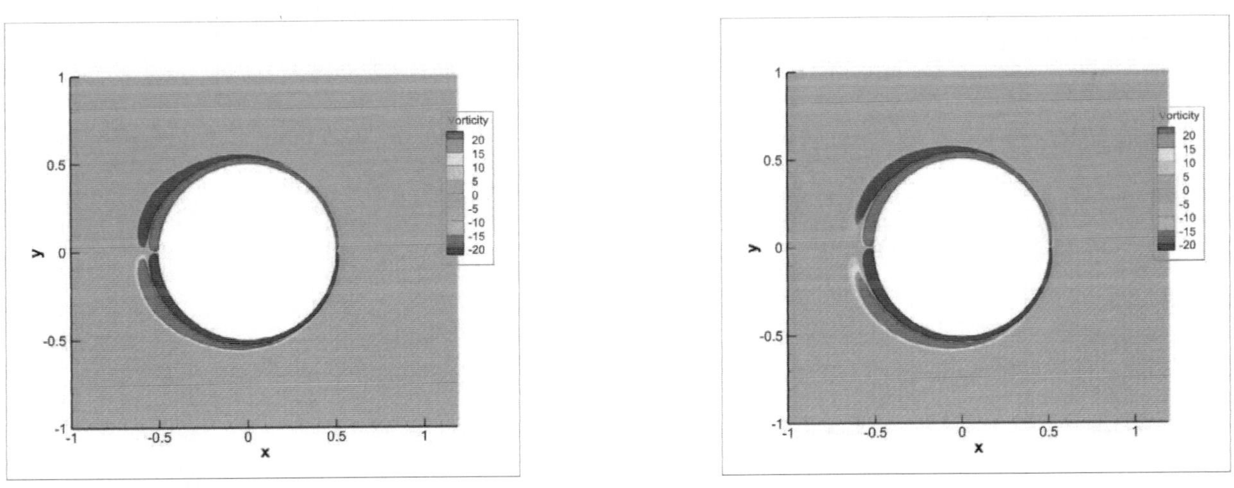

a) VISVE-laminar   b) VISVE-turbulent

Figure 10. Vorticity contours: Re=2985, KC=2; t/T=0.5

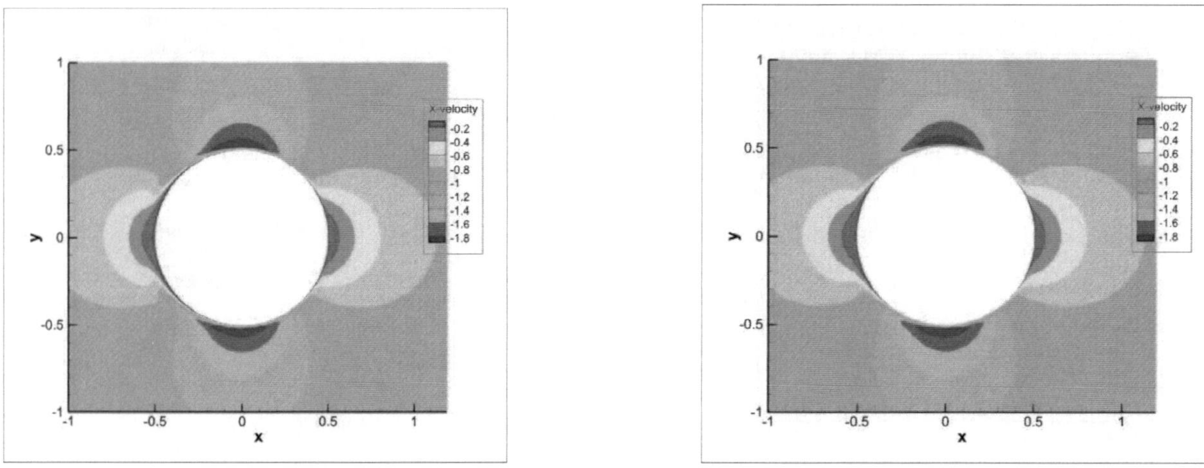

a) VISVE-laminar

b) VISVE-turbulent

Figure 11. Velocity contours: Re=2985, KC=2; t/T=0.5

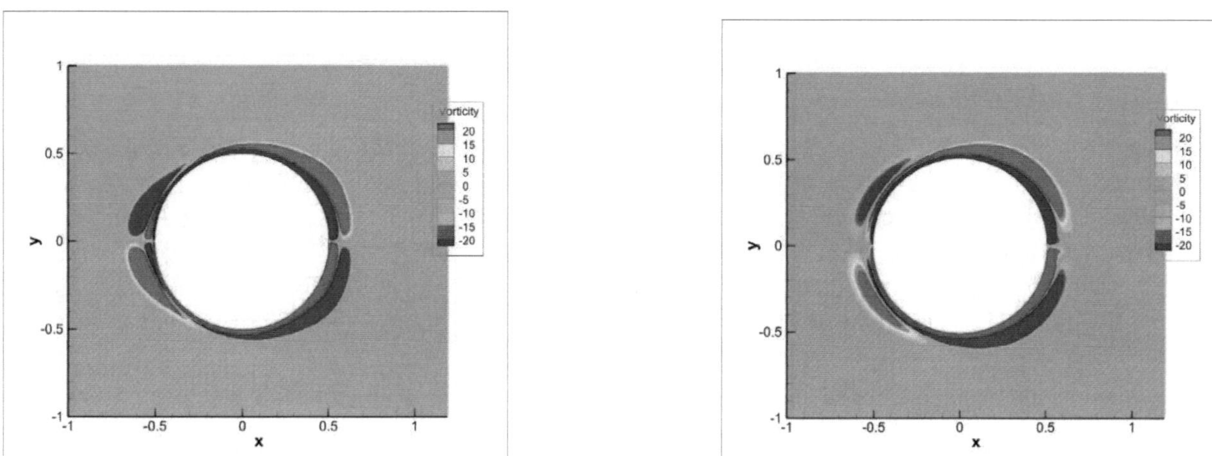

a) VISVE-laminar

b) VISVE-turbulent

Figure 12. Vorticity contours: Re=2985, KC=2; t/T=1

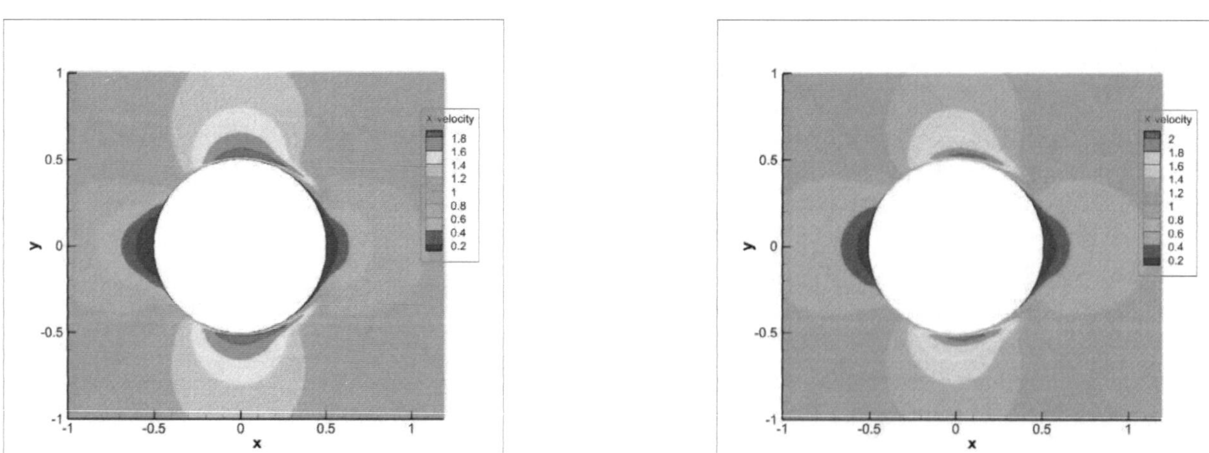

a) VISVE-laminar

b) VISVE-turbulent

Figure 13. Velocity contours: Re=2985, KC=2; t/T=1

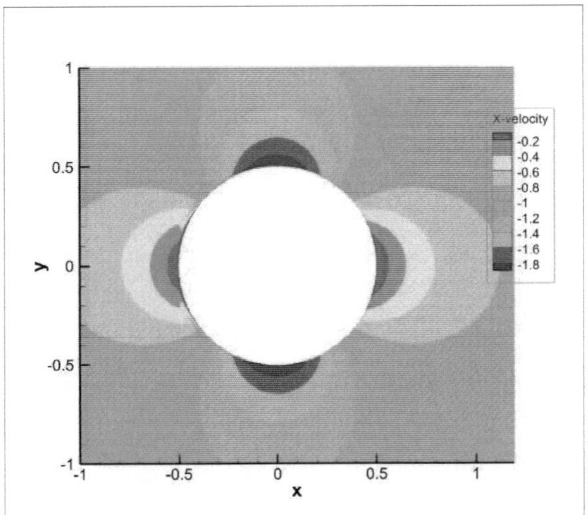

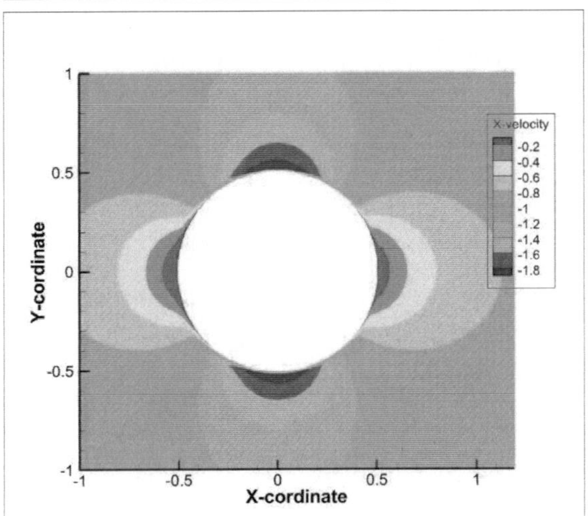

a) VISVE-laminar

b) VISVE-turbulent

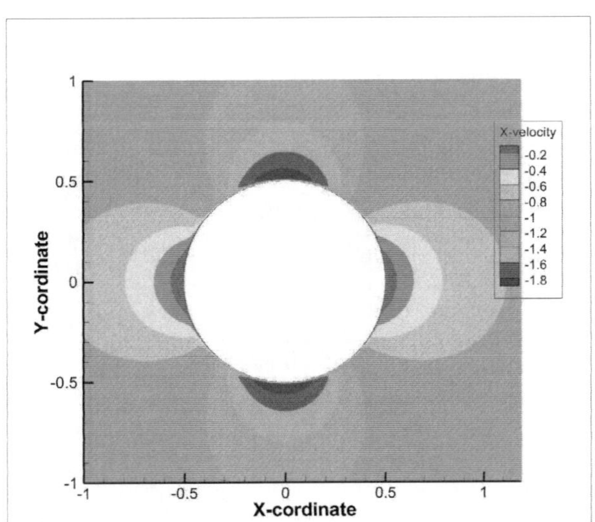

c) RANS

Figure 14. Velocity contours: Re=40000, KC=2, t/T=0.5

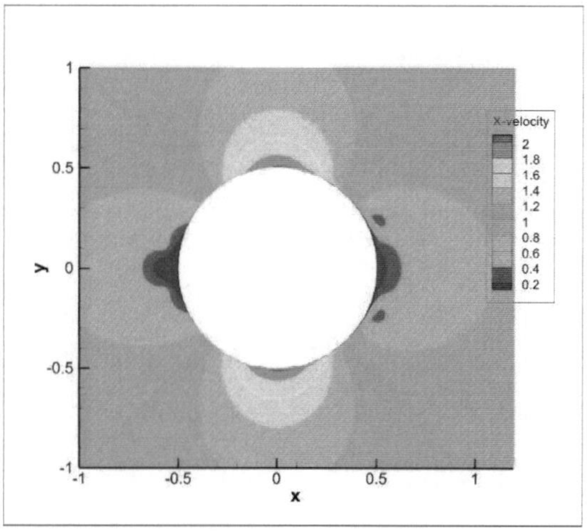

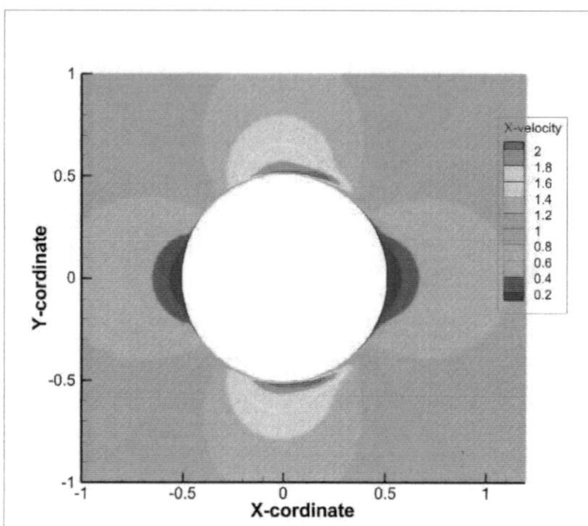

a) VISVE-laminar

b) VISVE-turbulent

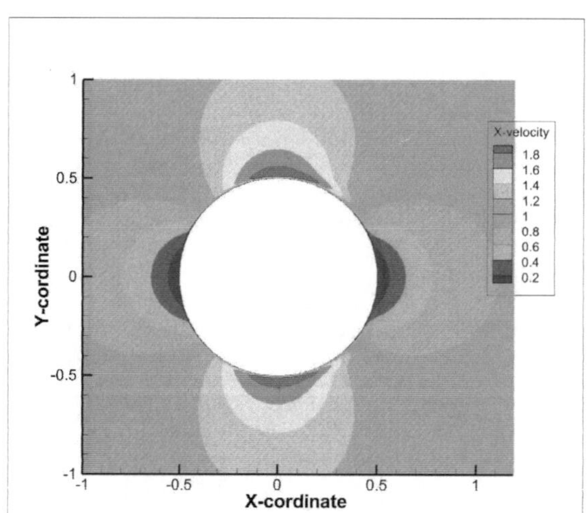

c) RANS

Figure 15. Velocity contours: Re=40000, KC=2, t/T=1

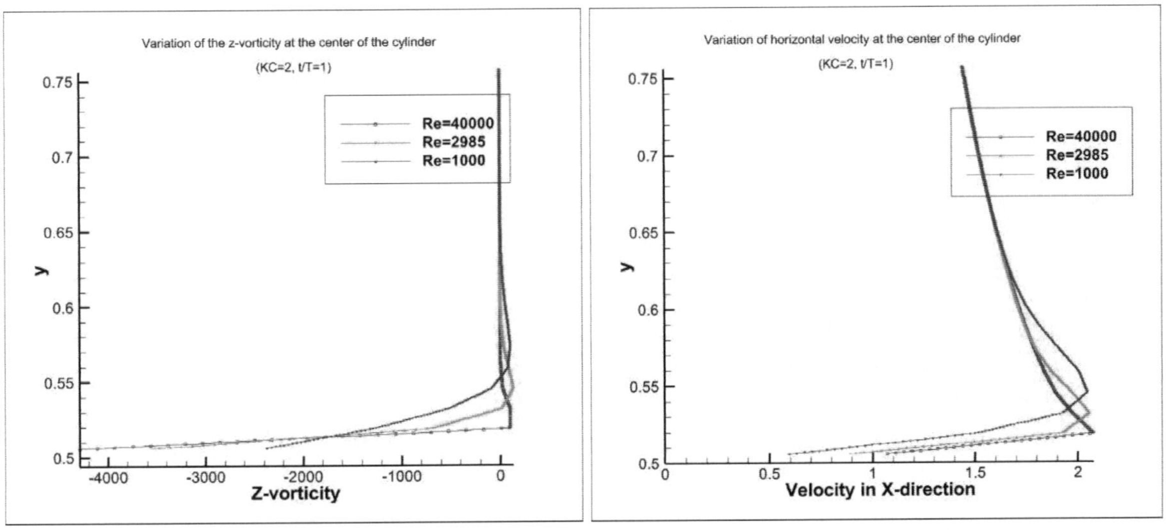

Figure 16. The vorticity and velocity profile for Re= 1,000, 2,985, 40,000

3. Effects of different KC numbers

The effects of different KC numbers were investigated in this section. Figure 17 and 18 show the comparisons of turbulent VISVE and RANS for Re=40, 000, KC=6, at t/T =0.5. Figure 19 and 20 show the same condition at t/T=1. Compared with KC=2, a larger KC number has a longer period and needs a longer computational domain. At half of the period (Figure 17), the differences were acceptable. In Figure 19, the turbulent VISVE has more diffusion in front of the cylinder, compared with RANS.

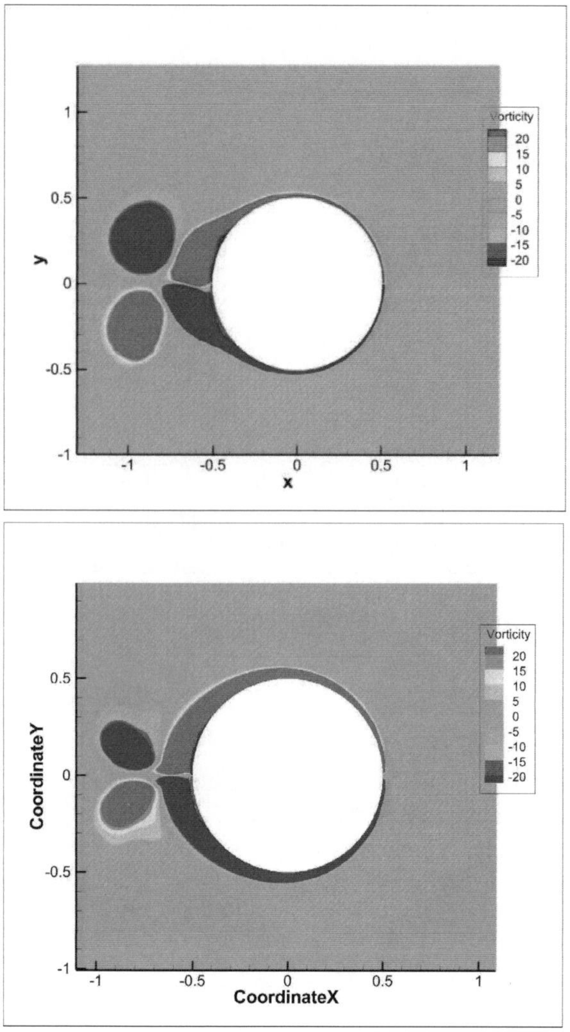

a) VISVE-turbulent b) RANS

Figure 17. Vorticity contours: Re=40000, KC=6, t/T=0.5

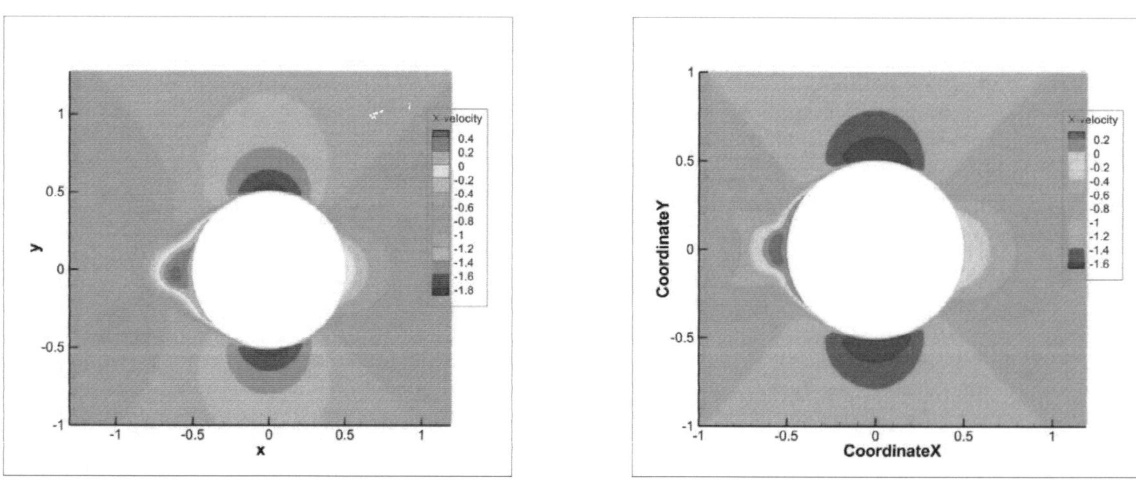

a) VISVE-turbulent b) RANS

Figure 18. Velocity contours: Re=40000, KC=6, t/T=0.5

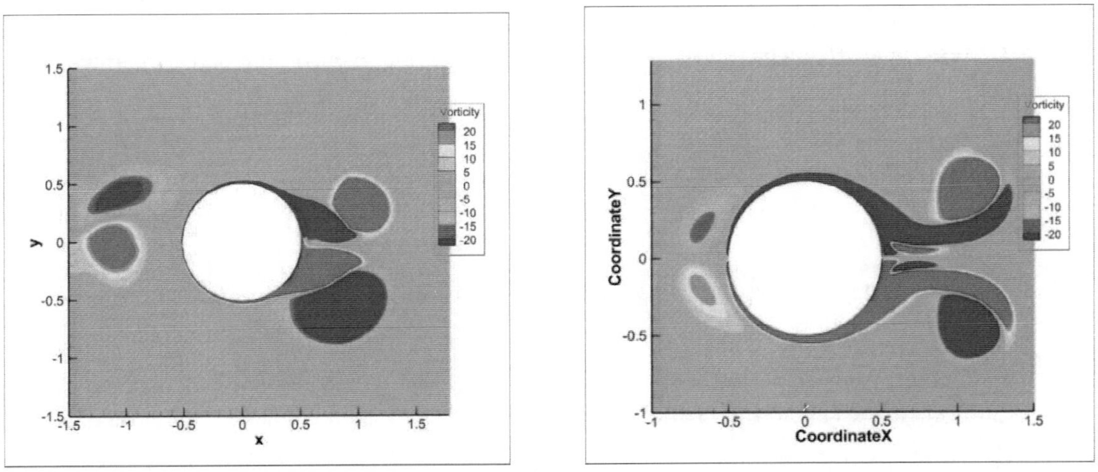

a) VISVE-turbulent

b) RANS

Figure 19. Vorticity contours: Re=40000, KC=6, t/T=1

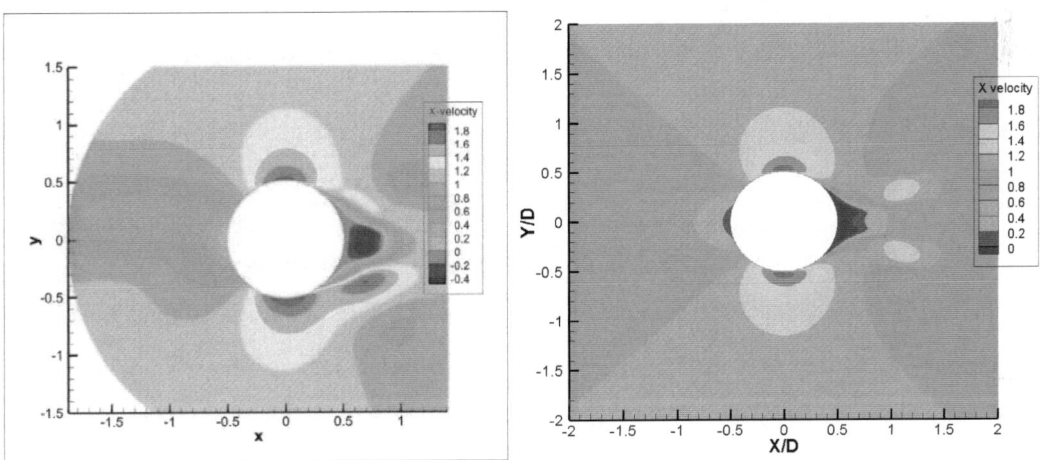

a) VISVE-turbulent

b) RANS

Figure 20. Velocity contours: Re=40000, KC=6, t/T=1

CONCLUSIONS

The $k - \omega$ SST turbulence model was implemented in the VISVE solver to include the effect of turbulence. It provides a newly developed turbulent computational model that can solve the turbulent flow independently and have the potential to be applied in many ocean engineering problems.

The turbulent VISVE solver was applied to the oscillating flow around a cylinder at different Reynolds numbers and Keulegan–Carpenter (KC) numbers. The results were compared with those from a RANS solver. The results show that the inclusion of the turbulence model does not make a difference at smaller Reynolds numbers (Re=1000 and Re=2980), but it improves the results compared with laminar VISVE at a higher Reynolds number (Re=40,000). For KC=2, the results of turbulent VISVE agree well with RANS. However, for KC=6, the turbulent VISVE has more diffusion ahead of the cylinder than RANS.

ACKNOWLEDGEMENTS

Support for this research was provided by the U.S. Office of Naval Research (Grant Numbers N00014-18-1-2276 and N00014-21-1-2488; Dr. Ki-Han Kim) and by Phases VIII and IX of the "Consortium on Cavitation Performance of High Speed Propulsors".

REFERENCES

Agishtein, M.E., Migdal, A.A., 1989. Dynamics of vortex surfaces in three dimensions: Theory and simulations. Physica D: Nonlinear Phenomena 40, 91–118.

Airiau, C., Giovanni, A., 1999. Vorticity evolution on a separated wavy wall flow, in: ESAIM: Proceedings, EDP Sciences. pp. 1–11.

Buttke, T., 1992. Lagrangian numerical methods which preserve the hamiltonian structure of incompressible fluid flow, in" vortex flows and related numerical methods". Vortex Flows and Related Numerical Methods 395, 39–57.

Cazalbou, J., Spalart, P.R., Bradshaw, P., 1994. On the behavior of two-equation models at the edge of a turbulent region. Physics of Fluids 6, 1797–1804.

Chang, Y.C., Hou, T., Merriman, B., Osher, S., 1996. A level set formulation of eulerian interface capturing methods for incompressible fluid flows. Journal of computational Physics 124, 449–464.

Chorin, A.J., 1980. Vortex models and boundary layer instability. SIAM Journal on Scientific and Statistical Computing 1, 1–21.

Cortez, R., 1998. On the accuracy of impulse methods for fluid flow. SIAM Journal on Scientific Computing 19, 1290–1302.

Davidson, L., Peng, S.H., 2003. Hybrid les-rans modelling: a one-equation sgs model combined with a k-ω model for predicting recirculating flows. International Journal for Numerical Methods in Fluids 43, 1003–1018.

Ferziger, J.H., Peric, M., Street, R.L., 2002. Computational methods for fluid dynamics. volume 3. Springer.

Gopalan, H., Heinz, S., Stöllinger, M.K., 2013. A unified rans–les model: Computational development, accuracy and cost. Journal of Computational Physics 249, 249–274.

Hertwig, D., Efthimiou, G.C., Bartzis, J.G., Leitl, B., 2012. Cfd-rans model validation of turbulent flow in a semi-idealized urban canopy. Journal of Wind Engineering and Industrial Aerodynamics 111, 61–72.

Huyer, S.A., Grant, J.R., 2000. Solution of two-dimensional vorticity equation on a lagrangian mesh. AIAA journal 38, 774–783.

Iliopoulos, K., Kinnas, S.A., 2021. Visve, a vorticity based model applied to 2-d hydrofoils in cavitating conditions, in: Proceedings of the 11th International Symposium on Cavitation, CAV2021, May 10-13, Daejon, Korea.

Jones, W., Launder, B.E., 1972. The prediction of laminarization with a two-equation model of turbulence. International journal of heat and mass transfer 15, 301–314.

Kinnas, S.A., 2020. Viscous vorticity equation (visve) for turbulent 2-d flows with variable density and viscosity. Journal of Marine Science and Engineering 8, 191.

Knio, O.M., Ghoniem, A.F., 1991. Three-dimensional vortex simulation of rollup and entrainment in a shear layer. Journal of Computational Physics 97, 172–223.

Knio, O.M., Ghoniem, A.F., 1992. Vortex simulation of a three-dimensional reacting shear layer with infinite-rate kinetics. AIAA journal 30, 105–116.

Lara, J., Losada, I., Guanche, R., 2008. Wave interaction with low-mound breakwaters using a rans model. Ocean engineering 35, 1388–1400.

Leonard, A., 1975. Numerical simulation of interacting three-dimensional vortex filaments, in: Proceedings of the Fourth International Conference on Numerical Methods in Fluid Dynamics, Springer. pp. 245–250.

Leonard, A., 1980. Vortex methods for flow simulation. Journal of Computational Physics 37, 289–335.

Leonard, B.P., 1979. A stable and accurate convective modelling procedure based on quadratic upstream interpolation. Computer methods in applied mechanics and engineering 19, 59–98.

Marshall, J., Grant, J., Gossler, A., Huyer, S., 2000. Vorticity transport on a lagrangian tetrahedral mesh. Journal of Computational Physics 161, 85–113.

Menter, F.R., 1994. Two-equation eddy-viscosity turbulence models for engineering applications. AIAA journal 32, 1598–1605.

Moin, P., Kim, J., 1982. Numerical investigation of turbulent channel flow. Journal of fluid mechanics 118, 341–377.

Moser, R.D., Kim, J., Mansour, N.N., 1999. Direct numerical simulation of turbulent channel flow up to re = 590. Physics of fluids 11, 943–945.

Li, Z. and Kinnas, S.A., 2017, February. VISVE, a Vorticity Based Model Applied to Unidirectional and Alternating Flow around a Cylinder. In SNAME 22nd Offshore Symposium. OnePetro.

Napolitano, M., Catalano, L., 1991. A multigrid solver for the vorticity-velocity navier-stokes equations. International journal for numerical methods in fluids 13, 49–59.

Osher, S., Sethian, J.A., 1988. Fronts propagating with curvature-dependent speed: Algorithms based on hamilton-jacobi formulations. Journal of computational physics 79, 12–49.

Ould-Salihi, M.L., Cottet, G.H., El Hamraoui, M., 2001. Blending finite-difference and vortex methods for incompressible flow computations.

SIAM Journal on Scientific Computing 22, 1655–1674.

Peaceman, D.W., Rachford, Jr, H.H., 1955. The numerical solution of parabolic and elliptic differential equations. Journal of the Society for industrial and Applied Mathematics 3, 28–41.

Ploumhans, P., Winckelmans, G., Salmon, J.K., Leonard, A., Warren, M., 2002. Vortex methods for direct numerical simulation of three-dimensional bluff body flows: application to the sphere at re= 300, 500, and 1000. Journal of Computational Physics 178, 427–463.

Robert, A., 1981. A stable numerical integration scheme for the primitive meteorological equations. Atmosphere-Ocean 19, 35–46.

Rogers, M.M., Moser, R.D., 1994. Direct simulation of a self-similar turbulent mixing layer. Physics of Fluids 6, 903–923.

Shur, M.L., Spalart, P.R., Strelets, M.K., Travin, A.K., 2008. A hybrid rans-les approach with delayed-des and wall-modelled les capabilities. International journal of heat and fluid flow 29, 1638–1649.

Stock, M., 2006. Summary of vortex method literature.

Tian, Y., 2014. Leading edge vortex modeling and its effect on propulsor performance. Ph.D. thesis. Ocean Engineering Group, Dept. of Civil, Architectural and Environmental Engineering, UT Austin.

Tian, Y., Kinnas, S.A., 2015. A viscous vorticity method for propeller tip flows and leading edge vortex, in: Fourth international symposium on marine propulsors (smp15), Austin, Texas, USA.

Wilcox, D.C., 1993. Comparison of two-equation turbulence models for boundary layers with pressure gradient. AIAA journal 31, 1414–1421.

Wilcox, D.C., 1998. Turbulence modeling for CFD. volume 2. DCW industries La Canada, CA.

Wu, C., Kinnas, S.A., 2019. A 3-d viscous vorticity equation (visve) method applied to flow past a hydrofoil of elliptical planform and a propeller, in: Proceedings of the 6th International Symposium on Marine Propulsors, SMP19, Rome, Italy, pp. 26–30.

Wu, C., Kinnas, S.A., 2020. Laminar and turbulent flow past a hydrofoil predicted by a distributed vorticity method, in: International Conference on Offshore Mechanics and Arctic Engineering, American Society of Mechanical Engineers. p. V008T08A008.

Wu, C., Kinnas, S.A., 2021a. Flow past a rotating cylinder predicted by a compact eulerian viscous vorticity method under non-inertial rotating frame. Ocean Engineering 230, 108882.

Wu, C., Kinnas, S.A., 2021b. Parallel implementation of a viscous vorticity equation (visve) method in 3-d laminar flow. Journal of Computational Physics 426, 109912.

Wu, C., Kinnas, S.A., Li, Z., Wu, Y., 2019. A conservative viscous vorticity method for unsteady unidirectional and oscillatory flow past a circular cylinder. Ocean Engineering 191, 106504.

Xing, L., 2018. VISVE, a vorticity based model applied to 2-D hydrofoils in backing and cavitating conditions. Master's thesis. Ocean Engineering Group, Dept. of Civil, Architectural and Environmental Engineering, UT Austin.

Yao, H., Kinnas, S.A., 2019. Coupling viscous vorticity equation (visve) method with openfoam to predict turbulent flow around 2-d hydrofoils and cylinders, in: Proceedings of the Twenty-ninth (2019) International Ocean and Polar Engineering Conference, ISOPE, Hawaii, USA, pp. 3442–3451.

SNAME
THE INTERNATIONAL COMMUNITY FOR MARITIME AND OCEAN PROFESSIONALS

FATIGUE ASSESSMENT FOR STEEL LAZY-WAVE RISERS (SLWR)

SONG LIU[1], XIAOCHUAN YU[2], SHANKAR BHAT[3], KEITH WANG[3]

1.COOEC; 2. UNIVERSITY OF NEW ORLEANS; 3. INDEPENDENT CONSULTANT

Proceedings of the 27th Offshore Symposium, February 22nd, 2022, Houston, Texas
Texas Section of the Society of Naval Architects and Marine Engineers

ABSTRACT

Both in-place slug induced riser fatigue and VIV fatigue during pre-laid phases are investigated in this paper for steel lazy wave risers (SLWR) tied back to a deep draft semi-submersible in about 1100m water depth, as part of the full scale engineering study effort for lazy-wave riser design. These results are compared in parallel with fatigue contributions from WLF, VIM and VIV fatigue.

Dynamic analysis of risers is conducted including the effect of slugging with the vessel in its nominal position to eliminate environmental and vessel motion dynamic effects. Slugs are modelled as variation of mass and inertia, and centrifugal and Coriolis forces are included in the analysis. Rainflow counting method is used to calculate fatigue damage, which is similar to that used for wave loading fatigue analysis. Slug flow conditions are described in terms of lengths of slugs, slug velocity, slug frequency and duration, density of liquid slugs and/or gas bubbles over the field life and including the turn-down and start-up phases and late service life. Analysis results show that the slug fatigue damage contribution for lazy-wave risers can be significantly higher than that for conventional SCRs and it even could play a governing contribution for the buoyancy section.

In-plane VIV fatigue is investigated for SLWR in wet parked phase with long-term current bins used in the analysis, which is from in-field measurement data. It is found that the riser wet park VIV fatigue analysis can be significant depending on the current profiles, hog bend height and the lay tension.

Keywords: Steel Lazy-wave Riser, Slug Induced Fatigue, Vortex-Induced Vibration

INTRODUCTION

Steel catenary risers (SCRs) have been the most robust and preferrable riser configuration for deepwater field development due to their simplicity and lower material and installation cost (Mekha, 2001). Strength and fatigue performance of the SCRs are critical feasibility issues for SCR design. The in-place riser fatigue assessment generally covers fatigue damages from wave loading fatigue (WLF), vortex induced motion (VIM) fatigue and vortex induced vibration (VIV) fatigue. Most fatigue damages during riser installation come from pipelay phases, and some from the recovery/pull-in stages if risers are pre-laid on seabed prior to final hookup to the host platform.

Suited for higher-motion floating production unit (FPU), steel lazy-wave risers (SLWR) are gaining popularity in the past few years, as the buoyancy section decouples touchdown zone (TDZ), usually the most critical region regarding to strength and fatigue life evaluation for conventional SCRs, from FPU motions (Cheng and Cao, 2014; Lal et al., 2019). For the SLWR, however, there are other fatigue contribution sources need to be considered due to the design complicity in configuration, caused by internal fluid density variations.

Slugging flow induced riser fatigue can be significant for lazy-wave configuration risers (Jeong et al., 2020; Campbell et al., 2021; Cooper et al., 2009). Slugging flow is featured by liquid slug with distributed gas bubbles between two gas pockets (Figure 1). Slugs can be generated by different mechanisms in a riser/pipeline, which includes slugging produced by transient effects, production start-up and blow-down, or pigging operations. The transient slugging could be caused by the elevations in the pipeline (Terrain slugging) or gas flowing at a fast rate over a slower flowing liquid phase (hydrodynamic slugging). In particular for risers, liquids could accumulate at the bottom of the riser until sufficient pressure is generated behind it to push the liquids over the top of the riser, overcoming the static head, which is called riser-based slugging. Slugs formed by terrain slugging, hydrodynamic slugging or riser-based slugging are periodical in nature.

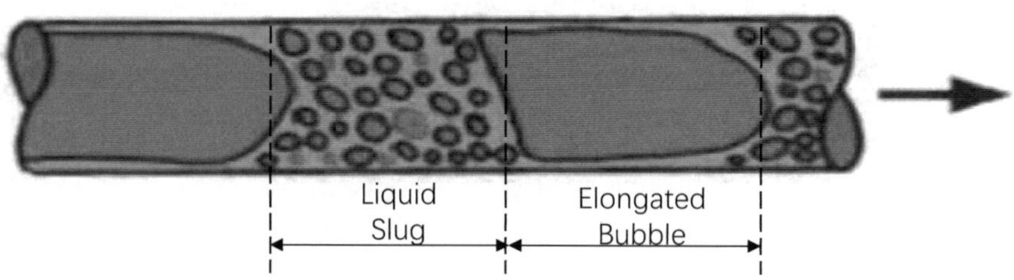

Figure 1: Slugging Flow Regime

Additional fatigue damage contribution could come from the wet-storage phase for SLWRs after they are pre-laid on the seabed prior to FPU hookup and before final pull-in stage. Depending on the wet-storage configuration of the SLWR and elevation of the hog bend section, the in-field background or bottom current could induce significant VIV fatigue damages which need to be considered for overall riser fatigue assessment.

In this study, both the slug-induced fatigue and VIV fatigue during SLWR wet parking phase are analyzed and evaluated for typical deepwater lazy-wave risers, based on past engineering effort and design input for deepwater field development projects.

STEEL LAZY-WAVE RISER DESIGN DATA

This section summarizes the configuration and other properties pertaining to the fatigue design and analysis of the 16 inch gas export and 10 inch production SLWRs investigated in this paper. Contents for the production SLWRs are mainly condensate during normal operation.

The host platform will be a moored semisubmersible production facility. The nominal water depth at the host location and riser touch down area is about 1,123 m. The risers are to be connected to the semisubmersible porches using flexible Joints (FJs).

Configuration of the SLWRs is shown in Figure 2 with section lengths summarized in Table 1, which includes the buoyancy section, strakes coverage length etc. Note that the risers are designed as almost fully covered by VIV suppression strakes with straked coverage length of (L_1+L_2+L_B+L_3). Key data of the steel riser line pipes are summarized in Table 2.

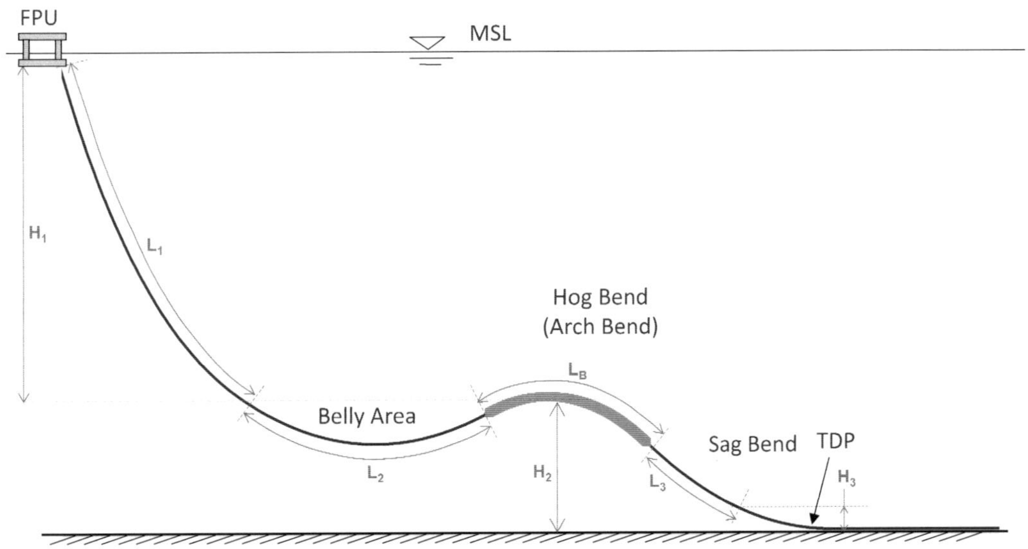

Figure 2: Riser Configuration Sketches

Table 1: SLWR Section Lengths and Configuration Parameters

Parameters	16 inch Gas Export SLWR	10 inch Production SLWR
L_1 (m), Top section length	1000	800
L_2 (m), Belly section length	670	870
L_B (m), Buoyancy section length	430	430
L_3 (m)	200	366
H_1 (m)	875	750
H_2 (m)	220	345
H_3 (m)	25	25
Departure angles (deg)	18	18
Touchdown Point (TDP) water depth (m)	1123	1123

Table 2: Riser Line Pipe Key Data

SLWRs	Unit	16-inch Export	10-inch Production
OD	inch	16	10.75
Wall thickness	inch	1.1 (top 1000m) / 0.75 (others)	1.024
Design Pressure	barg	201	462
Steel Grade	-	DNV Gr 450	DNV Gr 450
Line Pipe Type	-	Seamless	Seamless

The process of SLWR configuration design is much more complicity than that for conventional SCRs, as the content density will significantly impact the elevation and shape of the lazy-wave configuration. The configuration design need to cover all the design category and operational scenarios, for which the content density in each section can spread in a huge range. As an example, for the 10 inch production SLWR, riser configuration are plotted in Figure 3 for void/flooded installation condition, mean operating and shut-in scenarios.

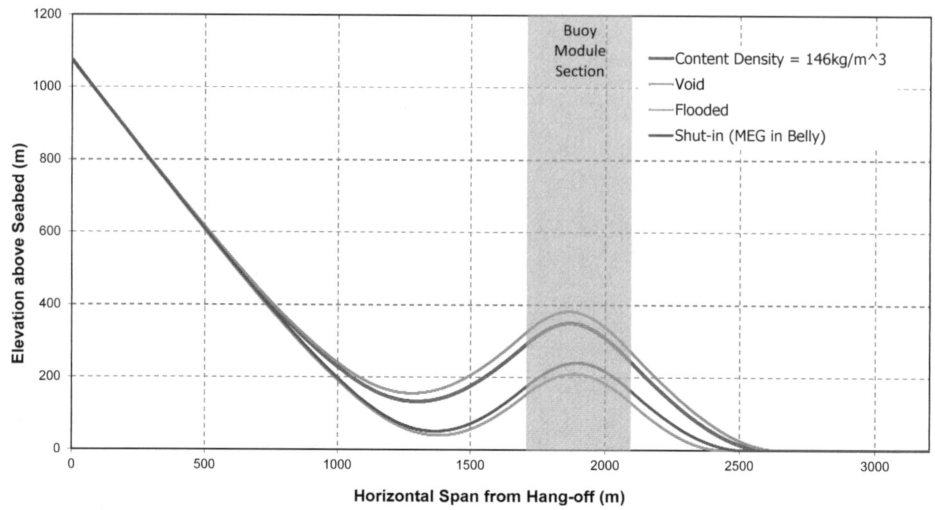

Figure 3: Production SLWR Configurations at Different Operational/Installation Conditions

Slugging Flow Data for Slug-Induced Fatigue Assessment

The characteristics of the slugging flow are calculated by the transient multiphase fluid flow simulation. Following slug/bubble parameters are extracted from the simulation results to represent the slug flow regimes and characteristics:

- Slug density / Bubbles density
- Slug length / Bubble length
- Flow velocity (assuming same velocity for Slug / Bubble)
- Slug / Bubble period or frequency.

Sufficient amount of flow assurance simulation results is required in order to derive reliable slugging parameters over the expected service life. In general in the early production years, the internal flow is more close to steady flow (no slugging effect) because of the high production rate. Slugging flow in the middle life years usually has higher density and longer slug length. During shut-in and re-start operation stages, the slug event can be continuous with significant slug density variation.

Table 3 summarizes the slug/bubble data for the 10 inch production SLWR, based on statistics results derived from transient multi-phase flow analysis. Note that the slugging flow is conservatively assumed to be continuous for the last 1.5 years of the 25 years design life due to low production rate decreases. Due to complicity and uncertainty in slugging flow simulation and slug prediction, Table 3 includes further statistical information such as median and maximum slug length, average and maximum slug velocities etc.

Table 3: Production SLWR Slug Data

	10 inch SLWR		
Turndown Slug Type	Type 1	Type 2	Type 3
FPS Pressure (bara)	60	60	40
Length of liquid slug (m)			

Median size	47	36	25
Maximum slug size	387	212	56
Density of liquid slug (kg/m^3)	573	922	980
Density of gas bubble (kg/m^3)	81	76	49
Average Slug velocity (m/s)	2.1	2.3	2.1
Max Slug velocity (m/s)	15	15	15
Slug Frequency (1/hr)			
Average slug frequency	9	3	3
Frequency for max slug size	0.25	0.1	0.1
Number of slug events over field service life	0-23.5years: 96 hours per year 23.5-25years: Continuous slugging events assumed.		

Slug Induced Fatigue Analysis Methodology

In this study, slugging flow is modelled as variation of mass and inertia due to the flow of fluid within the risers. Additional load on riser pipes, such as the Centrifugal and Coriolis forces are included in the analysis, illustrated in Figure 4.

It is demonstrated by previous study of slug induced fatigue in combination with environmental load and corresponding vessel motions (Jeong et al., 2020), that the slugging flow will have big effect for mild environmental conditions, while will have negligible impact for extreme environmental conditions such as the 100-year return-period cases. For simplification, in this paper the slugging dynamic analysis is conducted with the vessel fixed at its nominal operation position and without environmental conditions.

Parameter studies are performed for median and maximum slug lengths. Both average and maximum slug velocities are considered. Median slug lengths associated with high occurrence frequencies, and maximum slug lengths associated with lower frequencies are applied as input during the analysis. A total of at least 10 slugging cycles are simulated for each slug type investigated.

The fatigue calculation methodology used for slugging fatigue damage is similar to that used for wave loading fatigue analysis. Rainflow counting method is used to calculate fatigue damage. For each slug type, the maximum slugging fatigue damages from the first 23.5 years and last 1.5 years are combined. Note that the slugging fatigue life is conservatively calculated from the parameter studies and slug type resulting in the worst fatigue damage.

Slugging fatigue analysis is conducted using Orcaflex v9.5. Note that flow friction effect can be modelled implicitly through velocity changes along the riser sections, which is not applicable in this study. As an example, Figure 5 shows the content density profiles along the riser length for Slug type #1 with median slug length modelled.

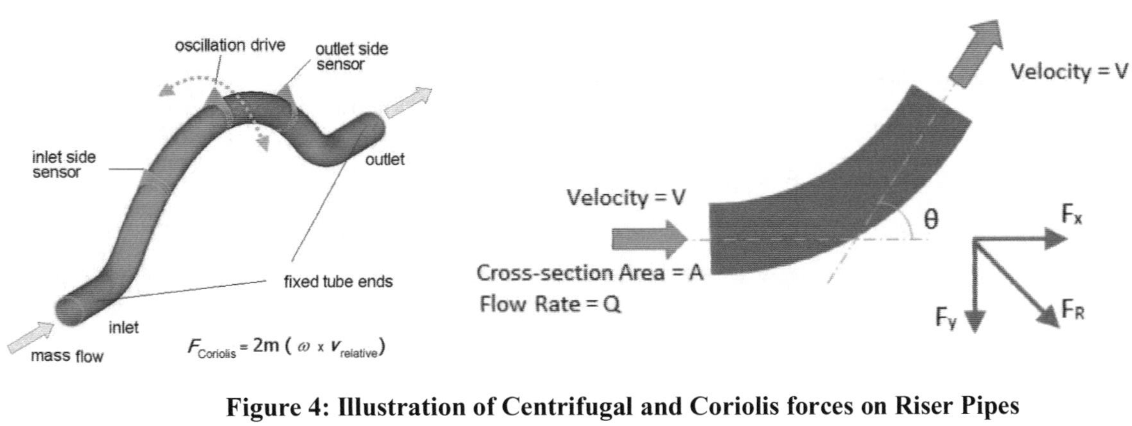

Figure 4: Illustration of Centrifugal and Coriolis forces on Riser Pipes

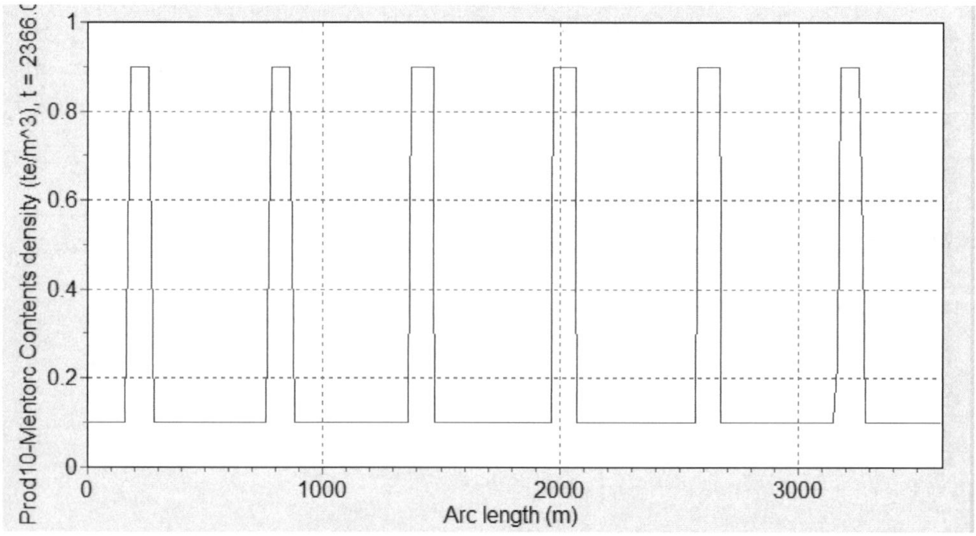

Figure 5: Fluid Density Profile along the Riser Length: Slug #1 (Medium Slug Length as Example)

SLUG INDUCED FATIGUE ANALYSIS RESULTS

The slugging fatigue lives for the 10-inch production SLWR are summarized in Table 4 and Table 5, for median slug length and max slug length respectively. Fatigue damages are derived for the critical hotspot at pipe inner diameter (ID) using BS 7608 E-curve with a knockdown factor of 2.0. Stress concentration factor of 1.2 is applied at both weld cap and weld root.

The results show that the most critical slugging fatigue regions are the nominal touchdown zone (TDZ) and buoyancy region where the curvature is at the highest.

It is shown by the slug-induced fatigue analysis results that slug length and velocity, in conjunction with other parameters such as slug frequencies, slugging/bubble density and bubble length (derived by other parameters), have big impact on slug fatigue results. In general, for the maximum slug length, due to the very low frequency, fatigue damages are much lower than the median slug length which occurs in higher frequencies, which proves that slug-induced fatigue is caused mainly by small riser dynamic fluctuations due to internal flow mass variation. Table 5 shows that the fatigue damage increases with higher slug velocity but not in proportion to the speed, while for long slug length, the increases in slug velocity may cause a fluctuation and thus fatigue damage decreases.

Table 4: Slugging Analysis Results for Median Slug Length

Slug Type	Slug Length	Slug Velocity	Fatigue Damage (E/2 at ID)		Critical Location
	Median/Max	Average / Max	0-23.5 years	23.5-25 years	
Type #1	Median	Average	0.08%	0.49%	TDZ
	Median	Maximum	0.12%	0.67%	Buoy Region
Type #2	Median	Average	0.09%	0.50%	TDZ
	Median	Maximum	0.12%	0.70%	Buoy Region
Type #3	Median	Average	0.02%	0.13%	TDZ
	Median	Maximum	0.08%	0.47%	Buoy Region

Table 5: Slugging Analysis Results for Maximum Slug Length

Slug Type	Slug Length	Slug Velocity	Fatigue Damage (E/2 at ID)		Critical Location
	Median/Max	Average / Max	0-23.5 years	23.5-25 years	
Type #1	Maximum	Average	0.033%	0.19%	TDZ
	Maximum	Maximum	0.030%	0.17%	TDZ
Type #2	Maximum	Average	0.015%	0.09%	TDZ
	Maximum	Maximum	0.014%	0.08%	TDZ
Type #3	Maximum	Average	0.008%	0.05%	TDZ
	Maximum	Maximum	0.004%	0.03%	TDZ

The slugging fatigue life results are combined with the other fatigue damage sources to calculate the combined fatigue life, which is shown in presented in Table 6 for the 10 inch SLWR. S-N curve of BS 7608 C curve with a knockdown factor of 2.5 is selected all weld caps. For weld root, E curve is used for a knockdown factor of 2.0 except for first onshore weld. As shown by the comparison, for the TDZ region, the slugging fatigue damage can be more severe compared to the VIM and VIV fatigue damage, and comparable with the wave loading fatigue damage.

Note when the factored combined fatigue life is calculated by equation below,

$$\text{Factored Fatigue Life} = (1-\beta)/\,(SF_{WLF}*DAM_{WLF} + SF_{VIV}*DAM_{VIV} + SF_{VIM}*DAM_{VIM} + SF_{Slug}*DAM_{slug})$$

Where β is the allowable fatigue damage during installation, taken as 0.1 in industry general practice. A robust riser design should return factored fatigue life at least larger than the design life of 25 years. Safety factors (SF) of 10, 20, 10 are used for WLF, VIV, VIM fatigue contributions respectively as general accepted industry practice. A safety factor of 10 is also assumed for slug induced fatigue, which is an assumption without validation, which may not be conservative.

Table 6: Combined Fatigue Lift Summary – 10'' SLWR

Fatigue Critical Region	Weld Root or Cap	SN Curve	Unfactored Fatigue Life (Years)				Factored Combined Fatigue Life (Years)
			WLF	VIM	VIV	Slugging	
TDZ	Weld Root	BS7608 E/2	1,440	4,700	49,000	3,050	61.1
	Weld Cap	BS7608 C/2.5	8,110	17,400	182,000	3,423	269.5
Buoyancy Region	Weld Root	BS7608 E/2	11,440	987,000	370,000	11,364	497
	Weld Cap	BS7608 C/2.5	93,500	8,500,000	3,200,000	9,260	760
First Offshore Weld	Weld Root	BS7608 E/2	1,070	177,000	10,000,000	1,000,000	96.1
	Weld Cap	BS7608 C/2.5	5,650	878,000	50,000,000	1,000,000	504

Pre-Laid VIV Fatigue Analysis for SLWR

Riser vortex-induced vibration (VIV) fatigue during wet parking period is investigated for 16 inch gas export SLWR in the study. The riser is assumed to be pre-laid on seabed in empty condition for at least 6 months.

As one of the key differences during riser installation comparing with conventional SCRs, the required SLWR lay tension is governed not only by factors such as sagbend pipe strain, seabed stability, but also dictated by the pre-laid riser configuration, i.e., the buoyancy section elevation. The SLWR pre-laid configuration need to be pre-determined based on bending strain criteria in accidently flooded contingency condition, and in addition, to limit the VIV fatigue damage during the wet parking period.

The bending strain at riser touchdown area will increase significantly if the riser is flooded accidentally (due to a leak for example) during the wet storage phase. Figure 6 shows the 16 inch gas export riser configuration in empty and flooded conditions, with pre-lay tensions of 35mT respectively.

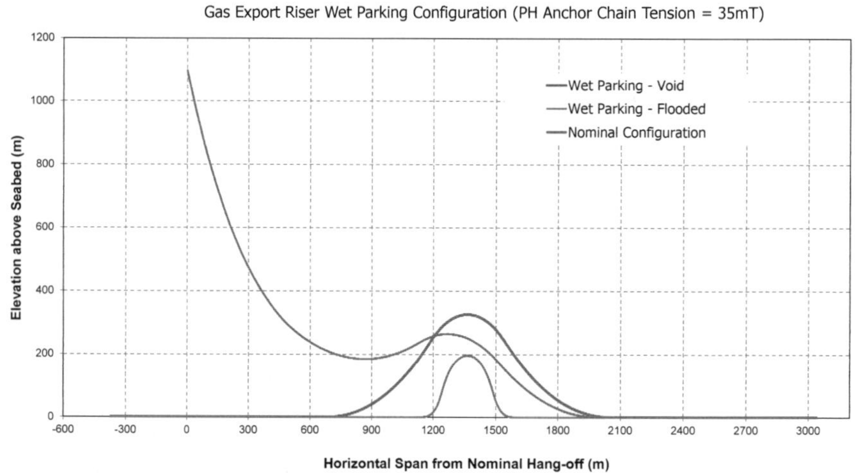

Figure 6: Gas Export SLWR Pre-laid Configurations at Pre-laid Tension of 35 mT

In-plane VIV fatigue analysis is performed using Shear7 4.3 for 16 inch gas export SLWR in wet parked phase. Both long-term current bins (9557 current profiles measured in 2 years) and 10year extreme current.

Table 7 and Figure 7 present the wet storage VIV analysis results using the same S-N curves as defined. A safety factor of 20 is imposed in the VIV fatigue results. The wet park time frame is assumed to be 6 months. The analysis results show that the VIV fatigue damage during the 6 months' pre-laid phase could accumulate to as much as 5.90% of the overall allowable fatigue for 35mT empty pre-laid tension, for which most of the total fatigue budget for installation phase (10%) could be consumed and leave very little fatigue budget for other installation operations.

Table 7: Gas Export SLWR Pre-Laid VIV Fatigue Analysis Results

Pre-laid Tension, mT	VIV Fatigue Damage during Pre-laid	Total Fatigue Budget for Installation Phase
35.0	5.90%	10%

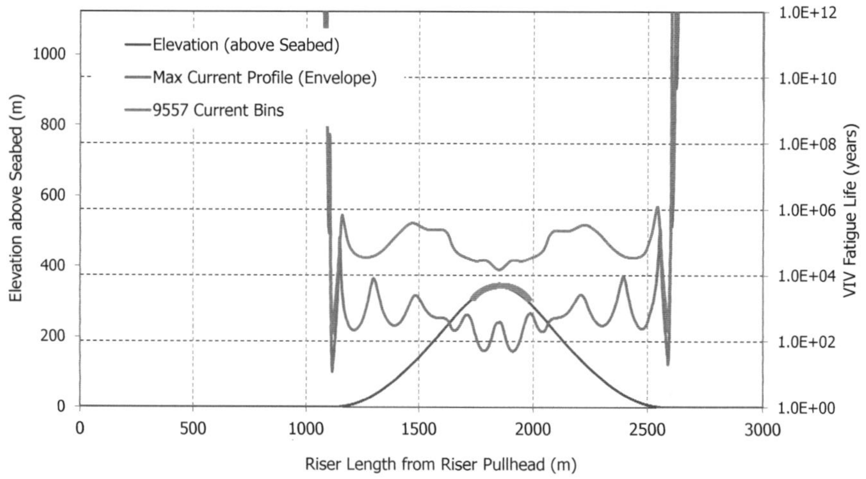

Figure 7: Gas Export SLWR Pre-Laid VIV Fatigue Life Profiles

CONCLUSION

In this study, SLWR fatigue assessment is carried out for slug-induced fatigue and VIV fatigue during wet storage phase, which are two of the key additional fatigue contributions in comparison with conventional SCRs. From the results of the analysis, the following conclusions can be drawn:

- Slugging flow could cause significant SLWR fatigue damages. Parameter studies show that slug length and velocity, in conjunction with other parameters such as slug frequencies, slugging/bubble density, have big impact on slug fatigue results.
- The results show that the most critical slugging fatigue regions are the nominal Touchdown Down Zone (TDZ) and Buoyancy region where the curvature is at the highest, while the impact on the top hang-off region is negligible. For the TDZ region, the slugging fatigue damage can be more severe compared to the VIM and VIV fatigue damage, and even comparable relative to the wave loading fatigue damage.
- Reliable and more realistic transient multi-phase flow simulation and data processing are critical for SLWR fatigue assessment. Safety factor for slug-induced fatigue need to be validated by experimental or field data.
- VIV fatigue damage during the pre-laid phase of SLWRs can be significant and exceeding the installation fatigue budget, i.e., currently 10% accepted by the industry practice. This fatigue damage contribution needs to be analyzed to pre-determine the pre-laid configuration and the lay tensions.

REFERENCES

Campbell, B., Agarwal,P., Curtis,C. and Yang, G., Singha,A., Casstevens, K. and Gokcal, G.E. (2021). Slugging fatigue assessment for steel lazy wave Risers, Offshore Technology Conference, OTC-30922-MS.

Cheng, J. and Cao, P. (2014). Steel lazy wave riser design for semi-submersible in harsh environment, ASME 2014 33rd International Conference on Ocean, Offshore and Arctic Engineering, OMAE2014, San Francisco, California, USA, OMAE2014-24283.

Cooper, P., Burnett, C. and Nash, I. (2009). Fatigue design of flowline systems with slug flow. ASME 2009 28th International Conference on Ocean, Offshore and Arctic Engineering - Honolulu, Hawaii, USA, OMAE2009-79308. (doi:10.1115/omae2009-79308)

Jeong, H., Jang, B.S., Kim, J.D., Park, G., Choi, J. and Lee, D. (2020). A study on effects of slug flow on dynamic response and fatigue damage of risers. Ocean Engineering, 217, 107965. (doi:10.1016/j.oceaneng.2020.107965)

Lal, M., Wang, F., Lu, X. and Sebastian, A. (2019). Strength and fatigue performance of steel lazy wave risers with change in configuration parameters. ASME 2019 38th International Conference on Ocean, Offshore and Arctic Engineering-Glasgow, Scotland, UK, OMAE 2019-95135. (doi:10.1115/OMAE2019-95135)

Mekha, B. (2001), New Frontiers in the Design of Steel Catenary Risers for Floating Production Systems, J. Offshore Mech. Arct. Eng., 123(4), 153-158.

Vásquez M., Arthur, J., Avila, J. and Pablo, J. (2019). A parametric analysis of the influence of the internal slug flow on the dynamic response of flexible marine risers. Ocean Engineering, 174, 169–185. (doi:10.1016/j.oceaneng)

AUTHOR INDEX

Society of Naval Architects and Marine Engineers (SNAME) Texas Section
99 Canal Center Plaza, Suite 310
Alexandria, VA 22314
USA

CURRAN ASSOCIATES INC.
proceedings
.com

ISBN 978-1-7138-5750-1
9 781713 857501

16th Asia Pacific Bioinformatics Conference (APBC 2018)

Yokohama, Japan
15 – 17 January 2018

BioMed Central
The Open Access Publisher